Plant Physiology

A TREATISE

EDITED BY

F. C. STEWARD

Professor Emeritus
Cornell University
Ithaca, New York

Volume VIII: Nitrogen Metabolism

Coedited by

F. C. STEWARD   and   R. G. S. BIDWELL

Charlottesville, Virginia   *Department of Biology*
Dalhousie University
Halifax, Nova Scotia, Canada

1983

ACADEMIC PRESS, INC.

(Harcourt Brace Jovanovich, Publishers)

Orlando   San Diego   San Francisco   New York   London
Toronto   Montreal   Sydney   Tokyo   São Paulo

ACADEMIC PRESS, INC.
Orlando, Florida 32887

United Kingdom Edition published by
ACADEMIC PRESS, INC. (LONDON) LTD.
24/28 Oval Road, London NW1 7DX

Library of Congress Cataloging in Publication Data

Main entry under title:

Plant physiology.

   Includes bibliographies and indexes.
   Contents: v. 1. A. Cellular organization and respiration.   B. Photosynthesis and chemosynthesis. --
v. 2. Plants in relation to water and solutes. -- [etc.]
-- v. 8. Nitrogen metabolism.
   1. Plant physiology--Collected works.   I. Steward,
F. C. (Frederick)
QK711.P58   581.1   59-7689
ISBN 0-12-668608-4 (v. 8)

PRINTED IN THE UNITED STATES OF AMERICA

83 84 85 86   9 8 7 6 5 4 3 2 1

Contents

CHAPTER ONE

Developments in Basic and Applied Biological Nitrogen Fixation *by* Y. OKON AND R. W. F. HARDY

CHAPTER TWO

Nitrogen Metabolism *by* D. J. DURZAN AND F. C. STEWARD

Contributors to Volume VIII

Numbers in parentheses indicate the pages on which the authors' contributions begin.

D. J. DURZAN (55), *Department of Pomology, University of California, Davis, California 95616*

R. W. F. HARDY (5), *Central Research and Development Department, Experimental Station, E. I. du Pont de Nemours and Company, Wilmington, Delaware 19898*

R. C. HUFFAKER (267), *Plant Growth Laboratory, Department of Agronomy and Range Science, University of California, Davis, California 95616*

Y. OKON (5), *Department of Plant Pathology and Microbiology, Faculty of Agriculture, The Hebrew University of Jerusalem, Rehovot 76100, Israel*

J. S. PATE (335), *Department of Botany, University of Western Australia, Nedlands, Western Australia 6009, Australia*

F. C. STEWARD (55), *Charlottesville, Virginia 22901*

Foreword: The New Volumes

The first volume of "Plant Physiology: A Treatise" appeared in 1959, though planning for it commenced as early as 1957. The treatise was arranged in three sections: Cell Physiology, Nutrition and Metabolism, Growth and Development. As it developed, its planned six volumes extended to eleven tomes, the last of which appeared in 1972. Over this time span change was inevitable although, in a measure, the later volumes could update the first.

The decision to add additional volumes to the treatise (Volumes VII *et seq.*) was not made either hastily or lightly. The new volumes will summarize salient developments and thus epitomize the present status of the subject. In so doing, they will add to the historical record of Volumes I through VI in the belief that the continued use of the treatise has owed much to its historical perspective. This perspective, set initially in brief historical sketches, has permeated the work throughout. Not all topics dealt with in Volumes I–VI have been selected for updating. For our arbitrary selections or omissions the editors assume responsibility.

An early injunction to authors, variously fulfilled according to their individual styles, to write comprehensively but in a narrative text that, as necessary, rises above the mass of detail still applies. Thus the new volumes do not discard but will build on the existing ones, for the progress of science is always indebted to its past.

Therefore, as much as possible of the format of the treatise as published will be preserved. This is done even though the restless mood of the times clamors for the new look, and change often seems a virtue. But natural science, like nature, should surely be conservative, and any look that emerges will be seen in the content of the new chapters and the points of view summarized by the authors. Meanwhile editors and authors will strive to preserve that essential continuity with the past that is the basis of scholarship and is essential to an understanding of how and when progress was made.

Organization may survive, but, inevitably, authors change. Many who have served the treatise well no longer survive; others have moved into retirement, albeit with varying degrees of maintained professional activity. The present group of authors represents a blend of some who are new to the task and others who have previously served in this capacity. The plan is to add new volumes that cover selected topics in the general areas of cell biology, metabolism, and growth and development, invoking the aid of three coeditors. Each coeditor assumes, with the editor, primary responsibility for one area while maintaining a general vigilance for the integrity of the treatise as a whole. Moreover, coeditors bring not only their own expertise but experience in teaching of plant physiology. Indeed, to promote the continued usefulness of the treatise to teachers, research workers, and students is part of our collective aims.

But in the final analysis, the value of the new volumes will rest on the ability of the invited authors to add to the existing content of Volumes I–VI and to present an overview of the status of the subject as it now moves through the turbulent last quarter of the twentieth century and anticipates the problems of the twenty-first.

F. C. Steward

Preface to Volume VIII

Volume VIII, like Volume VII, emerges on a vibrant Plant Physiological Contemporary Scene (cf. Volume VII, pp. xi–xiv) in which topics that were first summarized in earlier volumes are now updated. Volume VIII is therefore a companion to Volume VII, and, indeed, it was somewhat arbitrarily detached from it to comprise a group of closely knit chapters which relate to topics first summarized in Volumes II, III, and IVA.

Volume VII dealt comprehensively with metabolism from the standpoint of the compounds of carbon as they move in plants from carbon dioxide through the events of photosynthesis to carbohydrates and then, via respiration, to carbon dioxide again. Between these extremes, involving the passage of electrons and protons from and to water, lie the vast array of reactions which constitute the carbon metabolism and the steps by which the energy intially fixed in photosynthesis is made available for all metabolic purposes. This entire metabolic scene, with its reversible and interlocking cycles, its innumerable enzyme-mediated steps by which energy changes are negotiated reversibly via the phosphorylated compounds as specific coenzymes, needs the overview of Volume VII to be comprehended. But the other "face" of the "metabolic coin" requires the equally comprehensive overview of the compounds of nitrogen and their metabolism as attempted in 1965 in Volume IVA, Chapter 4, and as supplemented in this volume.

The organization of Volume VIII follows the established pattern of the treatise, with its indexes to subjects, to plant names, and to authors. Chapter 2 has appendixes to the text which, together with earlier ones in Volume IVA, Chapter 4, now cover the occurrence in plants of a large number of nonprotein nitrogen compounds. The many citations in these appendixes to the discovery and identification of these compounds were not repeated in the bibliography to the chapter unless the text specifically required this be done.

The editors are again indebted to Dr. W. J. Dress for the Index to Plant Names and for his oversight of the names used in the text. For assistance with the Subject Index we are indebted to Mrs. Shirley Bidwell. Again it is a pleasure to express our appreciation to the staff of Academic Press for their understanding of and help with the problems which Volume VIII interrelated with Volume VII presented.

The Editors

Note on the Use of Plant Names

The policy has been to identify by its scientific name, whenever possible, any plant mentioned by a vernacular name by the contributors to this work. In general, this has been done on the first occasion in each chapter when a vernacular name has been used. Particular care was taken to ensure the correct designation of plants mentioned in tables and figures which record actual observations. Sometimes, when reference has been made by an author to work done by others, it has not been possible to ascertain the exact identity of the plant material originally used because the original workers did not identify their material except by generic or common name.

It should be unnecessary to state that the precise identification of plant material used in experimental work is as important for the enduring value of the work as the precise definition of any other variables in the work. "Warm" or "cold" would not usually be considered an acceptable substitute for a precisely stated temperature, nor could a general designation of "sugar" take the place of the precise molecular configuration of the substance used; "sunflower" and "*Helianthus*" are no more acceptable as plant names, considering how many diverse species are covered by either designation. Plant physiologists are becoming increasingly aware that different species of one genus (even different varieties or cultivars of one species) may differ in their physiological responses as well as in their external morphology and that experimental plants should therefore be identified as precisely as possible if the observations made are to be verified by others.

On the assumption that such common names as lettuce and bean are well understood, it may appear pedantic to append the scientific names to them—but such an assumption cannot safely be made. Workers in the United States who use the unmodified word "bean" almost invariably are referring to some form of *Phaseolus vulgaris;* whereas in Britain *Vicia faba*, a plant of another genus entirely, might be implied. "Artichoke" is another such name that comes to mind, sometimes used for *Helianthus tuberosus* (properly, the Jerusalem artichoke), though the true artichoke is *Cynara scolymus.*

By the frequent interpolation of scientific names, consideration has also been given to the difficulties that any vernacular English name alone may present to a reader whose native tongue is not English. Even some American and most British botanists would be led into a misinterpretation of the identity of "yellow poplar," for instance, if this vernacular American name were not supplemented by its scientific equivalent *Liriodendron tulipifera,* for this is not a species of *Populus* as might be expected, but a member of the quite unrelated magnolia family.

When reference has been made to the work of another investigator who, in his published papers, has used a plant name not now accepted by the nomenclature authorities followed in the present work, that name ordinarily has been included in parentheses, as a synonym, immediately after the accepted name. In a few instances, when it seemed expedient to employ a plant name as it was used by an original author, even though that name is not now recognized as the valid one, the valid name, preceded by the sign =, has been supplied in parentheses: e.g., *Betula verrucosa* (= *B. pendula*). Synonyms have occasionally been added elsewhere also, as in the case of a plant known and frequently reported on in the literature under more than one name: e.g., *Pseudotsuga menziesii* (*P. taxifolia*); species of *Elodea* (*Anacharis*).

Having adopted these conventions, their implementation rested first with each contributor to this work; but all outstanding problems of nomenclature have been referred to Dr. W. J. Dress of the L. H. Bailey Hortorium, Cornell University. The authority for the nomenclature now employed in this work has been the Bailey Hortorium's "Hortus Third." For bacteria Bergey's "Manual of Determinative Bacteriology" and Skerman, McGowan, and Sneath's "Approved List of Bacterial Names" and for fungi Ainsworth and Bisbee's "Dictionary of the Fungi" have been used as reference sources; other names have been checked where necessary against Engler's "Syllabus der Pflanzenfamilien." Recent taxonomic monographs and floras have been consulted where necessary. Dr. Dress's work in ensuring consistency and accuracy in the use of plant names is deeply appreciated.

The Editors

PREAMBLE TO CHAPTERS ONE THROUGH FOUR

It is not possible to conceive of life as we know it without the distinctive naturally occurring compounds of nitrogen and the specificities and asymmetries they convey to the environment *in vivo* through which metabolism moves. This stress on nitrogen and its compounds should not overlook the essentiality of sulfur, which operates, however, over a more restricted metabolic range or of the compounds of phosphorus, so important in energetics (Volume VII) and in heredity.

With respect to nitrogen, and as in the case of carbon compounds, there are several levels of interpretation to be comprehended. The first is the descriptive biochemistry by which compounds and their reactions are recognized. The second is the way the feasible reactions are assembled into metabolic pathways and cycles that give meaningful purpose and direction to the course of metabolism. Finally, the ongoing fate of nitrogen compounds needs to be seen, even as do form and morphogenesis, as the inevitable concomitant of ontogeny, growth, and development. Volume VIII attempts this overview in the following ways.

Chapter 1 begins with the nitrogen of the air, which is only directly made availabe by the still remarkable events of nitrogen fixation made feasible in the few free-living organisms that enjoy this high degree of autotrophy for nitrogen, and especially as it occurs in the happy symbiosis of those higher plants that are hosts to the nitrogen-fixing organisms. Chapter 1 gives an account, biochemically far advanced from that possible in 1963 (Volume III, Chapter 5), of the agriculturally important mechanisms by which it is now known that nature brings about the conversion of atmospheric nitrogen, initially to a biologically useful reduced form (i.e., ammonia), but eventually into plant protein. Nature does this without the fanfare of the nitrogen-fixing industry, but it does require highly sophisticated arrangements that have only recently yielded to investigation. Notable among these are the very elaborate proteins with their localized centers of iron and molybdenum which are the essential catalysts that nature has elaborated for the necessary biochemical conversion. In Chapter 1, this topic is carried through the biology and biochemistry of nitrogen-fixing associations to the point of its practical application in agriculture.

For most purposes, plants in nature and in agriculture draw on nitrate or ammonia as their source of organic nitrogen, and earlier, as in Volume IVA, the roles of nitrate and ammonia were presented as aspects of inorganic nutrition. Using intermediates of carbon metabolism (keto acids) as substrates, there are a number of "ports of entry" for exogenous nitrogen into organic combinations involving carbon compounds at different levels of complexity with one to five carbons. Chapter 2 first traces the modern knowledge of the reactions by which these processes occur and the ways in which the resultant nitrogenous metabolites "fan out," as it were, into the very large number of nonprotein nitrogen compounds which plants are known to make and harbor in their cells. As in the case of carbon, one needs to comprehend the molecules that the subsequent nitrogen metabolism entails, the reactions and metabolic schemata that ensue, and, finally, the ways in which simple compounds give rise to amino acids and then to proteins. Because of its complexity, both in terms of the structure of the end products (proteins) and the means by which they are fabricated, the last phase merits a separate discussion which supplements that in Volume IVA, Chapter 4, Section VIII. As knowledge progresses, however, the constant wonder is that nature ever managed to utilize and control such elaborate chemical situations to provide for the synthesis of the proteins that are so universally needed for the life of cells.

But the metabolism of nitrogen-containing compounds—from nitrate or ammonia to protein, even with its sequel of breakdown and turnover to return nitrogen to the usable pools and to release carbon frameworks to be reintroduced to the respiratory cycle—is not all that the term nitrogen metabolism implies. Therefore, Chapter 2 also considers nitrogen metabolism not in the abstract, but in the context of various actual situations in which cells divide and grow and specific plants develop and mature and reproduce. Here the obtrusive events are those which illustrate how plants as they develop provide in their organs morphogenetically very different internal environments that control and greatly vary the fate of the nitrogen compounds. Notably, in these aspects, nutrition and the diurnal and seasonal periodicities in the external physical environments play the causal roles.

The proteins merit special discussion in Chapter 3, which analyzes their metabolism and turnover and provides a critical examination of the methods used for their investigation. Specific proteins (the enzymes ribulosebisphosphate carboxylase and nitrate reductase and seed protein) are selected as the type cases to illustrate the fate of special proteins through their formation and turnover. The mechanisms and biological settings of protein degradation, as distinct from breakdown during protein turnover, are also covered in detail.

In 1959, Chapters 5 and 6 of Volume II dealt with movements and circulation within the plant body of organic and inorganic solutes, respectively. Chapter 4 in this volume now undertakes, with reference to selected plants, a comprehensive survey of what happens to metabolites (particularly nitrogenous ones) as they arise from nutrients from the inorganic world to move through the ramifications of metabolism to enter the transport system (xylem and phloem) and to circulate, both physically and metabolically, through the plant body to developing organs. From the use of isotopic tracers, schemata have been drawn that trace metabolic events from legume nodules throughout the course of the biochemistry and development of the specific plants. From these analyses, models are developed for the study of processes of transport of metabolites throughout the plants in question and of the mechanisms by which metabolites are partitioned and transported to specific destinations in the plant body.

Aspects of the roles of nitrogen compounds in general and of proteins in particular which impinge upon growth and development are, however, still outside the scope of this volume as outlined above. Many details of plant nutrition of cells and organs as they grow are still outstanding to update the comprehensive Volume III. Also, when growth and metabolism subside, as in the onset of dormancy in seeds and organs of perennation, their specific storage proteins present their peculiar physiological problems. Plant storage proteins were historically very important even as today they have great nutritional and economic significance.

Thus Volume VIII presents a comprehensive overview of those aspects of metabolism which involve nitrogen as the element that best illustrates the breadth and scope of the metabolism of plants throughout their development. Since the constituent chapters are so coherent, however, they need no further preamble than that given above.

F.C.S.

CHAPTER ONE

Developments in Basic and Applied Biological Nitrogen Fixation[1]

Y. OKON AND R. W. F. HARDY

[1]Abbreviations used in this chapter: ADP, adenine diphosphate; AF, activating factor; ATP, adenosine triphosphate; CoA/CoASH, coenzyme A, oxidized/reduced; FAD, flavin adenosine dinucleotide; FeMoCof, molybdenum–iron protein cofactor; $FvdH^-/FvdH_2$, flavodoxin semiquinone/hydroquinone; GDH, glutamate dehydrogenase; GOGAT, glutamate synthetase; GS, glutamine synthetase; $NAD^+/NADH$, nicotine adenine dinucleotide, oxidized/reduced; $NADP^+/NADPH$, nicotine adenine dinucleotide phosphate, oxidized/reduced; *nif* gene, gene coding for nitrogenase; PHB, poly-β-hydroxybutyrate; P_i, inorganic phosphate.

Plant Physiology
A Treatise
Vol. VIII: Nitrogen Metabolism

5

I. Economic Importance of Biological Nitrogen Fixation

According to FAO estimates (1979), in order to sustain the world population in the year 2000 it will be necessary to increase agricultural production by 60%. The role that effective fertilizer uses can play in increasing agricultural production is fully recognized. A clear relationship has been established between the increasing yields of cereals, especially in developed countries over the last 50 years, and the introduction of high-yielding varieties, better pest control, and the increase in fertilizer consumption (i.e., nitrogen, phosphorus, potassium). It is generally assumed that an input of 1 kg of fertilizer produces up to 10 kg of additional cereal, at least for the initial applications of fertilizer (21).

Prices have had definite influence on the growth of fertilizer consumption during the last few years. For the first time in 30 years, fertilizer use decreased in 1974–1975 as a result of sharp increases in fertilizer prices (21). The most common nutrient limiting the production of agricultural crops is nitrogen. The earth's atmosphere is 80% molecular nitrogen; it contains 4×10^{15} metric tons of this element. Nevertheless, plants can utilize nitrogen only in the combined mineral form (fixed nitrogen), such as ammonium (NH_4^+) or nitrate (NO_3^-). Microorganisms such as bacteria and fungi are involved in the recycling of nitrogen in nature, that is, the nitrogen cycle (13, 28, 32).

Up to the nineteenth century, crop yields obtained in cultivated fields were generally low. With the advent of modern agriculture, the natural nitrogen cycle, chiefly the process of biological nitrogen fixation, was no longer able to provide for the nitrogen needs of the plants. Nitrogen fertilizer produced by the Haber–Bosch process ($>50 \times 10^6$ metric tons were produced in 1980) seemed capable of supplying these increasing demands, during the first part of this century. For both more developed and less developed countries, however, capital and energy costs by the Haber–Bosch process have become significant. Anhydrous ammonia, the form in which most nitrogen fertilizers are applied to soil, currently costs about $120/ton to produce in the United States. But as natural gas is deregulated in the marketplace to its petroleum-energy-equivalent prices, the energy cost of fertilizer is projected to soar to $400/ton by 1990. Transportation, when available, and application also add to the cost.

Moreover, in more developed countries, fertilizer use is inefficient. It is estimated that only 50% of the applied nitrogen fertilizer is utilized by plants, with most of the remainder lost by either denitrification or leaching (32). Nitrate concentration has increased in water reservoirs in the vicinity of heavily fertilized fields, with potential pollution hazards.

The increasing demand for fixed nitrogen in modern agriculture could be solved by the enhancement and extension of nitrogen fixation. Agriculturally important legumes are estimated to account for about one-half (80 × 10^6 metric tons/year) of all nitrogen fixed by biological systems. Although legumes have had a major role in food production throughout history, the total world area currently cultivated with these plants is approximately 15% of the area used for cereal and forage grasses, the main source of food in the modern world (32).

The production of meat, alcohol, and sugar depends in part on the availability of cereal and forage grasses. In order to obtain high crop yield, especially when using highly productive cultivars, it is necessary to apply nitrogenous fertilizer in large amounts. For example, a crop of sweet corn (*Zea mays*) in Israel is usually fertilized with 240 kg of nitrogen per hectare in order to obtain a yield of 20–25 tons of fresh ears/ha; irrigated wheat (*Triticum*) fertilized with 120 kg of nitrogen per hectare yields 6–7 tons of grain/ha.

If biological nitrogen fixation could replace the application of fertilizer nitrogen for the production of high-yield crops, even in part, there would be an obvious and immediate benefit for both more and less developed countries. The purpose of this chapter is to review the current status and achievements of the field of nitrogen fixation and to assess the importance of the recent developments and new knowledge to the reaping of this benefit through the development of agriculture.

II. Mechanisms of Nitrogen Fixation

A. INTRODUCTION

Biological nitrogen fixation is a property of several prokaryotic organisms belonging to different taxonomic groups (Table I and Fig. 1). These bacteria carry in their genome the nitrogen-fixing *nif* genes coding for the synthesis of the enzyme nitrogenase, which is capable of reducing the nitrogen molecule to ammonia. This reaction is expressed by

$$N{\equiv}N \xrightarrow{6\,e^- + 6\,H^+} 2\,NH_3$$

This process has several prerequisites:

1. All organisms capable of fixing nitrogen must possess a source of appropriate reducing power and a system to transport electrons to

TABLE I

BEST-STUDIED BACTERIA INVOLVED IN NITROGEN FIXATION[a]

Organisms	Metabolism when fixing nitrogen, habitats, morphology	Biochemical, physiological, and morphological features	Economic importance and use in agriculture
Azotobacteraceae *Azotobacter vinelandii* *A. chroococcum* *A. beijerinckii* *A. paspali* *Azomonas* *Beijerinckia* *Derxia*	Heterotrophic, aerobic, or micro-aerophilic Soil, water, rhizosphere leaf surfaces *Azotobacter* form cysts, produce slime—large capsules, PHB granules Gram-negative, cell shape variable	Nitrogenase and electron transport to nitrogenase well characterized Very high respiration rates Membrane energization producing reductant for nitrogenase Complex of nitrogenase with FeS protein protect nitrogenase from oxygen destruction	Proposed benefit to crops not confirmed Hormonal effect on root and plant growth Specific association of *A. paspali* with *Paspalum* roots
Bacillaceae *Bacillus polymyxa* *Clostridium pasteurianum*	Heterotrophic, anaerobic Soil rhizosphere and decomposing plant material Gram-positive rods with endospores	Nitrogenase, hydrogenase, and electron transport for nitrogenase well characterized	Marginal benefit to agriculture Probably add combined nitrogen to soil when decomposing cellulose *B. polymyxa* reported in association with roots of some cultivars of wheat
Enterobacteriaceae *Klebsiella pneumoniae*	Heterotrophic, anaerobic, or microaerophilic Soil, water, digestive tract of animals Gram-negative short rods	Most used in genetics and regulation studies of nitrogen fixation Nitrogenase, FeMoCof active site of FeMo protein well characterized *nif* gene mapped	Associated with rhizosphere of grasses Benefit to agriculture unknown
Azospirillum brasilense *A. lipoferum*	Heterotrophic, microaerophilic Soil and rhizosphere Gram-negative, cell shape variable, PHB granules	In very close association with roots of grasses Symbiosis?	Potential use in increasing yield of grasses and savings of nitrogen fertilizers Inoculation benefits crops

8

Rhizobiaceae *Rhizobium* I. Fast growers *R. leguminosarum* *R. phaseoli* *R. trifolii* *R. meliloti* II. Slow growers *R. japonicum* *R. lupini*	Heterotrophic, microaerophilic Soil and roots (nodules) of legumes Gram-negative rods, PHB granules Cells differentiating into bacteroids inside plant cells	*Rhizobium*–legume symbiosis Leghemoglobin regulates oxygen supply to the bacteroid Plant lectins involved in bacterial plant specificity Hydrogenase recycling hydrogen evolution from nitrogenase reaction increases effectiveness Genetics and regualtion of nitrogenase studied Free nitrogen fixation demonstrated Physiology of the legume–*Rhizobium* symbiosis very much studied Increases in P_{CO_2} increases legume yield Carbon and energy sources to the nodule is the factor limiting nitrogen fixation	Hormonal effect on roots and plant growth Fix nitrogen in association with roots Very important in agriculture Legume crops are benefited by inoculation with infective and effective strains Efforts to improve inoculants made, collection of rhizobia available
Endophytes Actinomycetes	Heterotrophic, microaerophilic Soil? Roots of trees	Symbiosis with nonleguminous wood trees (*Alnus* and *Myrica*) Endophytes have been isolated in pure culture Active nitrogenase cell-free preparations	Potentially important in forestation, wood production

(Continued)

TABLE I (Continued)

Organisms	Metabolism when fixing nitrogen, habitats, morphology	Biochemical, physiological, and morphological features	Economic importance and use in agriculture
Phototrophic bacteria	Phototrophic, anaerobic	Nitrogenase and electron transport well characterized in *R. rubrum*	Economic importance unknown
Rhodospirillaceae	Aquatic habitats		Involved in red tides
I. *Rhodospirillum rubrum*	I. Former Athiorhodaceae	Inactive Fe protein of nitrogenase activated by a protein called activating factor	
Chromatiaceae	Organic substrates serve as hydrogen sources	Organisms used for studies of bacterial photosynthesis (no photolysis of water)	
II. *Chromatium*	II, III. Former Thiorhodaceae		
Chlorobiaceae	Sulfide serving as hydrogen source		
III. *Chlorobium*			
Blue-green algae, Cyanophyceae	Phototrophic, aerobic, or microaerophilic	Vegetative cells donate photosynthate to heterocyst, specializing in fixing nitrogen	Very important in enhancing crop of rice in paddy soils
Filamentous, heterocystous	Contain chlorophyll, as in higher plants	Symbiosis with algae (lichens) and water ferns (*Azolla*)	Especially the *Azolla–Anabaena azollae* symbiosis
Anabaena, Nostoc	Aquatic and terrestial habitats		Successfully used as green manure in rice fields
Filamentous, nonheterocystous			
Oscillatoria			
Unicellular			
Gloeocapsa			

a Data taken from references 1, 14, 32, 40, 41, 46, 52, 79, 82.

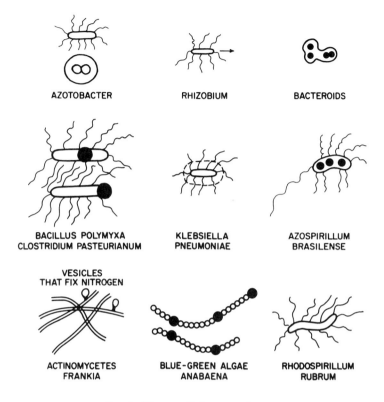

AZOTOBACTER   RHIZOBIUM   BACTEROIDS

BACILLUS POLYMYXA   KLEBSIELLA   AZOSPIRILLUM
CLOSTRIDIUM PASTEURIANUM   PNEUMONIAE   BRASILENSE

VESICLES
THAT FIX NITROGEN

ACTINOMYCETES   BLUE-GREEN ALGAE   RHODOSPIRILLUM
FRANKIA   ANABAENA   RUBRUM

FIG. 1. Nitrogen-fixing organisms.

nitrogenase. The redox potential needed for reducing N≡N to NH_3 has to be below -0.4 V.

2. Nitrogenase must be protected from oxygen damage, as the purified enzyme is easily destroyed by oxygen.
3. A control mechanism for enzyme-synthesis activity has to be present because fixation of nitrogen is a costly energy process for the cell. Up to the equivalent of 25 molecules of ATP are used for the reduction of one molecule of nitrogen. When ammonia levels increase in the cytoplasm the synthesis of nitrogenase is repressed, but when the bacterial cell is starved for combined nitrogen the *nif* gene is derepressed, permitting expression of nitrogenase.

Many reviews, proceedings of meetings, and books on biological nitrogen fixation have been written since 1970 (8–10, 20, 24, 29, 31, 32, 34, 37, 48–51, 54, 62, 64, 69, 72, 75). For the history of nitrogen fixation, the reader is referred to Burris (12). Apart from obvious economic reasons, interest in this subject is increasing, probably because biological nitrogen fixation can be considered as a general topic in biological sciences. It can be studied by many research disciplines at many levels:

1. Chemical understanding of how a molecule (i.e., the enzyme nitrogenase or its active site) is capable of reducing the very stable nitrogen molecule to form ammonia at normal atmospheric pressures and temperatures, as compared with the industrial Haber–Bosch process, in which high temperatures (~500°C), high pressures (~500 atm), hydrogen, and a metal catalyst are needed
2. Biochemical and enzymologic studies, as the enzyme nitrogenase is both highly complex and peculiar, the reaction taking place only in the presence of two different proteins—one smaller protein containing iron and the other, larger, protein containing both iron and molybdenum
3. Physiology of nitrogen-fixing bacteria, such as the electron transport chain to nitrogenase, as well as carbon and energy metabolism, ammonia incorporation, and the mechanism for the protection of nitrogenase from oxygen damage
4. Interactions between nitrogen-fixing microorganisms and plants, such as the *Rhizobium*–legume symbiosis
5. Physiology of the plant when associated with nitrogen-fixing bacteria, particularly in relation to the cost in photosynthetic energy of fixing nitrogen
6. Application of free or plant-associated biological nitrogen fixation in order to produce more food and save nitrogen fertilizers in modern agriculture
7. Genetics and regulation of nitrogen fixation and investigation of the *nif* gene coding for nitrogenase with the aim of eventually transferring this property to other prokaryotic or eukaryotic organisms, thereby minimizing the use of costly chemical fertilizer

The discovery in the 1960s that nitrogen fixation can be indirectly measured by the acetylene reduction assay (reduction of C_2H_2 to C_2H_4 by nitrogenase followed by gas chromatographic separation of C_2H_4 and its assay by the sensitive flame ionization technique) greatly enhanced the study of systems that fix nitrogen (11, 30, 33).

B. Biochemistry

The enzyme nitrogenase is composed of two proteins—one containing molybdenum and iron atoms, called the Mo–Fe protein, or component

I, and the other containing iron, called the Fe protein, component II, or, occasionally, nitrogenase reductase (9, 22, 23, 87). Both are soluble proteins. Nitrogenase complexes isolated from a wide variety of diazotrophs are remarkably similar, and the two components interact with each other to different extents. Mo–Fe protein from organism A will interact with Fe protein from organism B, and vice versa (Table II) (9, 22, 23, 87). Antiserum prepared against the purified protein of one nitrogen-fixing organism can react specifically with the same protein isolated from another (9).

The Mo–Fe protein has a molecular weight of 200,000–250,000 and contains 2 molybdenum atoms, 28–34 nonheme iron atoms, and 26–28 acid-labile sulfides. It is probably composed of two copies each of two subunits ($\alpha_2\beta_2$), each having a molecular weight of ~60,000 (Table III) (22, 23).

A cofactor of low molecular weight called Mo–Fe cofactor (FeMoCof) has been purified from the Mo–Fe protein (68). It contains all molybdenum and one-half the iron. FeMoCof restores nitrogenase activity in cell-free preparations of mutants lacking catalytically active Mo–Fe protein. Purified FeMoCof can reduce C_2H_2 to C_2H_4 using borohydride as reductant. FeMoCof is probably the active catalytic site of nitrogenase (68). The understanding of FeMoCof nitrogen-fixing activity may yield important clues to the design of new catalysts for abiological nitrogen fixation.

The Fe protein has a molecular weight of 55,000–65,000 and is composed of two copies of a single subunit. It contains four nonheme iron atoms and four acid-labile sulfides (Table IV) (9, 22, 23). The amino acid sequences of the Fe proteins of *Anabaena, Clostridium, Klebsiella,* and *Azotobacter* have been determined (Fig. 2) (44, 76). Interestingly, the amino acid sequences around five cysteines, located in the NH_2-terminal two-thirds of the protein, are highly conserved in all four species (Fig. 2).

The Fe protein is isolated in its active form from most diazotrophs. However, the Fe proteins from *Rhodospirillum rubrum* and *Azospirillum brasilense* were inactive as isolated (42). The enzyme from both organisms can be activated by a membrane-bound protein called activating factor (AF) in the presence of a divalent metal (Mg^{2+}, Mn^{2+}) and ATP(42). ATP is not hydrolyzed during activation. Activation involves the removal of an adenine-like molecule from the Fe-protein precursor. It has long been known that NH_4^+ will rapidly turn off nitrogenase activity *in vivo* in *R. rubrum,* whereas in other diazotrophs NH_4^+ inhibits the synthesis of new nitrogenase but without an immediate effect on existing nitrogenase activity. The AF of *R. rubrum* is probably involved in the rapid turning on and off of its nitrogenase activity.

TABLE II

NITROGENASE ACTIVITY OBTAINED BY CROSSING Fe PROTEIN AND Mo–Fe PROTEIN PURIFIED FROM DIFFERENT NITROGEN-FIXING BACTERIA[a,b]

Source of Fe protein	Azotobacter vinelandii	Klebsiella pneumoniae	Rhodospirillum rubrum	Azospirillum brasilense	Chromatium vinosum	Rhizobium japonicum	Bacillus polymyxa	Clostridium pasteurianum
Av	=	++++	+++	+	++	++++	+++	(−)
Kp	++++	=	+	++	+	++++	+++	+
Rr	++++	++++	=	++	++++	++	++	(−)
Ab	+++	++++	++++	=	++++	+	+	(−)
Cv	++++	++	++	+	=	ND	+	(−)
Rj	++++	++++	++++	+	++	=	+++	(−)
Bp	+	+++	+++	+	+	++++	=	++
Cp	(−)	+	(−)	(−)	(−)	(−)	++	=

[a] From Emerich et al. (23).

[b] Data expressed as percent activity of that observed in homologous crosses.

[c] Symbols: =, homologous crosses 100% activity; ++++, 75–100% activity; +++, 50–74% activity; ++, 25–49% activity; +, 1–24% activity; (−), 0% activity.

14

TABLE III

PROPERTIES OF THE Mo–Fe PROTEINS[a]

Property	Organism							
	Azotobacter chroococcum	*Azotobacter vinelandii*	*Klebsiella pneumoniae*	*Rhizobium japonicum*	*Rhizobium lupini*	*Rhodospirillum rubrum*	*Bacillus polymyxa*	*Clostridium pasteurianum*
Molecular weight	227,000	216,000	218,000	200,000	194,000	234,000	215,000	210,000
Subunit composition	1 type: 60,000	1 type: 56,000	2 types: 59,600 51,300	1 type: 50,000	1 type: 57,000	1 type: 58,000	1 type: 60,000	2 types: 60,000 51,000
Metal and sulfide content (mol/mol protein)								
Mo	1.90	1.54	2.01	1.3	1.0	0.9	1.80	1.95
Fe	23.00	24.00	32.50	29.0	29.0	20.0	29.40	22.24
S_2	20.00	20.00	18.00	26.0	26.0	10.1	18.80	22.24
Electron paramagnetic resonance spectra	4.29	4.30	4.30	—	—	—	4.30	4.29
	3.65	3.65	3.73	—	—	—	3.65	3.77
	2.01	2.01	2.02	—	—	—	2.01	2.01

[a] From Emerich and Evans (22).

Sequence alignment (nitrogenase iron protein sequences). Rows labelled *An*, *Cp*, *Kp*, *Av*; position numbers above.

Block 1 (positions 10, 20, 30)

An: H₂N-Met-Thr-Asp-Glu-Asn-Ile-Arg-Gln-Ile-Ala-Phe-Tyr-Gly-Lys-Gly-Gly-Ile-Gly-Lys-Ser-Thr-Thr-Ser-Gln-Asn-
Cp: H₂N-Met-Arg-Gln-Val-Ala-Ile-Tyr-Gly-Lys-Gly-Gly-Ile-Gly-Lys-Ser-Thr-Thr-Gln-Asn-
Kp: H₂N-Thr-Met-Arg-Gln-Cys-Ala-Ile-Tyr-Gly-Lys-Gly-Gly-Ile-Gly-Lys-Ser-Thr-Thr-Gln-Asn-
Av: H₃N-Ala-Met-Arg-Gln-Cys-Ala-Met-Ile-Tyr-Gly-Lys-Gly-Gly-Ile-Gly-Lys-Ser-Thr-Thr-Gln-Asn-

Block 2 (positions 30, 40, 50)

An: Thr-Leu-Ala-Ala-Met-Ala-Glu-Met-Gly-Gln-Arg-Ile-Met-Ile-Val-Gly-Cys-Asp-Pro-Lys-Ala-Asp-Ser-Thr-Arg-
Cp: Leu-Thr-Ser-Gly-Leu-His-Ala-Met-Gly-Lys-Thr-Ile-Met-Val-Val-Gly-Cys-Asp-Pro-Lys-Ala-Asp-Ser-Thr-Arg-
Kp: Leu-Val-Ala-Ala-Leu-Ala-Glu-Met-Gly-Lys-Lys-Val-Met-Ile-Val-Gly-Cys-Asp-Pro-Lys-Ala-Asp-Ser-Thr-Arg-
Av: Leu-Val-Val-Met-Ile-Val-Gly-Cys-Asp-Pro-Lys-

Block 3 (positions 60, 70)

An: Leu-Met-Leu-His-Ser-Lys-Ala-Gln-Thr-Thr-Val-Leu-His-Leu-Ala-Ala-Glu-Arg-Gly-Ala-Val-Glu-Asp-Leu-Glu-
Cp: Leu-Leu-Leu-Gly-Gly-Leu-Ala-Gln-Lys-Ser-Val-Leu-Asp-Thr-Leu-Arg-Glu-Glu-Gly-Glu----Glu-Asp-Val-Glu-
Kp: Leu-Ile-Leu-His-Ala-Lys-Ala-Gln-Asn-Thr-Ile-Met-Glu-Met-Ala-Ala-Glu-Val-Gly-Ser-Val-Glu-Asp-Leu-Glu-

Block 4 (positions 80, 90, 100)

An: Leu-His-Glu-Val-Met-Leu-Thr-Gly-Phe-Arg-Gly-Val-Lys-Cys-Val-Glu-Ser-Gly-Gly-Pro-Glu-Pro-Gly-Val-Gly-
Cp: Leu-Asp-Ser-Ile-Leu-Lys-Glu-Gly-Tyr-Gly-Gly-Ile-Arg-Cys-Val-Glu-Ser-Gly-Gly-Pro-Glu-Pro-Gly-Val-Gly-
Kp: Leu-Glu-Asp-Val-Leu-Gln-Ile-Gly-Tyr-Gly-Asp-Val-Arg-Cys-Ala-Glu-Ser-Gly-Gly-Pro-Glu-Pro-Gly-Val-Gly-
Av: Cys-Val-Glu-Ser-Gly-Gly-Pro-Glu-Pro-Gly-Val-Gly-

Block 5 (positions 110, 120)

An: Cys-Ala-Gly-Arg-Gly-Ile-Ile-Thr-Ala-Ile-Asn-Phe-Leu-Glu-Glu-Asn-Gly-Ala-Tyr-Gln-Asp----Leu-Asp-Phe-
Cp: Cys-Ala-Gly-Arg-Gly-Ile-Ile-Thr-Ser-Ile-Asn-Met-Leu-Glu-Gln-Leu-Gly-Ala-Tyr-Thr-Asp-Asp-Leu-Asp-Tyr-
Kp: Cys-Ala-Gly-Arg-Gly-Val-Ile-Thr-Ala-Ile-Asn-Phe-Leu-Glu-Glu-Glu-Gly-Ala-Tyr-Glu-Asp-Asp-Leu-Asp-Phe-
Av: Cys-Ala-Gly-Arg-Gly-

Block 6 (positions 130, 140)

An: Val-Ser-Tyr-Asp-Val-Leu-Gly-Asp-Val-Val-Cys-Gly-Gly-Phe-Ala-Met-Pro-Ile-Arg-Glu-Glu-Lys-Ala-Gln-Glu-
Cp: Val-Phe-Tyr-Asp-Val-Leu-Gly-Asp-Val-Val-Cys-Gly-Gly-Phe-Ala-Met-Pro-Ile-Arg-Glu-Gly-Lys-Ala-Gln-Glu-
Kp: Val-Phe-Tyr-Asp-Val-Leu-Gly-Asp-Val-Val-Cys-Gly-Gly-Phe-Ala-Met-Pro-Ile-Arg-Glu-Asn-Lys-Ala-Gln-Glu-
Av: Asp-Val-Val-Cys-Gly-Gly-Phe-Ala-Met-Pro-Ile-Arg-

Fig. 2. Amino acid sequence of the Fe protein of *Anabaena* (*An*), *Clostridium pasteurianum* (*Cp*), *Azotobacter vinelandii* (*Av*), and *Klebsiella pneumoniae* (*Kp*). Conserved residues are enclosed in boxes. Conserved cysteines are encircled. After Mevarech *et al.* (44) and Sundaresan and Ausubel (76).

180

An: Ile - Tyr - Ile - Val - Thr - Ser - Gly - Glu - Met - Met - Ala - Met - Tyr - Ala - Ala - Asn - Asn - Ile - Ala - Arg - Gly - Ile - Leu - Lys - Tyr -
Cp: Ile - Tyr - Ile - Val - Ala - Ser - Gly - Glu - Met - Met - Ala - Met - Ala - Leu - Tyr - Ala - Ala - Asn - Asn - Ile - Ser - Lys - Gly - Ile - Gln - Lys - Tyr -
Kp: Ile - Tyr - Ile - Val - Cys - Ser - Gly - Glu - Met - Met - Ala - Met - Ala - Met - Tyr - Ala - Ala - Asn - Asn - Ile - Ser - Lys - Gly - Ile - Val - Lys - Tyr -
Av: Ile - Tyr - Ile - Val - Cys - Ser - Gly - Glu -

190

An: Ala - His - Ser - Gly - Val - Arg - Leu - Gly - Gly - Leu - Ile - Cys - Asn - Ser - Arg - Lys - Val - Asp - Arg - Glu - Asp - Glu - Leu - Ile -
Cp: Ala - Lys - Ser - Gly - Val - Arg - Leu - Gly - Gly - Ile - Ile - Cys - Asn - Ser - Arg - Lys - Val - Ala - Asn - Glu - Tyr - Glu - Leu - Leu -
Kp: Ala - Lys - Ser - Gly - Lys - Val - Arg - Leu - Gly - Gly - Leu - Ile - Cys - Asn - Ser - Arg - Gln - Thr - Asp - Arg - Glu - Asp - Glu - Leu - Ile -
Av: Leu - Gly - Gly - Leu - Ile - Cys - Asn - Ser - Arg -

200 / 210

An: Met - Asn - Leu - Ala - Glu - Arg - Leu - Asn - Thr - Gln - Met - Ile - His - Phe - Val - Pro - Arg - Asp - Asn - Ile - Val - Gln - His - Ala - Glu -
Cp: Asp - Ala - Lys - Phe - Ala - Lys - Glu - Leu - Gly - Ser - Gln - Leu - Ile - His - Phe - Val - Pro - Arg - Ser - Pro - Met - Val - Thr - Lys - Ala - Glu -
Kp: Ile - Ala - Leu - Arg - Glu - Lys - Leu - Gly - Thr - Gln - Met - Ile - His - Phe - Val - Pro - Arg - Asp - Asn - Ile - Val - Gln - Arg - Ala - Glu -

230 / 240

An: Leu - Arg - Arg - Met - Thr - Val - Ala - Glu - Tyr - Ala - Pro - Asp - Ser - Asn - Gln - Gly - Gln - Glu - Tyr - Arg - Ala - Leu - Ala - Lys - Lys -
Cp: Ile - Asn - Lys - Gln - Thr - Val - Ile - Glu - Tyr - Asp - Pro - Thr - Cys - Glu - Gln - Ala - Glu - Tyr - Arg - Glu - Leu - Ala - Arg - Lys -
Kp: Ile - Arg - Arg - Met - Thr - Val - Ile - Glu - Tyr - Asp - Pro - Ala - Cys - Lys - Gln - Ala - Asn - Glu - Tyr - Arg - Thr - Leu - Ala - Gln - Lys -
Av: Tyr - Asp - Pro - Lys - Ala - Lys - Gln - Ala - Asp - Glu -

250 / 260 / 270

An: Ile - Asn ----- Asn - Asp - Lys - Leu - Thr - Ile - Pro - Thr - Pro - Met - Glu - Met - Asp - Glu - Glu - Ala - Leu - Lys - Ile - Glu - Tyr -
Cp: Val - Asp - Ala - Asn - Glu - Leu - Phe - Val - Ile - Pro - Lys - Pro - Met - Thr - Gln - Glu - Arg - Leu - Glu - Glu - Ile - Leu - Met - Gln - Tyr -
Kp: Ile - Val - Asn - Asn - Thr - Met - Lys - Val - Val - Pro - Thr - Pro - Cys - Thr - Met - Asp - Glu - Leu - Glu - Ser - Leu - Met - Glu - Met - Phe -
Av: Glu - Glu - Leu - Leu - Glu - Met - Glu - Phe -

280 / 290

An: Gly - Leu - Leu - Asp - Asp - Thr - Lys - His - Ser - Glu - Ile - Ile - Gly - Lys - Pro - Ala - Glu - Ala - Thr - Asn - Arg - Ser - Cys - Arg - Asn - COOH
Cp: Gly - Leu - Met - Asp - Leu - COOH
Kp: Gly - Ile - Met - Glu - Glu - Asp - Thr - Ser - Ile - Gly - Lys - Thr - Ala - Ala - Glu - Glu - Asn - Ala - Ala - COOH
Av: Gly - Ile - Met - Glu - Glu - Val - Glu - Asp - Glu - Ser - Ile - Val - Lys - Thr - Ala - Glu - Glu - Val - COOH

TABLE IV

PROPERTIES OF THE Fe PROTEINS[a]

Property	Organism						
	Azotobacter chroococcum	Azotobacter vinelandii	Klebsiella pneumoniae	Rhizobium lupini	Rhodospirillum rubrum	Bacillus polymyxa	Clostridium pasteurianum
Molecular weight	64,000	64,000	66,800	65,000	60,000	55,000	56,000
Subunit composition	1 type: 30,800	1 type: 33,000	1 type: 34,600	1 type: 32,000	1 type: 30,000	1 type: 31,000	1 type: 27,500
Metal and sulfide composition (mol/mol protein)							
Fe	4.00	3.45	4.00	3.20	3.50	2.80	4.50
S_2	3.90	2.95	3.95	—	2.30	3.20	4.00
Electron paramagnetic resonance spectra	2.05	2.05	2.05	—	—	2.04	2.04
	1.94	1.94	1.94	—	—	1.94	1.94
	1.87	1.88	1.86	—	—	1.88	1.88

[a] From Emerich and Evans (22).

A combination of the Fe protein and the Mo–Fe protein in the presence of Mg^{2+}-ATP and an appropriate electron donor such as reduced ferredoxin or flavodoxin *in vivo* or $Na_2S_2O_4$ *in vitro* catalyzes the reduction of substrates including $N_2 \rightarrow NH_3$, $2\,H^+ \rightarrow H_2$, as well as the hydrolysis of ATP. In addition to $C_2H_2 \rightarrow C_2H_4$, other nonbiological substrates including cyanide, azide, isocyanate, and N_2O are also reduced.

Highly purified Mo–Fe and Fe proteins have been obtained from most diazotrophic types by using various techniques for protein purification under strictly anaerobic conditions (22, 23). The half-lives of the Mo–Fe and Fe proteins in air are 10 min and less than 1 min, respectively.

C. MECHANISMS OF NITROGENASE ACTIVITY

Ferredoxin and flavodoxin are probably the natural electron carriers for the reduction of the Fe protein *in vivo* in most diazotrophs. Their physiological significance still remains in doubt because in some cases only low nitrogenase activities have been obtained in reconstituted electron-transport chains. In *in vitro* reactions, $Na_2S_2O_4$ reduces the Fe protein.

Electron paramagnetic resonance studies and kinetic studies have helped elucidate the mechanism of nitrogenase activity (Fig. 3) (22, 23).

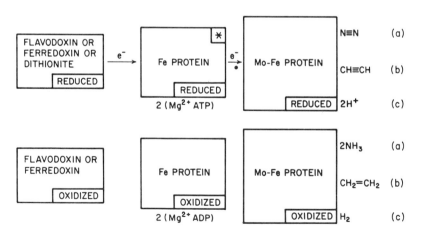

FIG. 3. Reaction scheme for nitrogenase function (see text for explanation). After Emerich and Evans (22) and Emerich *et al.* (23). (a) Six bindings and dissociations, six one-step electrons transferred. (b) or (c) Two bindings and dissociations, two one-step electrons transferred. Asterisk indicates that two or one Fe proteins may bind to one Mo–Fe protein.

The data so far obtained indicate that reduced Fe protein binds to MgATP, creating a complex with the Mo–Fe protein. The complex may be composed of one or two Fe proteins bound to one Mo–Fe protein. Dissociation of the two proteins occurs between electron-transfer events (23). Evidence that dissociation occurs between electron transfers is implied by the observation that the tight binding complex formed between *Azotobacter vinelandii*, Mo–Fe proteins, and *Clostridium pasteurianum* Fe protein dissociates very slowly and cannot support substrate reduction (23). The oxidized Fe protein (now combined with MgADP) dissociates and becomes reduced again (now combined with MgATP). It recombines randomly with another nitrogenase until all the electrons needed for the reduction of the substrate (e.g., six for nitrogen) are accumulated. Apart from H^+, substrates such as $N\equiv N$ or $HC\equiv CH$ are believed to be bound to the same site in the Mo–Fe protein.

D. HYDROGENASE AND THE RECYCLING OF NITROGENASE-DEPENDENT HYDROGEN EVOLUTION

Many nitrogen-fixing bacteria evolve hydrogen when they fix nitrogen, but not when they are grown on NH_4^+. When electrons and ATP are available to nitrogenase but nitrogen is not, the electrons combine with protons to yield hydrogen. However, even when nitrogen is present, some of the electrons are evolved as hydrogen.

It has been proposed that the hydrogen evolved serves to protect the nitrogenase from inhibition by oxygen. However, anaerobes also evolve hydrogen. Hydrogen formation appears to be a consequence of the nature of the nitrogenase active site, and no nitrogenase has evolved with an active site capable of eliminating this wasteful side reaction.

In vivo, electrons and ATP are wasted during nitrogenase activity. About 50% of the electrons are lost as hydrogen under normal bacterial growth conditions and 30% in legumes (22, 25, 60, 65, 66).

All studied nitrogen-fixing bacteria possess hydrogenase. Evidence has accumulated indicating that hydrogen evolved from the nitrogenase reaction may be recycled *in vivo* through a process that captures the energy of gaseous hydrogen (65). Both reversible and unidirectional hydrogenases exist in nitrogen-fixing organisms. The hydrogenase from soybean (*Glycine max*) root-nodule bacteroids has a molecular weight near 60,000 and appears to have properties similar to other hydrogenases, that is, an iron–sulfur center as a catalytic site, as evidenced by its low-temperature electron paramagnetic resonance spectrum.[2]

[2]For a discussion of iron–sulfur centers and their characterization by electron paramagnetic reasonance, see Volume VII, Chapter 2. (Eds.)

There are several explanations for the role of hydrogenase activity *in vivo* (25, 65):

1. It may remove inhibitory hydrogen from nitrogenase active site, but hydrogen would not accumulate sufficiently to inhibit nitrogen reduction in bacteria.
2. It may enhance respiratory protection for nitrogenase by scavenging oxygen. It has been observed that nitrogenase activity is protected by hydrogenase-mediated hydrogen uptake in *Rhizobium japonicum* bacteroids, that exogenous hydrogen decreases the oxygen sensitivity of *Azotobacter chroococcum* nitrogenase, and that exogenous hydrogen also augments nitrogenase activity in *Anabaena*.
3. Hydrogenase can provide ATP and/or reducing power for nitrogenase activity. Although hydrogen-dependent ATP generation has been observed in several diazotrophs, it is considered a minor function of recycled hydrogen.

E. GENETICS OF NITROGEN FIXATION

As in all biological systems, the ability to produce nitrogenase depends on both the occurrence of the structural genes for the enzyme and the expression of the information of the structural genes through enzyme synthesis. The genes for nitrogenase are designated *nif*. In addition to the nitrogenase enzymes, supplementary genetic information for the following syntheses is required to enable a diazotroph to fix nitrogen: (a) proteins to take up and incorporate iron, molybdenum, and sulfur into the Mo–Fe and Fe proteins; (b) ferredoxins or flavodoxins; (c) enzymes to remove product ammonia as amino acids; (d) molecules to regulate expression of nitrogenase, depending on the availability of and need for fixed nitrogen; (e) molecules to protect nitrogenase from oxygen; and (f) several other molecules required for nitrogen fixation but the function of which has not yet been identified. Obviously, the synthesis of new nitrogen-fixing crops will not be simple.

Studies of the genetics of nitrogen fixation were initiated in 1971, when several genes crucial to nitrogen fixation were identified on a small segment of the chromosome of the diazotroph *Klebsiella pneumoniae* (69). With the aid of a bacteriophage, the *nif* genes were transferred from a normal *Klebsiella* to a nonfixing mutant *Klebsiella*. Nitrogen fixation was thereby restored in the mutant. Subsequently, conjugation was used to transfer the *nif* locus from *Klebsiella* to the closely related but non-nitrogen-fixing *Escherichia coli* to produce the first example of a synthetic diazotroph (8, 9, 62). More recently, the entire *Klebsiella nif* locus has

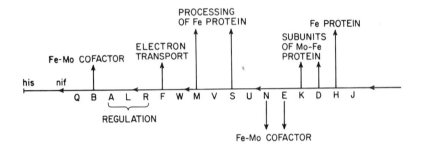

FIG. 4. Order, transcriptional direction, and function of *Klebsiella pneumoniae nif* genes. From Brill (9).

been mapped using Mu-induced point and deletion mutations. Seventeen genes have been identified by complementation analysis (9). A physical map of the *nif* region has been constructed by restriction analysis (9), and the genetic and physical maps have been correlated through the use of transposons (9).

Figure 4 is a diagrammatic representation of the *nif* genes in *Klebsiella*, as well as their transcriptional relationships (9). All *nif* mutations cluster near the *his* operon in the chromosome of *K. pneumoniae*. The genes are organized into seven distinct operons, with transcription of all the operons in the same direction, that is, toward *his* genes. The *nif* region is ~24 kilobases long. In *Anabaena*, the *nif* genes are rearranged relative to those in *Klebsiella* (35). Identification and cloning of the *nif* DNA of unrelated bacteria have been possible because of the high homology retained among *nif* genes of all tested species (9, 35, 44, 76).

A better understanding of the molecular genetics of nitrogen fixation may make it possible to transfer the ability to fix nitrogen to organisms such as higher plants that do not now have such ability, thereby making them self-sufficient for their nitrogen needs. This approach may provide the ideal technology for nitrogen fixation but is perhaps the most speculative.

Plasmids have been used to transfer the ability to fix nitrogen to the closely related species *Escherichia coli*, *Salmonella typhimurium*, and *Klebsiella aerogenes* (= *K. pneumoniae*) and to mutants of the unrelated *Azotobacter vinelandii* that had lost the ability to synthesize either the Mo–Fe or Fe protein. Another bacterium, *Agrobacterium tumefaciens*, a plant pathogen responsible for crown galls, accepts and maintains the *nif*-containing plasmids, but does not fix nitrogen. The results are similar for *Erwinia herbicola*, another plant pathogen. In another case, the *nif* plasmids would not enter the bacterium *Pseudomonas aeruginosa*. The reasons for failure are unknown.

Even though an organism contains *nif,* it will not synthesize nitrogenase when an adequate amount of fixed nitrogen such as ammonia is available to meet the needs of the organism. A regulatory system prevents synthesis, thereby saving the high energy cost of fixing nitrogen when fixed nitrogen is available. The details of this regulatory system are complex and have not yet been unraveled, but evidence suggests that glutamine synthetase, which synthesizes glutamine from glutamate and ammonia, regulates nitrogenase synthesis.

Genetic manipulation and chemical treatments have produced diazotrophs that continue to fix nitrogen in the presence of fixed nitrogen. Such organisms, when further genetically manipulated to prevent the incorporation of ammonia into amino acids, synthesize and excrete ammonia. It has been suggested that such organisms may be useful as microbial nitrogen-fixing factories, but their large carbohydrate requirement for nitrogen fixation probably precludes their application. However, this limitation may not apply to photosynthetic diazotrophs such as blue-green algae.

III. Metabolic Aspects

A. Ammonia Metabolism[3]

Biological nitrogen fixation can be considered a situation in which the bacterial cell is starved for nitrogen. Because the process of nitrogen fixation is very costly in energy, the ammonia produced has to be assimilated into nitrogenous compounds of the cell (i.e., amino acids and nucleic acids) efficiently (8, 67).

Studies with *Escherichia coli* and *Klebsiella pneumoniae* growing on nitrogen-deficient medium showed that ammonia is assimilated by way of two enzymes: glutamine synthetase (GS), EC 6.3.1.2 [see Eq. (1)], and glutamate synthetase (GOGAT), EC 2.7.1.5.3 [see Eq. (2)].[4]

$$\text{L-Glutamate} + \text{NH}_4^+ + \text{ATP} \rightarrow \text{L-glutamine} + \text{ADP} + \text{P}_i \quad (1)$$

$$\text{L-Glutamine} + \text{2-oxoglutarate} + \text{NAD(P)H} \rightarrow 2 \text{ glutamate} + \text{NAD(P)} \quad (2)$$

---

[3]For further discussion of ammonia metabolism, see Chapter 2. Translocation and distribution of nitrogenous products of nitrogen fixation are described in Chapter 4. (Eds.)

[4]These enzymes and their associations with other metabolic systems are described in detail in metabolic charts 1 and 8, Chapter 2. (Eds.)

In bacteria growing in a medium supplied with ammonia, glutamate dehydrogenase (GDH), EC 1.4.1.3 [see Eq. (3)], is the most active enzyme for ammonia assimilation.

$$\text{2-Oxoglutarate} + \text{NH}_4{}^+ + \text{NAD(P)H} \rightarrow \text{L-glutamate} + \text{NAD(P)}^+ \qquad (3)$$

The main advantage of the GS–GOGAT pathway resides in its much higher affinity for the substrate. In *Klebsiella pneumoniae* K_m $\text{NH}_3 \simeq 0.3$ mM in Eq. (1), K_m glutamate $\simeq 1$ mM in Eq. (2), and K_m $\text{NH}_3 \simeq 12.0$ mM in Eq. (3). The GS–GOGAT pathway has been demonstrated to take place in all nitrogen-fixing organisms studied, growing on nitrogen as sole nitrogen source. When the pool of ammonia increases in the cytoplasm, it is assimilated by GDH. The GS–GOGAT pathway also functions in higher plants (67).

Glutamine synthetase of *Klebsiella pneumoniae* is deadenylated in its active (biosynthetic) form. When ammonia concentration increases, the enzyme gains adenylyl groups (67).

Adenylated GS has been shown to regulate, in some unknown way, the repression of nitrogenase synthesis, whereas the deadenylated form is involved in derepression of nitrogenase synthesis (8, 67, 69). Nitrogen-fixing bacteria growing in the presence of ammonia with specific inhibitors of GS and GOGAT, such as methionine sulfoximine or methionine sulfone, continue to exhibit nitrogenase activity, and ammonia is excreted into the medium (9). Similarly, mutants of *Klebsiella pneumoniae* defective in GS are capable of fixing nitrogen in the presence of ammonia. In the case of *Rhizobium* living symbiotically as bacteroids in legume nodules, GS–GOGAT activity is very low. The ammonia produced is not utilized by the bacteriod but is excreted into the cytoplasm of the host cells, where it is then incorporated by their GS–GOGAT pathway. The main end product suggested is asparagine, which is translocated to the shoot of the legume through the vascular tissues (67). In free-living rhizobia grown under conditions of nitrogen fixation and under low partial pressure of oxygen ammonia is excreted into the medium, and, as in the bacteroids, it does not regulate nitrogenase synthesis. However, at higher P_{O_2} levels, ammonia is assimilated by the GS–GOGAT pathway, and increasing ammonia concentration does repress nitrogenase synthesis (67).

The ureides allantoin and allantoic acid have been known for many years to be major storage and translocation forms of nitrogen in such plants as cowpea, pea, and soybeans. These ureides have been found as the dominant nitrogen compound in the xylem sap of nodulated soybeans and cowpeas (43, 89).

The site of allantoin synthesis is the root nodule, where the enzymes that synthesize allantoin from guanine → hypoxanthine → xanthine are found (soybean, cowpea). Allantoin accumulation was suppressed in nonnodulating soybean genotypes or in soybeans grown on combined nitrogen. It has been suggested that the relative ureide content of xylem sap can be used as an indicator of nitrogen fixation in soybeans, but more work remains to be done in order to establish this hypothesis and its application to other legumes (43, 89).

B. CARBON METABOLISM IN FREE-LIVING BACTERIA

The carbon metabolism of nitrogen-fixing bacteria does not differ from that of non-nitrogen-fixing organisms belonging to the same taxonomic groups. However, they have special requirements for ATP and electrons to support nitrogenase activity as well as for the carbon skeleton for assimilation of the fixed ammonia (Table V) (39, 90).

In anaerobic diazotrophs, such as *Clostridium pasteurianum* and *C. butyricum*, the fermentative metabolism readily supplies ATP and electrons from substrate-level reactions in which ADP is phosphorylated in a tightly coupled manner by direct enzymatic transfer from phosphorylated sugars or organic acids, for example, the pyruvate phosphoroclastic cleavage reaction:

$$\text{Pyruvate} + \text{CoASH} + \text{Fd}_{ox} \rightarrow \text{CH}_3\text{-COSCoA} + \text{Fd}_{re} + \text{CO}_2$$

$$\text{CH}_3\text{-COSCoA} + \text{H}_2\text{PO}_4^- \xrightarrow{\text{phosphate transacetylase}} \text{CH}_3\text{COPO}_4\text{H}^- + \text{CoASH}$$

$$\text{CH}_3\text{COPO}_4\text{H}^- + \xrightarrow{\text{ADP ATP:acetate phosphotransferase}} \text{CH}_3\text{COO}^- + \text{ATP}$$

Because substrate-level phosphorylation is relatively inefficient, large quantities of reduced carbon must be oxidized to provide sufficient ATP, especially for nitrogen fixation. The electron surplus obtained can be disposed of either by the reduction of oxidized compounds, by transhydrogenation of pyridine nucleotides (NAD^+/NADH), or by release of gas such as hydrogen. This process is called fermentation (Volume VII, Chapter 4).

The electrons produced by the pyruvate phosphoroclastic cleavage reaction can be transferred directly to nitrogenase via ferredoxin, as has been demonstrated in nitrogenase cell-free preparations.

The butyric acid fermentation of glucose can then be summarized by the reaction

$$\text{C}_6\text{H}_{12}\text{O}_6 + 3\ \text{ADP}^{3-} + 3\ \text{P}_i + 3\ \text{H}^+ \rightarrow$$
$$\text{CH}_3\text{CH}_2\text{CH}_2\text{COO}^- + 2\ \text{CO}_2 + 2\ \text{H}_2 + 3\ \text{ATP}^{4-} + 3\ \text{H}_2\text{O}$$

Given the heavy demand for ATP in nitrogen fixation (15–25 ATP/mol per nitrogen molecule fixed), at least five molecules of glucose

TABLE V

CARBON METABOLISM AND NITROGEN FIXATION IN FREE-LIVING BACTERIA

Type of metabolism	Source of ATP for nitrogenase	Source of reductant for nitrogenase	Limiting factors	Excess of reductant[a]	Nitrogen-fixation efficiency in pure cultures[b]
Anaerobic fermentation	Substrate level	Pyruvate, substrate level, transhydrogenation	Rate of ATP production	Butyric acid, hydrogen gas	5–10
Anaerobic photosynthetic	Cyclic photophosphorylation, cytochrome involved	Weakly reduced inorganic or organic substrates	Rate of reductant production	CO_2 fixation	?
Aerobic heterotrophic	TCA cycle oxidative phosphorylation, cytochrome involved	TCA cycle, pyruvate	Need for protection of nitrogenase from oxygen damage	Membrane energization, oxygen reduction	10–20
Microaerophilic heterotrophic	TCA cycle oxidative phosphorylation, cytochrome involved	TCA cycle, pyruvate	Slower rates of ATP and reductant production	PHB	20–40
Aerobic photosynthetic	Oxidative photophosphorylation, cytochrome involved	H_2O photolysis	Nitrogen fixation only in heterocysts	Carbon dioxide fixation	?

[a] Energy is also consumed for cell maintenance and biosynthesis.
[b] Milligrams of nitrogen fixed per gram carbon substrate consumed.

must be fermented with ATP, not electron supply, limiting nitrogen fixation in anaerobic fermentative bacteria, possibly explaining the low efficiency (5–10 mg of nitrogen fixed per gram of C substrate consumed) of nitrogen fixation by anaerobic bacteria.

Anaerobic photosynthetic nitrogen-fixing bacteria possess a cytochrome system for photoelectron transport linked to ATP formation and reductant generation. In these organisms, electrons originating from weakly reducing electron donors, such as sulfur, sulfide, or thiosulfate, or organic substrates, such as alcohols or hydroxybutyrate, can be energized by absorption of light by a porphyrin (bacteriochlorophyll), leading to formation of reduced ferredoxin or reduced NADPH. In this process, oxygen is not evolved. ATP can be synthesized from ADP and P_i by cyclic photophosphorylation in which no additional electron input is made. In this system, in contrast to fermentative ATP formation, reductant rather than ATP is the factor limiting nitrogen fixation (39).

Aerobic heterotrophic nitrogen-fixing bacteria (Azotobacter) are characterized by a vigorous rate of oxygen consumption. Pyruvate is fully oxidized by the TCA cycle to carbon dioxide, ATP is formed by oxidative phosphorylation, and partially reduced keto acids are supplied. In aerobic nitrogen fixers, provided an oxidizable substrate is present in sufficient quantity, neither ATP nor reductant would be limiting for nitrogen fixation. However, large amounts of energy are spent in order to prevent nitrogenase destruction by oxygen. The efficiency of nitrogen fixation, 10–20 mg of nitrogen fixed per gram of C substrate consumed, is much lower than expected.

Repression of nitrogenase synthesis by oxygen has been observed in *Klebsiella pneumoniae*, *Rhizobium* 32H1, and *Azotobacter chroococcum*. It may be a general phenomenon to ensure that wasteful synthesis of nitrogenase is avoided. The mechanism of oxygen repression is unknown. Pulse-labeling techniques were used to show that oxygen stress *in vivo* not only destroyed existing nitrogenase, but slowed down the rate of synthesis of the enzyme as well.

Microaerophilic bacteria (free-living *Rhizobium*, *Azotobacter*, *Beijerinckia*, *Azospirillum*), which possess no protective mechanism against the oxygen damage of nitrogenase, produce ATP, reductant, and intermediates like true aerobes. Under conditions of low P_{O_2}, the excess of reduced NAD(P)H produced is probably disposed of by synthesis of PHB; a large fraction of the cell dry weight (up to 70%) is composed of PHB. When growing under vigorous aeration in the presence of ammonia, PHB content is reduced to less than 1.0% of the cell dry weight (5, 52). PHB may serve as a source of energy when the substrate in the medium is depleted (65, 90).

It has been found that *Azotobacter, Derxia, Rhizobium,* and *Azospirillum* are capable of autotrophic growth with hydrogen as the energy source in the absence of an organic carbon substrate. Hydrogen-dependent carbon dioxide fixation and high activities of hydrogen-uptake hydrogenase and ribulose-1,5-bisphosphate carboxylase were found in bacterial cells under autotrophic growth conditions (75).

Aerobic photosynthetic blue-green algae produce reductant, ATP, and intermediates by oxidative phosphorylation and photophosphorylation as in higher plants. Nitrogen fixation occurs in most photosynthetic aerobes in specialized differentiated cells, the heterocysts.

C. PROTECTION OF NITROGENASE FROM OXYGEN
   DAMAGE

Obligatory anaerobic nitrogen-fixing bacteria such as *Clostridium pasteurianum* and faculative anaerobes such as *Klebsiella pneumoniae* and *Bacillus polymyxa* fix nitrogen only under strict anaerobic conditions, so there is no problem in protecting their nitrogenase from oxygen damage. There has also been great interest in the means by which aerobic nitrogen-fixing bacteria are able to support vigorous nitrogen fixation under high oxygen partial pressures and at the same time maintain a very reduced environment for nitrogenase activity (61, 62, 65, 90).

The high respiration rates of cultures of *Azotobacter* are assumed to scavenge excess oxygen from the vicinity of nitrogenase. A conformational protection and compartmentation of nitrogenase has also been proposed, as cell-free preparations of *A. vinelandii* or *A. chroococcum* containing the membrane fraction are less oxygen sensitive. Nitrogen-fixing cells subjected to excessive aeration were observed to "switch off" activity and, after aeration was reduced, to "switch on" the activity. This occurred without *de novo* protein synthesis. Another proposed mechanism was that the ATP concentration or its ratio to another nucleotide was involved in the mechanism of "switching on and switching off" nitrogenase activity (61, 65).

In more recent work, Veeger *et al.* (84) proposed that membrane energization—the need for an intact chemiosmotic gradient in intact membranes, rather than ATP—is the determining factor in aerobic nitrogen fixation.

The proposed mechanism is summarized in Fig. 5 (84). Respiration produces a proton-driving force that lowers the pH at the same site where flavodoxin semiquinone (FvdH) binds to the membrane (Fig. 5, I). Under conditions of pH 4.5–5.0, the potential of $FvdH^-$ and $FvdH_2$

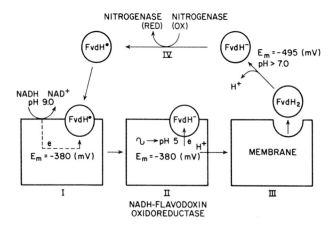

Fig. 5. Proposed scheme for membrane energization and for electron transport from NADH to nitrogenase in *Azotobacter*. From Veeger *et al.* (84).

is approximately -380 mV. This potential is low enough to permit reduction of flavodoxin by NAD(P)H situated in another site of the membrane at pH 9.0. This reduction is catalyzed by NADH-flavodoxin oxidoreductase (Fig. 5, I → II → III).

Dissociation of the hydroquinone ($FvdH_2$) obtained from the membrane (Fig. 5, III) and its subsequent deprotonation in the cytoplasm at pH > 7 decreases the potential (~460 mV). Reduced flavodoxin hydroquinone is then capable of reducing nitrogenase in the cytoplasm (Fig. 5, IV). NAD(P)H-flavodoxin oxidoreductase is an FAD-containing protein capable of reducing flavodoxin to both hydroquinone for electron donation to nitrogenase in the absence of energized membranes (84).

The oxygen stability of the *Azotobacter vinelandii* complex is achieved by formation of a tight, high molecular weight ($1-2 \times 10^6$) stoichiometric complex of the two components of nitrogenase with a protein called FeS II. FeS II is a pink 2 Fe–2S protein with a molecular weight of 24,000 (84). It seems to contain more than one peptide. In this complex, the formation of which is promoted by Mg^{2+} and by oxidized flavodoxin, the three proteins are in the oxidized state, resistant to oxygen, hence they switch off. Reduction of the complex, possibly by reduced flavodoxin, leads to nitrogenase activity, that is, they switch on. It lowers the molecular weight of the complex to 300,000, possibly by dissociation of the complex to the active free components of nitrogenase Fe and Mo–Fe proteins.

Another type of nitrogen-fixing organism is represented by the microaerophils, such as *Azospirillum, Azotobacter paspali,* and free-living *Rhi-*

zobium, which have been isolated from, and live in close association in, the surface and endorhizosphere of grasses. These aerobic organisms are capable of fixing nitrogen only at very low P_{O_2} levels (0.005–0.007 atm), probably because they lack mechanism(s) that protect the nitrogenase from oxygen damage (5, 52, 53, 55). In contrast to obligatory microaerophils (*Spirillum volutans* and *Campylobacter*), *Azospirillum* and *Azotobacter paspali* as well as free-living *Rhizobium* are capable of vigorous growth under atmospheric P_{O_2} in media supplied with combined nitrogen (55). In contrast, *Azospirillum* actively seeks microaerobic conditions even when growing on ammonia. Aerotaxis to low P_{O_2} (microaerophilic taxis) has been demonstrated in *Azospirillum*, but the physiological basis of this behavior is still unknown. It has been shown that some strains of *Azospirillum* produce red carotenoids (53). Those pigmented strains that also tend to form cell aggregates in liquid medium are capable of fixing nitrogen at slightly higher P_{O_2} levels, as compared with nonpigmented strains. Carotenoids have been proposed to have a role in protecting algal and bacterial cells from damage by free oxygen radicals and singlet oxygen (53). Another protective mechanism may be encountered in slimy nitrogen-fixing bacteria such as *Derxia gummosa*, which have large capsules that may restrict oxygen diffusion into the cell.

The more specialized means of nitrogenase protection from oxygen are encountered in heterocystous blue-green algae and in the *Rhizobium* legume symbiosis containing leghemoglobin.

D. CYANOPHYCEAE HETEROCYST

The blue-green algae (Cyanophyceae) are the only prokaryotic organisms that exhibit a higher form of photosynthesis. Some are capable of fixing nitrogen. The organisms are widely distributed in fresh water, marine, and terrestrial environments and occur in nitrogen-fixing association with a relatively diverse group of eukaryotic organisms (35, 58, 71, 73, 74, 88).

Nitrogen-fixing blue-green algae can be divided into three groups: filamentous forms with and without the ability to differentiate heterocysts and unicellular forms. Generally, all heterocystous forms can fix nitrogen, whereas only some non-heterocystous forms and one unicellular species (*Gloeocapsa*) have this ability. The most-studied nitrogen-fixing genera are the filamentous heterocyst-forming *Anabaena* and *Nostoc*.

The heterocyst (Fig. 6) differs morphologically from the vegetative cell. It is surrounded by a multilayered envelope exterior to the cell wall; this extra layer confers resistance to both lysozymal and mechanical

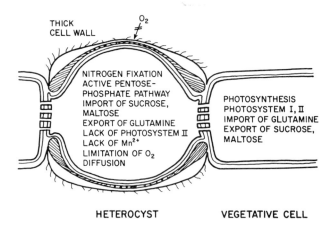

THICK
CELL WALL O_2

NITROGEN FIXATION
ACTIVE PENTOSE-
PHOSPHATE PATHWAY
IMPORT OF SUCROSE,
MALTOSE
EXPORT OF GLUTAMINE
LACK OF PHOTOSYSTEM II
LACK OF Mn^{2+}
LIMITATION OF O_2
DIFFUSION

PHOTOSYNTHESIS
PHOTOSYSTEM I, II
IMPORT OF GLUTAMINE
EXPORT OF SUCROSE,
MALTOSE

HETEROCYST   VEGETATIVE CELL

Fig. 6. Model for activities taking place in the heterocyst and vegetative cell of blue-green algae. From Haselkorn *et al.* (35).

stress. It also limits diffusion of gases into the cell. Physiologically, heterocysts differ from vegetative cells. They lack photosystem II activity, contain very little phycobiliprotein (the light-harvesting pigment for photosystem II in blue-green algae), and hold only 10% of the bound Mn^{2+} of vegetative cells, an important component of photosystem II (35, 74). All other components of the photosynthetic electron-transport chain are present in the heterocyst. In illuminated cultures, photosystem I and cyclic photophosphorylation activity in the heterocysts are the main source of ATP and reductant required for nitrogen fixation.

The carbon metabolism in heterocysts is also different from that of the vegetative cells. There is no carbon dioxide fixation and no ribulosebisphosphate carboxylase activity. On the other hand, glucose 6-phosphate and 6-phosphogluconate dehydrogenase are very active (35, 74). The form in which carbon is transported from vegetative cells has not been clearly demonstrated. Evidence has accumulated that disaccharides such as maltose enter heterocysts from vegetative cells, are metabolized by the pentose phosphate cycle, and may serve as an additional energy source for nitrogenase activity (35, 74).

Fixed nitrogen is incorporated in the heterocyst by the GS–GOGAT pathway. Glutamine is transported to vegetative cells. Ammonia or glutamine or its derivatives have a role in the regulation of both heterocyst differentiation and nitrogenase synthesis.

Nitrogenase and nitrogenase activity have been demonstrated to be restricted to the heterocyst. There, nitrogenase is protected from oxygen damage because there is no oxygen evolution (lack of photosystem

II), diffusion of oxygen is slowed by the multilayered envelope, and oxygen is rapidly scavenged by the high respiratory activity of the pentose phosphate pathway (35, 74).

IV. Nitrogen Fixation in Association with Plants

A. Free-Living Nitrogen Association

The amount of nitrogen fixed by free-living organisms in nature may be large (40, 41), but it has been so far very difficult to estimate because of the constantly changing environmental conditions at any given site under study. The number of free-living nitrogen-fixing organisms usually found in soil is low ($10^2–10^4$/g soil), so they must compete with other organisms ($10^7–10^9$/g soil) for carbon and energy sources, which are limiting.

Although the contribution of fixed nitrogen to crops by free-living organisms is considered marginal in modern agriculture, free-living nitrogen fixation is an important factor in supplying critical amounts of nitrogen for the maintenance of plant populations in natural grasslands. Also, when the C/N ratio of the soil is increased, there may be a selective enrichment of nitrogen-fixing populations, with a concomitant enhancement in their nitrogen-fixing activity.

The potential exploitation of biological nitrogen fixation in modern agriculture will therefore be confined to organisms living in association with the roots or other parts of plants where their populations can be selectively enriched and can compete successfully with other organisms for the supply of plant photosynthates. In this case, those plants that excrete considerable amounts of carbon compounds or that have a large turnover in root cells will be the most likely candidates to support nitrogen-fixing activities.

Associations that may be immediately exploited for the benefit of agriculture are the rhizobia–legume symbiosis, actinomycete–tree symbiosis, blue-green alga–water fern symbiosis, and the so-called associative symbiosis (or close associations) between roots of grasses and such bacteria as *Azospirillum*, *Azotobacter*, *Bacillus polymyxa*, and *Klebsiella pneumoniae*.

B. INTERACTIONS BETWEEN NITROGEN-FIXING BACTERIA AND THE SURFACE OF ROOTS

1. Rhizobium and Leguminous Roots

The high degree of specificity of *Rhizobium* species toward certain groups of leguminous species has been well established. Rhizobia live in soil as free organisms. The first step in nodule initiation would be a selective increase in the number of the specific bacteria near or on the root surface of the specific legume.

There has been increasing evidence to show that specific lectins are involved in the selection mechanism (3, 4, 17). The lectins are proteins that specifically bind to free glycoside radicals. The cell wall and capsule of rhizobia are composed of acidic heteropolysaccharides. The most studied specificity systems are those of the specific binding of the lectin trifoliin produced by clover to *Rhizobium trifolii* and of the soybean lectin to *R. japonicum* (17).

A model to explain the selective adherence of *Rhizobium trifolii* to clover (*Trifolium*) roots related to the early expression of host specificity has been proposed. The surface of the *R. trifolii* cell and the surface of the clover root hairs possess identical antigens that are cross-reactive (antibodies produced from injection of bacterial or plant antigens to rabbits cross-react with either the bacterial or plant antigen). The lectin trifoliin binds to the antigen forming a bridge between the bacteria and the root surface (Fig. 7). Experimental evidence to support this model has been presented (17). Clover root hairs or purified trifoliin extracted from clover roots preferentially adsorb infective *R. trifolii* or its purified cross-reacting antigen, which is encountered only in infective bacteria. Trifoliin is multivalent in binding and specifically agglutinates *R. trifolii*. The sugar 2-deoxyglucose specifically inhibits agglutination of *R. trifolii* by trifoliin and elutes trifoliin from intact clover roots or trifoliin-coated *R. trifolii*. The specific adsorptions can be suppressed *in vivo* in plants growing under high concentrations of nitrate or ammonium, suggesting some kind of regulation mechanism. Indeed, nodule formation and development are generally repressed in soils fertilized with combined nitrogen.

A similar recognition system has been demonstrated in the soybean *Rhizobium japonicum* symbiosis. Purified soybean lectin only binds to infective *R. japonicum*. This binding is reversed by N-acetylgalactosamine, a potent specific inhibitor of binding by soybean lectin (3, 4).

There is a much lesser degree of specificity in other systems, such as

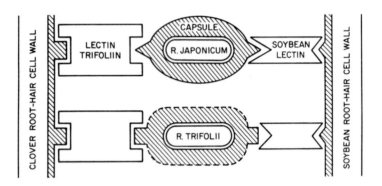

Fig. 7. Diagram showing lectins involved in specificity. After Dazzo (17). Cross-reacting antigens of (▷, top) *Rhizobium japonicum* capsule and soybean cell wall and (□, bottom) *R. trifolii* capsule and clover cell wall.

the *Rhizobium*–tropical legume combinations. Lectins extracted from tropical legumes such as concanavalin A are able to bind with surface determinants shared among many rhizobia.

Although the evidence for specific binding in clover *Rhizobium trifolii* is well documented, it is still far from showing that lectin–rhizobia binding combinations occur *in vivo* at the site of infection. Furthermore, it is not understood how those interactions, even if they do occur, are translated into triggering the very specific events of root curling and invagination that lead to the formation of the infection thread and ultimately of a successful nodule.

2. Azospirillum *and Roots of Grasses*

The association of free-living nitrogen-fixing *Azospirillum* species with roots of grasses has been studied since the 1970s (5, 19, 52, 80, 83). *Azospirillum lipoferum* was formally known as *Spirillum lipoferum*. Its isolation from the grass *Digitaria* in Brazil by J. Döbereiner and co-workers stirred wide interest in the study of grass–bacteria associations because of their potential ability to save nitrogen fertilizers in cereal crops.

Azospirilla have been isolated from many soils all over the world. These highly motile, gram-negative, half-twisted rods have a long polar flagellum and many shorter lateral flagella. The cell swells and contains PHB granules in old cultures. Two species, *Azospirillum brasilense* and *A. lipoferum*, have been characterized. The DNA base composition of the genus is 70–71% G + C (79). In contrast to rhizobia–legume symbiosis, the *Azospirillum*–grass interaction does not produce visible structures on roots. Direct and indirect evidence showing *Azospirillum* in close associa-

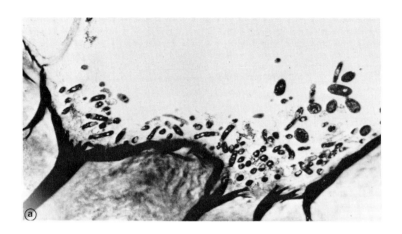

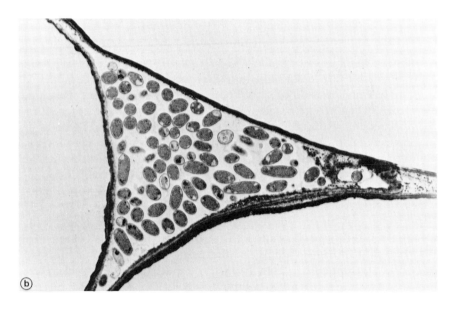

FIG. 8. *Azospirillum brasilense* in association with roots of grasses. (a) Embedded in mucigel layers (magnification 3200×). (b) Intercellular space of cortex cells (3200×). (c) Filling apparently dead cells (3200×). Courtesy of Drs. P. G. Heytler and R. R. Herbert, Experimental Station, E. I. du Pont de Nemours and Company, Wilmington, Delaware.

(Continued)

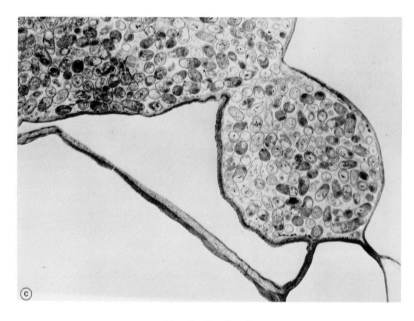

FIG. 8. (*Continued*)

tion with roots of grasses [millet (*Setaria italica*), guinea grass (*Panicum maximum*), maize (*Zea mays*), sorghum (*Sorghum bicolor*)] has accumulated. After surface sterilization of grass roots and their subsequent crushing, many azospirilla are released. Light and electron microscopy show azospirilla embedded in the mucigel layer, or mucilagenous sheath (Fig. 8a), between cortex cells (Fig. 8b), and filling apparently dead cells (Fig. 8c). Tetrazolium-reducing bacteria-like structures have been found in inter- and intracellular spaces of some plant cells in the cortex, endodermis, xylem, and stele (56). *Azospirillum* produces plant growth hormones (auxin, gibberellic acid) affecting production of more root hairs, lateral roots, and mucigel as compared with noninoculated plants (80). Adherent bacteria are associated with granular material of epidermal cells. Potassium nitrate (5 m*M*) in the plant growth medium inhibits *Azospirillum* attachment to root hairs. Proteinaceous material excreted from *Pennisetum* roots bind to *Azospirillum* cells and promotes their adsorption to root hairs. Millet root hairs have been shown to absorb azospirilla in significantly higher numbers than cells of other free nitrogen fixers (*Rhizobium, Azotobacter, Klebsiella*) (83).

Pectolytic enzymes, including pectin transeliminase and endopolygalacturonase, were detected in pure cultures of *Azospirillum brasilense* (83). Some degree of specificity between *Azospirillum brasilense* and roots of

C_3 plants (wheat) and *A. lipoferum* to roots of C_4 grasses (maize) has been proposed (2).

C. NITROGEN-FIXING BACTERIA INSIDE PLANTS

1. Legume–Rhizobium *Nodule Morphogenesis*

The process that begins on the root surface and culminates in the establishment of an effective nitrogen-fixing nodule has several states (Fig. 9) (4, 18, 22, 47, 85). Most of the knowledge available for understanding this process is based on light and electron microscopic studies. At the preinfection stage, rhizobial cells multiply and become attached to the infection site. Specific lectins are probably involved in specific attachments.

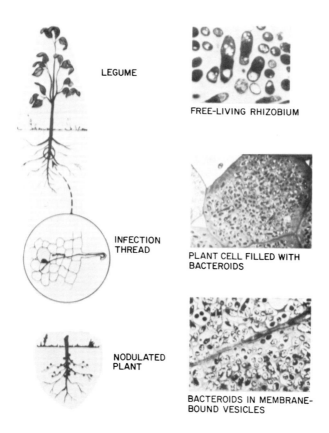

LEGUME

FREE-LIVING RHIZOBIUM

INFECTION
THREAD

PLANT CELL FILLED WITH
BACTEROIDS

NODULATED
PLANT

BACTEROIDS IN MEMBRANE-
BOUND VESICLES

FIG. 9. *Rhizobium*–legume symbiosis.

Substances excreted by the rhizobia cause branching, deformation, and curling of root hairs. The infection stage begins when the rhizobia enter at the base of a fold, possibly through a pore. Within the root hair cell, the rhizobia are encapsulated within an infection thread, consisting of an outer plant cell wall and an inner mucopolysaccharide matrix in which the rhizobia are embedded. Outer and inner cortical cells divide rapidly and increase in size. This stage is also called division of polyploid meristem.

The rhizobia are liberated from branches of the thread into the cytoplasm of cortical cells, where they multiply, enlarge, are surrounded by membrane envelopes, and become pleomorphic (bacteroids). The third stage is the functional nodule. The bacteroids are capable of fixing nitrogen, they excrete ammonia which is utilized by the plant cells, leghemoglobin mediates oxygen transport to the bacteroids, and the plant cells supply carbon and energy source to the bacteroids.

2. Leghemoglobin

The nitrogen-fixing *Rhizobium* bacteroids are usually located within membrane-bound vesicles or envelopes inside plant cells in the nodules. Leghemoglobin, a protein of plant origin, is encountered only in nodule tissue of legumes (22). Its presence within the membrane-bound vesicles and in contact with the bacteroids is generally accepted, but there have also been reports of leghemoglobin encountered outside the membrane and not in contact with the bacteroids. The principal role of leghemoglobin is to facilitate diffusion of oxygen from the nodule surface to the bacteroid. However, if leghemoglobin does not come into contact with the bacteroids, another mechanism of oxygen carrying protein within the membrane sacs must function to supply oxygen to bacteroids.

The rate constants for the formation ($1.18 \times 10^8 \ M^{-1} \ sec^{-1}$) and dissociation ($4.4 \ sec^{-1}$) of the leghemoglobin–oxygen complex indicate that it has the essential properties necessary for mediation and facilitation of oxygen diffusion. Leghemoglobin thus delivers oxygen for aerobic nitrogen fixation by the bacteroid, but at the same time, because of its high affinity for oxygen, it prevents local oxygen from reaching concentrations that can damage nitrogenase.

Genetic information for leghemoglobin synthesis is specified by the plant. The heme moiety is apparently synthesized by the bacterium. It is still not understood how rhizobia induce plants to synthestize globin messenger ribonucleic acid, which is found in nodule plant cytosol. There are more than one species of leghemoglobin, leghemoglobin a, and leghemoglobin C_2 (of soybean). They differ in their amino acids, and can be separated by charge differences. They may have different functions, because the ratio of leghemoglobin C_2 to leghemoglobin a changes during soybean development. It has been proposed that bacteroids contain two or more oxidase systems; one of these, which functions at low oxygen concentration maintained by the leghemoglobin system, efficiently produces ATP and raises the ATP/ADP ratio to a level advantageous to the nitrogen-fixing process. The other system functions at higher oxygen concentrations and may provide energy for cellular respiration and respiratory protection for nitrogenase.

The physiological and biochemical changes taking place either in the host or in the invading rhizobia are unknown. Evidence for bacteria-derived plant growth hormones and the involvement of hydrolytic enzymes in the morphogenetic changes is still indirect. Bacterial plasmids may code for infection phenotypes. New genetic approaches utilizing rhizobial mutants that stop the process at different stages of nodule development may help elucidate this complex interaction between a prokaryotic and a eukaryotic organism.

3. Nitrogen Fixation in Angiosperms

About 160 species in 15 genera among 7 families worldwide of angiosperms have been reported to have in their roots nodules containing actinomycetes capable of fixing nitrogen (7, 63, 82). So far few of these plants have been of direct agricultural importance. The alders (*Alnus* species) are good-size trees that produce timber. *Casuarina* species are useful in semitropical climates.

Root nodules induced by soil actinomycetes are morphologically and anatomically distinct from legume nodules. These modified, more branched lateral roots condense to form a spherical mass. Two general types of nodules are known, the *Alnus* type and the *Myrica* type (Fig. 10).

The actinomycete, surrounded by a polysaccharide layer produced by the host, occupies specific cortical cell layers. Nitrogenase develops in the endophyte. Active cell-free preparations of nitrogenase from endophytes in the nodules have been obtained, the enzyme having properties similar to those of other known nitrogenases.

The actinomycetes involved in nitrogen fixation have been isolated in pure culture; their involvement in nodule formation has been demonstrated according to Koch's postulates. The actinomycetes grow in culture typically as a filamentous mat. They are capable of developing sporangia and spores. The filaments lack the polysaccharide layer observed when inside the host. The best-characterized organisms are from the genus *Frankia*. Free-living *Frankia* grown aerobically in appropriate nutrient conditions showed nitrogenase activity associated with vesicles, a distinctive morphological structure formed both in effective nodules and in pure culture (81).

Isolated actinomycetes are able to infect a broad spectrum of plant

(a)   (b)

FIG. 10. Nodules formed by actinomycetes. (a) *Alnus* type. (b) *Myrica* type. From Torrey (82).

species. For example, an isolate from *Comptonia* was capable of nodulating plants of both the *Alnus* and *Myrica* groups, but not *Casuarina*. The cross-inoculation groups are therefore much broader than with *Rhizobium*.

Nitrogen-fixation rates in nodulated angiosperms are of the same order as in nodulated legumes. As woody dicotyledons are generally perennial plants, fixed nitrogen is more available to the soil on a long-term basis from decomposing leaf litter, or when the nodules and the roots die. Developments in the understanding of dicot–actinomycete symbiosis will permit more immediate exploitation of the system for wood and fiber production, soil improvement, reclamation of strip-mine areas, and revegetating roadsides, as well as for the improvement of forested ecosystems.

4. Nitrogen Fixation in Blue-Green Algae in Symbiosis with Plants

Symbiotic association of nitrogen-fixing blue-green algae occurs in symbiotic association with several bryophytes (liverworts), a pteridophyte (water fern), gymnosperms (cycads), and the angiosperm *Gunnera* (45, 57, 58). The algal symbiont is always a *Nostoc* or *Anabaena* species. It is often

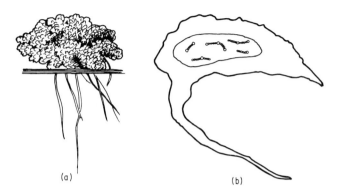

FIG. 11. *Azolla–Anabaena azollae* symbiosis. (a) *Azolla*. (b) Enlarged dorsal lobe, leaf cavity filled with *Anabaena azollae*. From Peters (58).

localized in a morphologically specialized structure of the host plant. The symbionts are also morphologically and physiologically distinct from the free-living species.

The *Azolla* (water fern)–*Anabaena azollae* symbiosis is of great economic importance. It has been studied in detail since 1970.

The six species of *Azolla* are widely distributed in temperate and tropical fresh waters. The fern has a branched floating stem with deeply bilobed leaves and true roots (Fig. 11). *Anabaena azollae* is entrapped in cavities formed in the dorsal aerial leaf lobes. As the leaf matures, the symbiont differentiates into heterocysts. As compared with free-living *Anabaena*, heterocyst frequency increases from 5–10% to about 25–30%. The leaf cavities are not directly connected to the vascular system of the fern. Nutrient exchanges occur probably through epidermal cells surrounding or extending into the cavity. The symbiont is capable of carbon dioxide fixation, and it can supply the total nitrogen needs of the host. Interestingly, nitrogenase is not repressed in the symbiont when *Azolla* is grown in combined nitrogen.

The *Azolla–Anabaena azollae* symbiosis is probably the ideal nitrogen-fixing system to be exploited in agriculture. Both the symbiont and the host are phototrophic organisms, possessing a very large light-harvesting surface, a factor limiting fixation by free-living nitrogen-fixing blue-green algae. The symbiont being protected inside the leaves from external adverse conditions fixes nitrogen efficiently, as shown by the high heterocyst frequency. At the same time, the symbiont is able to supply its own ATP and electrons by photosynthesis.

TABLE VI

Determinations of Metabolic Cost of Nitrogen Fixation[a]

Legume	Means of controlling nodulation	Measurement approach	Grams CH_2O per grams nitrogen
Clover	NO_3^-	Whole-plant balance	0.7
Pea	NO_3^-	Whole-plant balance	9.6
Cowpea	Development	Whole-plant balance	2.9
Clover	NO_3^-	Whole-plant balance	9.5
Pea	NO_3^-, age, light, defoliation	Gas flow	24–37
Lupine	NO_3^-	Whole-plant balance	2.1
Cowpea	Development	Balance + gas flow	11.6
Lupine	Development	Balance + gas flow	25.5
Pea	NO_3^-, age, light, temperature	Gas flow	8.5–23
Soybean Cowpea White clover	NO_3^-	Gas flow	15.8–17
Soybean	Genetic	Gas flow	
	Plant		10.0
	Bacteria		5.5

[a] Literature values compiled by P. G. Heytler, Central Research and Development Department, Experimental Station, E. I. du Pont de Nemours and Company, Wilmington, Delaware.

D. Efficiency of Nitrogen Fixation in Plant–Bacteria Associations

Efficiency in a biological system may be defined as the effectiveness with which energy is used in a particular process. The benefit of a nitrogen-fixing system to a plant and to agronomy is self-evident. Yet the symbiosis exacts a price from the metabolism of the plant. Maintenance of the bacteroids and nodules or of the rhizosphere bacteria in nonlegumes, increased translocation of metabolites, and energy requirements of enzymatic reduction of dinitrogen must all be supported by the plant's photosynthate pool, presumably decreasing the carbohydrate available for growth (27, 30, 36, 60).

Although early balance studies with legumes indicated negligible differences between energy expended on nitrogen fixation and that required for nitrate assimilation, later works have demonstrated higher costs for nitrogen fixation in a broad range of legumes (Table VI) (36). Evaluation of results is complicated because the values obtained depend on differences among species and even among cultivars, on the methodology used, on stages of plant ontogeny, and on establishing valid controls or reference populations.

A comparison of nodulating and nonnodulating soybean isolines demonstrated that a nodulated root system in a soybean may require four times the metabolic expenditure of a nonnodulated root. Of this about one-third was needed for maintenance; the remainder was used for nitrogenase activity at an efficiency of 5–7 g carbohydrate consumed per g nitrogen fixed (36).

Important agricultural objectives are to increase the efficiency and minimize the energy cost to the plant in order to obtain higher yields. It has been demonstrated that efficiency can be improved by increasing the P_{CO_2} in the air surrounding soybean plants (CO_2 fertilization), by eliminating the wasteful process of photorespiration,[5] or by using *Rhizobium* strains that actively take up hydrogen evolved from the nitrogenase reaction (in nodules) and use it as a further energy source.

The carbon dioxide–fixing enzyme ribulose-1,5-bisphosphate carboxylase was shown capable of using different substrates. The enzyme interacts with oxygen (as well as carbon dioxide), leading to the formation of glycolate, which in part is reoxidized to carbon dioxide without retention of any energy. This process is called photorespiration. Attempts to eliminate this apparently wasteful process by genetic or chemical means have not yet proved successful. However, artificially increasing the CO_2/O_2 ratio in the air does reduce photorespiration in plants.

Increasing the amount of photosynthate dramatically increases nitrogen fixation and seed yield in the soybean and other legumes. When the carbon dioxide content of air was increased from 300 to 1000 parts per million from 40 days of age to maturity, soybeans fixed five times as much nitrogen. This experiment demonstrated that soybeans could be made almost self-sufficient in nitrogen through nitrogen fixation. Thus the nitrogen and yield barriers of soybean can be broken; it is clear that the photosynthate supply in a *Rhizobium*–legume symbiosis limits nitrogen fixation.

It was demonstrated that hydrogenase-positive *Rhizobium* strains produced consistently significant increases in the yield of soybeans in the field as compared with hydrogenase-negative strains. Thus efficiency may be improved by using rhizobial strains that have a capacity to form nodules that recycle hydrogen (25).

Other improvements can be obtained by controlling the P_{O_2} level near

---

[5]For a discussion of this controversial viewpoint, see Volume VII, Chapter 3. It has yet to be shown that photorespiration can be eliminated, or that doing so would improve efficiency. (Eds.)

the nodules to the level that will permit highest respiration rates by the bacteroids without damaging their nitrogenase. Understanding and manipulation of specific recognition between plant lectins and rhizobia will permit the use of competitive, highly efficient *Rhizobium* strains.

C_3 and C_4 Photosynthetic Plants in Rhizosphere Association with Nitrogen-Fixing Bacteria

Many grasses possess the C_4 dicarboxylic acid cycle.[6] These grasses have been reported to utilize their available nitrogen more efficiently in producing dry matter and in fixing carbon dioxide than do grasses that possess the C_3 pathway, mainly because they respond to intense radiation and temperature more efficiently (5, 6). It has been suggested that C_4 grasses (mainly of tropical origin) may have the capacity to stimulate nitrogen fixation by bacteria in their root zones. Studies carried out with C_4 grasses, such as maize (*Zea mays*) and foxtail millet (*Setaria italica*) show that the highest acetylene reduction activities in intact plants inoculated with *Azospirillum* can be preferentially obtained at high light intensities and temperatures above 30°C. *Azospirillum*, the main nitrogen-fixing organism isolated from roots of grasses, utilizes C_4 compounds preferentially (malate and succinate). However, although malate or aspartate accumulate in leaves, there is no evidence that malate accumulates in the roots of C_4 grasses and, if it does, that it is capable of driving nitrogen fixation by *Azospirillum* species. Moreover, nitrogen fixation in C_4 grasses has not been properly compared with nitrogen fixation in association with C_3 grasses or other plants. More work has to be done to ascertain which non-legume plants are most capable of active nitrogen fixation in association with rhizosphere bacteria.

V. Utilization of Nitrogen-Fixing Systems in Agriculture

A. INOCULATION OF LEGUMES WITH *RHIZOBIUM*

Inoculation of seeds, plants, and soil with the correct strain of precultured *Rhizobium* has long been practiced (14, 16, 26). The most successful and widely used carrier for the inoculum is peat. A mixture of *Rhizobium* cells and carrier constitutes the inoculant.

[6]For a description of C_4 photosynthesis, see Volume VII, Chapter 3. (Eds.)

It is generally recognized that with successful nodulation only a very small amount of initial nitrogen fertilizer is needed and that *Rhizobium* can provide essentially all the nitrogen needs of the plant. Decomposing roots of the legume will also enrich the soil with combined nitrogen after the crop has been removed.

In the United States about 40% of the legume seeds planted are inoculated, soybeans constituting about 80% of the total. Brazil, Australia, Canada, India, and most of the European nations manufacture *Rhizobium* inoculants. However, many developing countries still lag in inoculant production and use. A successful *Rhizobium* inoculant must possess the following properties:

1. Nitrogen-fixing ability or effectiveness. The nodule formed must contain many bacteroids that actively fix nitrogen. High fixation activities are controlled by the right amount of photosynthate reaching the bacteroids, by utilizing the hydrogen evolved from the nitrogenase reaction as further energy source, by the right amount of oxygen being delivered by leghemoglobin for optimal bacteroid respiration rates, and further by prompt incorporation and transport by the plant of the ammonia produced. All of these requirements are governed by an optimal interaction between the symbiont and the host.

2. Competitiveness. This property implies the ability to produce nodules in a soil containing other highly infective rhizobia. Infective strains with higher effectiveness should be used for inoculation. Competitiveness is probably governed by selective recognition processes where lectins are involved.

3. Nodulating capability. It is important that the inoculum have a strong ability to nodulate over the range of temperature, pH, soil composition, and other factors under which the particular host plant will grow. The *Rhizobium* strain must be able to grow well in culture medium, in the carrier medium, and in the soil after the seed is planted. Furthermore, the organism should be able to survive in the soil from one season to another.

The methods for the production of *Rhizobium* inoculum, the different carriers for the cells, and the method for applying the inoculants in the field have been extensively studied. The main purpose is that inoculants should contain the largest possible number of viable cells at the time of use. After mixing with seeds or after direct application to soil, the number of viable cells should be sufficient to infect the developing legume root competitively. It is recommended that the strain of *Rhizobium* to be used and the method of inoculant preparation, composition, and application be adapted to local needs, usually after a local survey for selecting the best method(s) has been conducted. The survey should include preliminary laboratory tests of acetylene reduction rates in seedlings of different legume cultivars inoculated with different rhizobial strains. The most active systems will then be tested in greenhouse and field trials under local soil and environmental conditions for selection of the combination that produces the highest plant yield.

B. INOCULATION OF GRASSES WITH *AZOSPIRILLUM*

The potential benefit to grasses associated with nitrogen-fixing *Azospirillum* has received considerable attention since 1975 because grasses are the main crop for food production and because of the increasing cost of chemically produced nitrogen fertilizers (38, 70). Few studies have shown clear evidence that in this association the bacteria supplied significant amounts of combined nitrogen to the plants, and thereby increased growth. Nitrogen fixation in grasses associated with *Azospirillum* has been investigated mainly in detached roots or in soil cores containing the roots. Under these conditions acetylene-reduction tests have probably overestimated the actual activity *in situ*. In contrast, much lower activities have been generally reported in intact systems.

Increases in plant growth have been attributed to an increase in the absorbing surface of the root system caused by growth substances produced by the bacterium. However, inoculation experiments carried out in Israel under greenhouse conditions clearly demonstrated increases in plant dry weight and in the total nitrogen content of maize, sorghum, and other grasses (15). High acetylene reduction rates were obtained mainly at reproductive stages of grasses in intact plant–soil systems incubated under high temperatures (30–32°C).

Inoculation of maize with *Azospirillum brasilense* in the field caused increases of yield in Florida but not in Maryland or Wisconsin. In Israel, inoculation experiments carried out during 1979–1981 in commercial fields in different areas under diverse soil and environmental conditions resulted in significant increases (10–35% over controls) of commercially valuable yield components in several crops (Table VII). Highest yield increases attributable to *Azospirillum brasilense* inoculation were obtained at medium levels of initial nitrogen fertilizer. The yields achieved were equivalent to those from fully fertilized, noninoculated fields. These results present evidence that it may be possible to replace nitrogen fertilizer with *Azospirillum* inoculation (at least under conditions comparable to those of Israel).

It has not been unequivocally demonstrated whether the benefit in

TABLE VII

EFFECT OF INOCULATION WITH *AZOSPIRILLUM BRASILENSE* ON YIELD OF VARIOUS GRASSES
IN COMMERCIAL FIELDS OF ISRAEL

Plant species and location	Season	Commercial yield[a,b] fresh ears (ton/ha)	
Zea mays			
Sweet corn, cv. Jubilee,	Summer 1979	C 24.62 a	
Negev, irrigated, ini-		I 28.00 b	
tial nitrogen fertilizer	Summer 1980	C 22.06 a	
120 kg nitrogen per		I 27.77 b	
hectare			
cv. Rinat	Summer 1980	C 13.30 a	
		I 16.65 b	
cv. H-851, corn meal,	Summer 1979	C 7.42 a	
Jordan Valley, non-		I 8.20 b	
fertilized			

		Total plant dry weight (ton/ha)	Total nitrogen yield (kg/ha)
cv. H-Nanasi, forage,	Summer 1979	C 11.93 a	103.8 a
irrigated, nonfertilized		I 14.83 b	180.9 b
Sorghum bicolor			
cv. 6078, forage, non-	Spring 1980	C 9.48 a	90.0 a
fertilized, nonirri-		I 11.28 b	120.0 b
gated, Negev			

		Panicle weight (ton/ha)	
cv. H-226, grain, non-	Spring 1980	C 2.88 a	
fertilized, nonirri-		I 3.89 b	
gated, Negev			

		Total plant dry weight (ton/ha)	Total nitrogen yield (kg/ha)
Setaria italica			
Foxtail millet, forage,	Summer 1979	C 1.4 a	11.48 a
nonfertilized, irri-		I 2.1 b	26.20 b
gated, Negev			

(*continued*)

TABLE VII (Continued)

Plant species and location	Season	Plant fresh weight (ton/ha)	Seed yield (kg/ha)
Panicum miliaceum			
Jordan Valley, nonirri-	Summer 1979	C 10.31 a	2.80 a
gated, nonfertilized		I 11.68 b	3.17 b
		Grain yield (ton/ha)	
Triticum			
Wheat, cv. Miriam, initial	Winter 1979–1980	C 3.80 a	
nitrogen-fertilizer, 40		I 4.01 b	
kg/ha, nonirrigated,			
Negev			
cv. Inbar, Jordan Valley,	Winter 1978–1979	C 4.22 a	
nonfertilized, irrigated		I 4.68 b	
cv. Barkai, Jordan Valley	Winter 1979–1980	C 4.26 a	
		I 4.99 b	

[a] Number accompanied by the same letter for each experiment in each column does not differ significantly at $p = .05$.

[b] Fields were inoculated with *Azospirillum*–peat inoculant (I). Controls were treated with sterilized peat (C).

plant yield obtained in the field was the result of nitrogen supplied by biological nitrogen fixation or of plant growth substances produced by the bacterium. It is probable that in adequately fertilized fields the plants develop initially on the combined nitrogen present. However, the data strongly suggest that the associated bacteria benefit the plant by supplying biologically fixed nitrogen after nitrogen becomes limiting because of leaching, denitrification, and its uptake by the plants. Nevertheless, it seems unlikely that *Azospirillum* provides all the nitrogen needed by the plant.

C. *AZOLLA–ANABAENA AZOLLAE* SYMBIOSIS

Field studies carried out in Davis, California and in the Philippines have clearly demonstrated the potential use of the *Azolla–Anabaena azollae* system for replacing commercial fertilizer in rice cultivation (77, 78, 86). *Azolla* can be used both as a green manure during the fallow season and as a cover crop for rice (*Oryza sativa*). Death and decay of

Azolla are normally required before its nitrogen becomes available. *Azolla filiculoides* and *A. mexicana* occur in California, and *A. pinnata* occurs in the Philippines. Phosphorus supplementation is necessary to promote *Azolla* growth.

At Davis, exceptional populations of *Azolla filiculoides* can yield 10^5 kg of nitrogen per hectare (75% of the nitrogen requirement for rice). Used as green manure, the fern, containing 50–60 kg of nitrogen per hectare, increased rice yields 112% over those of unfertilized controls in California. In the Philippines, used as cover crop, *A. pinnata* increased rice yield by 14–40%.

VI. Conclusions

Enormous advances in the understanding of biological nitrogen fixation have occurred since 1970. However, apart from the principles of *Rhizobium*–legume technology, which were already known during the early twentieth century, no significant benefit to agriculture has been obtained so far from the accumulated knowledge of biological nitrogen fixation.

The obvious potential benefits that the exploitation of biological nitrogen fixation could bring to agriculture, as well as the increasing number of research groups all over the world working on the subject and the fast accumulating knowledge and methodology should soon translate our understanding of the process into successful practical applications.

References

1. Arnon, D. I., and Yoch, D. C. (1974). Photosynthetic bacteria. In "The Biology of Nitrogen Fixation" (A. Quispel, ed.), pp. 168–201. North-Holland, Amsterdam.
2. Baldani, V. L. D., and Döbereiner, J. (1980). Host plant specificity in the infection of cereals with *Azospirillum* spp. *Soil Biol. Biochem.* **12,** 433–439.
3. Bauer, W. D. (1980). Role of soybean lectin in the soybean–*Rhizobium japonicum* symbiosis. *In* "Nitrogen Fixation" (W. E. Newton and W. H. Orme-Johnson, eds.), Vol. 2, pp. 205–214. Univ. Park Press, Baltimore.
4. Bauer, W. D. (1981). Infection of legumes by *Rhizobium*. *Annu. Rev. Plant Physiol* **32,** 407–449.
5. van Berkum, P., and Bohlool, B. B., (1980). Evaluation of nitrogen fixation by bacteria in association with roots of tropical grasses. *Microbiol. Rev.* **44,** 491–517.
6. Black, C. C., Brown, R. H., and Moore, R. C. (1978). Plant photosynthesis. *In* "Limitations and Potentials for Biological Nitrogen Fixation in the Tropics" (J. Döbereiner, R. H. Burris, and A. Hollaender, eds.), pp. 95–110. Plenum, New York.
7. Bond, G. (1977). Some reflections on *Alnus*-type root nodules. *In* "Recent Developments in Nitrogen Fixation" (W. E. Newton, J. R. Postgate, and C. Rodriguez-Barrueco, eds.), pp. 531–537. Academic Press, New York.

50   Y. OKON AND R. W. F. HARDY

8. Brill, W. J. (1975). Regulation and genetics of bacterial nitrogen fixation. *Annu. Rev. Microbiol.* **29**, 109–129.
9. Brill, W. J. (1980). Biochemical genetics of nitrogen fixation. *Microbiol. Rev.* **44**, 449–467.
10. Burns, R. C., and Hardy, R. W. F. (1975). "Nitrogen Fixation in Bacteria and Higher Plants." Springer, New York.
11. Burris, R. H. (1974). Methodology. *In* "The Biology of Nitrogen Fixation" (A. Quispel, ed.), pp. 9–33. North-Holland, Amsterdam.
12. Burris, R. H. (1974). Biological nitrogen fixation, 1924–1974. *Plant Physiol.* **54**, 443–449.
13. Burris, R. H. (1980). The global nitrogen budget—science or seance. *In* "Nitrogen Fixation" (W. E. Newton, and W. H. Orme-Johnson, eds.), Vol. 1, pp. 7–16. Univ. Park Press, Baltimore, Maryland.
14. Burton, J. C. (1979). Rhizobium species. *In* "Microbial Technology" (H. J. Peppler, ed.), pp. 29–58. Academic Press, New York.
15. Cohen, E., Okon, Y., Kigel, J., Nur, I., and Henis, Y. (1980). Increase in dry weight and total nitrogen content in *Zea mays* and *Setaria italica* associated with nitrogen-fixing *Azospirillum* spp. *Plant Physiol.* **66**, 746–749.
16. Date, R. A., and Roughley, R. J. (1977). Preparation of legume seed inoculants. *In* "A Treatise on Dinitrogen Fixation" (R. W. F. Hardy, and A. H. Gibson eds.), Section 4, pp. 243–276. Wiley, New York.
17. Dazzo, F. B. (1980). Determinants of host specificity in the *Rhizobium*–clover symbiosis. *In* "Nitrogen Fixation" (W. E. Newton and W. H. Orme-Johnson, eds.), Vol. 2, pp. 165–188. Univ. Park Press, Baltimore.
18. Dilworth, M. J. (1980). Host and *Rhizobium* contributions to the physiology of legume nodules. *In* "Nitrogen Fixation" (W. E. Newton and W. H. Orme-Johnson, eds.), Vol. 2, pp. 3–32. Univ. Park Press, Baltimore.
19. Döbereiner, J., and Day, J. M. (1976). Associative symbioses in tropical grasses: characterization of microorganisms and dinitrogen-fixing sites. *Proc. Int. Symp. Nitrogen Fixation 1st*, pp. 518–538.
20. Döbereiner, J., Burris, R. H., and Hollaender, A. (eds.) (1978). "Limitations and Potentials for Biological Nitrogen Fixation in the Tropics." Plenum, New York and London.
21. Dudal, R. (1980). Fertilizers for food production in developing countries. *Proc. ISMA Annu. Conf. 1980*, pp. 1–5.
22. Emerich, D. W., and Evans, H. J. (1980). Biological nitrogen fixation with emphasis on the legumes. *In* "Biochemical and Photosynthetic Aspects of Energy Production" (A. San Pietro, ed.), pp. 117–145. Academic Press, New York.
23. Emerich, W. E., Hageman, R. V., and Burris, R. H. (1981). Interactions of dinitrogenase and dinitrogenase reductase. *Adv. Enzymol. Relat. Areas Mol. Biol.* **52**, 1–22.
24. Evans, H. J., and Barber, L. E. (1977). Biological nitrogen fixation for food and fiber production. *Science (Washington, D.C.)* **197**, 332–339.
25. Evans, H. J., Emerich, D. W., Ruiz-Argüeso, T., Maier, R. J., and Albrecht, S. L. (1980). Hydrogen metabolism in the legume–*Rhizobium* symbiosis. *In* "Nitrogen Fixation" (W. E. Newton and W. H. Orme-Johnson, eds.), Vol. 2, pp. 69–86. Univ. Park Press, Baltimore.
26. Franco, A. A. (1978). Contribution of the legume–*Rhizobium* symbiosis to the ecosystem and food production. *In* "Limitations and Potentials for Biological Nitrogen Fixation in the Tropics" (J. Döbereiner, R. H. Burris, and A. Hollaender, eds.), pp. 65–74. Plenum, New York and London.

27. Hardy, R. W. F. (1977). Rate-limiting steps in biological photoproductivity. *In* "Genetic Engineering for Nitrogen Fixation" (A. Hollaender, ed.), pp. 369–399. Plenum, New York.

28. Hardy, R. W. F. (1980). The global carbon and nitrogen economy. *In* "Nitrogen Fixation" (W. E. Newton, and W. H. Orme-Johnson, eds.), Vol. 1, pp. 3–5. Univ. Park Press, Baltimore, Maryland.

29. Hardy, R. W. F., Bottomley, F., and Burns, R. C. (eds.) (1979). "A Treatise on Dinitrogen Fixation," Sections 1 and 2. Wiley, New York.

30. Hardy, R. W. F., Criswell, J. G., and Havelka, U. D. (1977). Investigations of possible limitations of nitrogen fixation by legumes: (1) methodology, (2) identifications, and (3) assessment of significance. *In* "Recent Developments in Nitrogen Fixation" (W. E. Newton, J. R. Postgate, and C. Rodriguez-Barrueco, eds.), pp. 451–467. Academic Press, New York.

31. Hardy, R. W. F., and Gibson, A. H. (eds.) (1977). "A Treatise on Dinitrogen Fixation," Section 4. Wiley, New York.

32. Hardy, R. W. F., and Havelka, U. D. (1975). Nitrogen fixation research: a key to world food? *Science (Washington, D.C.)* **188,** 633–643.

33. Hardy, R. W. F., and Holsten, R. D. (1977). Methods for measurement of dinitrogen fixation. *In* "A Treatise on Dinitrogen Fixation" (R. W. F. Hardy, and A. H. Gibson, eds.), Section 4, pp. 451–486. Wiley, New York.

34. Hardy, R. W. F., and Silver, W. S. (eds.) (1977). "A Treatise on Dinitrogen Fixation," Section 3. Wiley, New York.

35. Haselkorn, R., Mazur, B., Orr, J., Rice, D., Wood, N., and Ripka, R. (1980). Heterocyst differentiation and nitrogen fixation in cyanobacteria (blue-green algae). *In* "Nitrogen Fixation" (W. E. Newton and W. H. Orme-Johnson, eds.), Vol. 2, pp. 259–278. Univ. Park Press, Baltimore.

36. Heytler, P. G., and Hardy, R. W. F. (1979). Energy requirement for N-fixation by Rhizobial nodules in soybeans. *Plant Physiol.* **63,** (Suppl.), 84.

37. Hollaender, A. (ed.) (1977). "Genetic Engineering for Nitrogen Fixation." Plenum, New York.

38. Kapulnik, Y., Sarig, S., Nur, I., Okon, Y., Kigel, J., and Henis, Y. (1981). Yield increases in summer cereal crops in Israeli fields inoculated with *Azospirillum*. *Exp. Agric.* **17,** 179–187.

39. Kennedy, I. R. (1979). Integration of nitrogenase in cellular metabolism. *In* "A Treatise on Dinitrogen Fixation" (R. W. F. Hardy, F. Bottomley, and R. C. Burris, eds.), Sections 1 & 2, pp. 653–690. Wiley, New York.

40. Knowles, R. (1977). The significance of symbiotic dinitrogen fixation by bacteria. *In* "A Treatise on Dinitrogen Fixation" (R. W. F. Hardy, and A. H. Gibson, eds.), Section 4, pp. 33–83. Wiley, New York.

41. Knowles, R. (1978). Free-living bacteria. *In* "Limitations and Potentials for Biological Nitrogen Fixation in the Tropics" (J. Döbereiner, R. H. Burris, and A. Hollaender, eds.), pp. 25–40. Plenum, New York and London.

42. Ludden, P. W. (1980). Nitrogen fixation by photosynthetic bacteria: properties and regulations of the enzyme system from *Rhodospirillum rubrum*. *In* "Nitrogen Fixation" (W. E. Newton, and W. H. Orme-Johnson, eds.), Vol. 1, pp. 139–156. Univ. Park Press, Baltimore, Maryland.

43. McClure, P. R., Israel, D. W., and Volk, R. J. (1980). Evaluation of the relative ureide content of xylem sap as an indicator of N_2 fixation in soybeans. Greenhouse studies. *Plant Physiol.* **66,** 720–725.

44. Mevarech, M., Rice, D., and Haselkorn, R. (1980). Nucleotide sequence of a cyanobac-

terial *nif* gene coding for nitrogenase reductase. *Proc. Natl. Acad. Sci. USA* **77,** 6476–6480.

45. Millbank, J. W. (1977). Lower plant associations. *In* "A Treatise on Dinitrogen Fixation" (R. W. F. Hardy, and W. S. Silver, eds.), Section 3, pp. 125–152. Wiley, New York.

46. Mulder, E. G., and Brotonegoro, S. (1974). Free-living heterotropic nitrogen-fixing bacteria. *In* "The Biology of Nitrogen Fixation" (A. Quispel, ed.), pp. 37–85. North-Holland, Amsterdam.

47. Newcomb, W. (1980). Control of morphogenesis and differentiation of pea root nodules. *In* "Nitrogen Fixation" (W. E. Newton and W. H. Orme-Johnson, eds.), Vol. 2, pp. 87–102. Univ. Park Press, Baltimore.

48. Newton, W. E., and C. J. Nyman, (eds.) (1976). *Proc. Int. Symp. Nitrogen Fixation 1st,* p. 717 (Vol. 1); p. 311 (Vol. 2).

49. Newton, W. E., and Orme-Johnson, W. H. (eds.) (1980). "Nitrogen Fixation," Vol. 1. Univ. Park Press, Baltimore, Maryland.

50. "Symbiotic Associations and Cyanobacteria," Vol. 2.

51. Newton, W. E., Postgate, J. R., and Rodriguez-Barrueco, C. (eds.) (1977). "Recent Developments in Nitrogen Fixation." Academic Press, New York.

52. Neyra, C. A., and Döbereiner, J. (1977). Nitrogen fixation in grasses. *Adv. Agron.* **29,** 1–38.

53. Nur, I., Okon, Y., Steinitz, Y. L., and Henis, Y. (1981). Carotenoid composition and function in nitrogen fixing bacteria of the genus *Azospirillum. J. Gen. Microbiol.* **122,** 27–32.

54. Nutman, P. S. (ed.) (1976). "Symbiotic Nitrogen Fixation in Plants." Cambridge Univ. Press, Cambridge.

55. Okon, Y., Cakmakci, L., Nur, I. and Chet, I. (1980). Aerotaxis and chemotaxis of *Azospirillum brasilense. Microb. Ecol.* **6,** 277–280.

56. Patriquin, D. G., and Döbereiner, J. (1978). Light microscopy observations of tetrazolium reducing bacteria in the endorhizosphere of maize and other grasses in Brazil. *Can. J. Microbiol.* **24,** 734–742.

57. Peters, G. A. (1977). The *Azolla–Anabaena azollae* symbiosis. *In* "Genetic Engineering for Nitrogen Fixation" (A. Holaender, ed.), pp. 231–258. Plenum, New York.

58. Peters, G. A. (1978). Blue-green algae and algal associations. *BioScience* **28,** 580–585.

59. Peters, G. A., Ray, T. B., Mayne, B. C., and Toia, R. E., Jr. (1980). *Azolla–Anabaena* association: morphological and physiological studies. *In* "Nitrogen Fixation" (W. E. Newton and W. H. Orme-Johnson, eds.), Vol. 2, pp. 293–310. Univ. Park Press, Baltimore.

60. Phillips, D. A. (1980). Efficiency of symbiotic nitrogen fixation in legumes. *Annu. Rev. Plant Physiol.* **31,** 29–49.

61. Postgate, J. R. (1974). Prerequisites for biological-nitrogen fixation in free-living heterotrophic bacteria. *In* "The Biology of Nitrogen Fixation" (A. Quispel, ed.), pp. 663–683. North-Holland, Amsterdam.

62. Postgate, J. R. (1978). "Nitrogen Fixation." Camelot Press, Southampton.

63. Quispel, A. (1974). The endophyles of the root nodules in non-leguminous plants. *In* "The Biology of Nitrogen Fixation" (A. Quispel, ed.), pp. 499–520. North-Holland, Amsterdam.

64. Quispel, A. (ed.) (1974). "The Biology of Nitrogen Fixation." North-Holland, Amsterdam.

65. Robson, R. L., and Postgate, J. R. (1980). Oxygen and hydrogen in biological nitrogen fixation. *Annu. Rev. Microbiol.* **34,** 183–207.

66. Schubert, K. R., and Evans, H. J. (1977). The relation of hydrogen reactions to nitro-
    gen fixation on nodulated symbionts. In "Recent Developments in Nitrogen Fixation"
    (W. E. Newton, J. R. Postgate, and C. Rodriguez-Barrueco, eds.), pp. 469–485. Aca-
    demic Press, New York.
67. Scott, D. B. (1978). Ammonia assimilation in N_2-fixing systems. In "Limitations and
    Potentials for Biological Nitrogen Fixation in the Tropics" (J. Döbereiner, R. H. Bur-
    ris, and A. Hollaender, eds.), pp. 25–40. Plenum, New York and London.
68. Shah, V. K. (1980). Iron–molybdenum cofactor of nitrogenase. In "Nitrogen Fixa-
    tion" (W. E. Newton, and W. H. Orme-Johnson, eds.), Vol. 1, pp. 237–747. Univ. Park
    Press, Baltimore, Maryland.
69. Shanmugam, K. T., O'Gara, F., Andersen, K., and Valentine, R. C. (1978). Biological
    nitrogen fixation. Annu. Rev. Plant Physiol. **79**, 263–276.
70. Smith, R. L., Bouton, J. H., Schank, S. C., Quesenberry, K. H., Tyler, M. E., Milam, J.
    R., Gaskins, M. H., and Litell, R. C. (1976). Nitrogen fixation in grasses inoculated
    with Spirillum lipoferum. Science (Washington, D.C.) **193**, 1003–1005.
71. Stewart, W. D. P. (1974). Blue-green algae. In "The Biology of Nitrogen Fixation" (A.
    Quispel, ed.), pp. 202–237. North-Holland, Amsterdam.
72. Stewart, W. D. P. (ed.) (1975). "Nitrogen Fixation by Free-Living Micro-Organisms."
    Cambridge Univ. Press, Cambridge.
73. Stewart, W. D. P. (1977). Blue-green algae. In "A Treatise on Dinitrogen Fixation" (R.
    W. F. Hardy, and W. S. Silver, eds.), Section 3, pp. 63–124. Wiley, New York.
74. Stewart, W. D. P. (1980). Some aspects of structure and function in N_2-fixing
    cyanobacteria. Annu. Rev. Microbiol. **34**, 497–536.
75. Stewart, W. D. P., and Gallon, J. R. (eds.) (1980). Nitrogen Fixation. Annu. Proc.
    Phytochem. Soc. Eur., No. 18, p. 453.
76. Sundaresan, V. and Ausubel, F. M. (1981). Nucleotide sequence of the gene coding for
    the nitrogenase iron protein from Klebsiella pneumoniae. J. Biol. Chem. **256**, 2808–2812.
77. Talley, S. N., and Rains, D. W. (1980). Azolla as a nitrogen source for temperate rice. In
    "Nitrogen Fixation" (W. E. Newton and W. H. Orme-Johnson, eds.), Vol. 2, pp.
    311–320. Univ. Park Press, Baltimore.
78. Talley, S. N., Talley, B. J., and Rains, D. W. (1971). Nitrogen fixation by Azolla in rice
    fields. In "Genetic Engineering for Nitrogen Fixation" (A. Hollaender, ed.), pp.
    259–282. Plenum, New York.
79. Tarrand, J. J., Krieg, N. R. and Döbereiner, J. (1978). A taxonomic study of the
    Spirillum lipoferum group, with description of a new genus, Azospirillum gen. nov., and
    two species, Azospirillum lipoferum (Beijerinck) comb. nov. and Azospirillum brasilense sp.
    nov. Can. J. Microbiol. **24**, 967–980.
80. Tien, T. M., Gaskins, M. H., and Hubbell, D. H. (1979). Plant growth substances
    produced by Azospirillum brasilense and their effect on the growth of pearl millet (Pen-
    nisetum americanum L.). Appl. Environ. Microbiol. **37**, 1016–1024.
81. Tjepkema, J. D., Ormerod, W., and Torrey, J. G. (1980). Vesicle formation and
    acetylene reduction activity in Frankia sp. CP11 cultured in defined nutrient media.
    Nature (London) **287**, 633–635.
82. Torrey, J. G. (1978). Nitrogen fixation by actinomycete-nodulated angiosperms. Bio-
    Science **28**, 586–592.
83. Umali-Garcia, M., Hubbell, D. H., Gaskins, M. H., and Dazzo, F. B. (1980). Association
    of Azospirillum with grass roots. Appl. Environ. Microbiol. **39**, 219–226.
84. Veeger, C., Laane, C., Scherings, G., Matz, L., Haaker, H., and Van Zeeland-Wolbers,
    L. (1980). Membrane energization and nitrogen fixation in Azotobacter vinelandii and
    Rhizobium leguminosarum. In "Nitrogen Fixation" (W. F. Newton, and W. H. Orme-
    Johnson, eds.), Vol. 1, pp. 111–137. Univ. Park Press, Baltimore, Maryland.

85. Vincent, J. M. (1980). Factors controlling the legume–*Rhizobium* symbiosis. *In* "Nitrogen Fixation" (W. E. Newton and W. H. Orme-Johnson, eds.), Vol. 2, pp. 103–130. Univ. Park Press, Baltimore.
86. Watanabe, I., Espinas, C. R., Berja, N. S., and Alimagno, B. V. (1977). Utilization of the *Azolla–Anabaena* complex as nitrogen fertilizer for rice. *IRRI Res. Pap. Ser.*, No. 110.
87. Winter, H. C., and Burris, R. H. (1976). Nitrogenase. *Annu. Rev. Biochem.* **45,** 409–426.
88. Wolk, C. P. (1980). Heterocysts, ^{13}N, and N_2-fixing plants. *In* "Nitrogen Fixation" (W. E. Newton and W. H. Orme-Johnson, eds.), Vol. 2, pp. 279–292. Univ. Park Press, Baltimore.
89. Woo, K. C., Atkins, C. A., and Pate, J. S. (1980). Biosynthesis of ureides from purines in cell-free systems from nodule extracts of cowpea [*Vigna unguiculata* (L.) Walp.]. *Plant Physiol.* **66,** 735–739.
90. Yates, M. G. (1977). Physiological aspects of nitrogen fixation. *In* "Recent Developments in Nitrogen Fixation" (W. E. Newton, J. R. Postgage, and C. Rodriguez-Barrueco, eds.), pp. 219–270. Academic Press, New York.

CHAPTER TWO

Nitrogen Metabolism[1]

D. J. Durzan and F. C. Steward

I. Introduction: Then and Now

Nitrogen metabolism of plants, like all current aspects of biology, has been affected by the almost explosive developments in science. This is

[1]Abbreviations used in this chapter: ADP, adenosine 5′-diphosphate; AMP, adenosine 5′-monophosphate; ATP, adenosine 5′-triphosphate; GMP, guanosine 5′-monophosphate; GOGAT, glutamate-2-oxoglutarate aminotransferase; GS, glutamine synthetase; GTP, guanosine 5′-triphosphate; NAD^+/NADH, nicotinamide adenine dinucleotide, oxidized or reduced; $NADP^+$/NADPH, nicotinamide adenine dinucleotide phosphate, oxidized or reduced; NAD(P)H, either NADH or NADPH; P_i, inorganic phosphate; PP_i, pyrophosphate.

Plant Physiology
A Treatise
Vol. VIII: Nitrogen Metabolism

apparent by the advances made since the earlier Chapter 4 in Volume IVA of this treatise was written. At the outset, however, one has to strike a balance between the flood of new information and the understanding it engenders.

Developments in methods, laboratory techniques, and instrumentation (particularly in the areas of biochemistry and enzymology) have given new dimensions to the study of plants. This is so in the means to isolate and identify plant constituents and to arrive at their precise molecular structures (92, 219). By 1965 the first fruits were already apparent from the use of two-dimensional paper chromatography and separations and purifications of substances on columns or acrylamide gels. But automation, miniaturization, and the use of sophisticated physical techniques to supplement, or even supplant, time-honored chemical means to prove molecular structures have transformed this field of endeavor. In no area is this more apparent than in the dramatic developments to identify protein and nucleic acid hydrolytic products and to sequence them in the molecule. Sophisticated miniaturized physical instrumentation to achieve their separations and purification and to arrive at structures have transformed operations that used to engage the skills of a few for many months, and they can now be carried out routinely, with almost incredible speed (e.g., 281). Also, the synthesis or production of natural products or their analogs, whether in the laboratory or on the industrial scale, have been revolutionized in the period.

Knowledge and understanding of enzymes have experienced an analogous development. Not only has the number of enzymes regarded as biocatalysts in cells vastly increased, but their uses in laboratory and even industrial practice to achieve specific and otherwise difficult conversions are now remarkable. Nowhere is this more apparent than in the new gene technology of splitting and splicing DNA and the reintroduction of the isolated genetic material into foreign cellular environments. The resultant techniques of so-called genetic engineering, which have produced such dramatic events as the bacterial synthesis of insulin, have achieved their great advances with bacteria. Nevertheless, one should still recognize that plant cells, with their complex organization, still provide the essential internal environment for the biochemical and metabolic events on which the life of higher plants, indeed of all life, depends. This is particularly true of the coordinated activities that constitute their use of and dependence on nitrogen.

Understanding of cells as the environment for metabolism has also advanced greatly (cf. Volume VII, Chapter 1), principally because of the applications of electron microscopy and autoradiography at high magnification to elucidate fine structure of cells in terms of their organelles, membranes, and ultimate particles (182). Particular organelles, long associated with vital functions (nuclei and chromosomes with heredity, nucleoli with the assembly or ribosomes, chloroplasts with fixation of light energy and carbon in photosynthesis, mitochondria with the oxidative release of energy in usable form as ATP, the endoplasmic reticulum and ribosome granules in their relations to protein synthesis) and the various membranes at the boundary surfaces of cells and their inclusions have now acquired an identity of their own (8, 9). Their properties are increasingly being studied in extracellular preparations (albeit these preparations rarely match their efficiency *in situ*). The organelles appear as more credible autonomous self-regulating entities because of their distinctive complements of DNA. However, having assigned in ever greater detail the particular steps of metabolism to this or that subcellular location, the problems of overall regulatory control become ever more formidable. One may trace in diagrams, with deceptive facility, the passage of a discrete amino acid via its specific transfer RNA to its genetically coded site on the surface of the protein-synthesis template and the eventual "peeling off" from that surface of a polypeptide chain with the correct linear array of amino acids. Nevertheless, even granted the coordinated steps involved in making one protein molecule, the synthesis of the great array of proteins (whether enzymes or structural) implies a seemingly incredible degree of chemical complexity and coordination; and yet it works. All of nitrogen metabolism presents the need to bring so many specific enzymes, with their functional groups, into the right juxtaposition with substrates, there to achieve what may seem to be relatively simple events but in a highly controlled way. (An example is the contrasted enzymatic machinery involved in the syntheses of asparagine or glutamine that only differ by one CH_2 group.)

But the problems of nitrogen metabolism are not only at the enzymatic or at the cellular level, for the metabolic performances of plants ramify into their highly organ-specific biochemistry. This degree of organ specificity may be achieved without irreversible loss of the genetic information that the mature living cells of specific organs received from the zygote. Biochemical differentiation thus becomes as evident a feature of morphogenesis as is the form within which it operates. Moreover, it is yet inexplicable.

A particular challenge, therefore, is to understand how growing regions transmit to their derivatives the specificity that is expressed in mature cells and organs without loss of the genetic totipotency of their living cells. In plants, more so than in higher animals, the regulatory control of morphogenesis (with its implications for metabolism and especially for nitrogen metabolism) is vested in responses, not only to nutrition (e.g., as by trace elements) but to environmental factors (length of day, fluctuations of temperature by day and night, the cycles of the seasons, etc. and, not to be overlooked, stresses in the economy of water). Therefore, the reductionist approach that has led to the great accumulation of factual minutiae about plants now needs to be balanced by the holistic vision of the integrated whole.

Plants are now increasingly important on the world scene. This is not only so through their historic role in the fixation of light energy and the regulation of the composition of the atmosphere. As the earth's green cover is persistently denuded, the consequences for climate and soil erosion are also painfully obvious. As plants are the basis of food production, their dependence on nitrogen (second only to their need for carbon) is again crucial because undernourished populations are usually more protein poor than carbohydrate poor. Much current preoccupation with biological nitrogen fixation, which even attempts to extend its scope into alternative hosts, reflects the almost desperate attempts to conserve and preserve natural reserves of nitrogen (otherwise dependent on restoration by industrial fixation) and to reverse the flow of the fertility of agricultural soils through the waters of the earth to the oceans. The natural nitrogen cycle, tacitly accepted since first enunciated, again becomes a potent topic of discussion (246). Also, as man contemplates movement into space, somewhat "far-out" discussions of space agriculture and of "balanced communities" on space platforms are now commonplace. But even if a cyclical balance with respect to carbon were achieved in space, the urgent need would remain to emulate the balanced terrestrial economy with respect to nitrogen that operates through the activities of plants.

Therefore, while this chapter essentially updates the earlier one in 1965, it also impinges on many topics in which events, in the interim, have moved apace.

II. The Extended Range of Nitrogen Compounds in Plants

The earlier 1965 chapter emphasized how far, in a few postwar years, the knowledge of amino compounds in plants had advanced. These advances stemmed from the search in the nineteenth and early twentieth centuries for evidence of all the amino acids of proteins occurring free in plants, with the added recognition of the two principal amides, asparagine (a compound of distinctively plant origin) and glutamine (of as much importance to bacteria, animals, and man as to plants). Even by 1965 a very large number of nonprotein nitrogenous compounds were known, but in the interim the subject has again proliferated. It is necessary, in order to see the subject in some perspective, to systematize the presentation. To this end several tables have been designed and are now presented.

Table I presents various named nitrogenous groups or configurations (with their structures), which serve as a guide to the nomenclature of their substituents (of which typical examples are given) as they occur in plants. Also, Table I lists the parent structures to which many other naturally occurring compounds in the appendixes relate.

Table II gives examples of the very varied occurrences of the different nonprotein-nitrogen compounds. These range from γ-aminobutyric acid (not recognized in higher plants until 1949 but now known to be virtually universal) to some well-identified substances that occur widespread, but not universally, or that occur infrequently in "often highly specific but seemingly unpredictable situations." Still other substances occur infrequently and apparently at random, and there are others that are still regarded as extremely rare in plants.

Table III shows how some hypothetical or more recently discovered or identified substances may be related to a parent structure, or configuration, by group substitutions or homologous relationships.

Table IV recognizes that the vast array of nitrogen compounds inevitably covers a wide range of functions, either established or hypothetical. Examples are listed of compounds to which such functions have been assigned in the plants in which they occur.

As the search for ever more nitrogenous substances in plants becomes more and more sophisticated, both with respect to the lower limits of their detection and the diversity of substances found, conceptions of their biological significance are broadened to include relationships between organisms. Parasitism, symbiosis, commensalism, disease susceptibility, and resistance (whether involving insects, fungi, or bacteria) or even allergies are all biological phenomena in which minute amounts of unusual nitrogen metabolites may play a regulatory role, but it would extend this chapter too far to pursue these topics to finality. Tables V and VI, however, show how far-reaching these considerations may become, both in terms of the organisms involved and the chemical substances concerned. Table V directs attention to predators on the seeds of legumes and the potentially toxic chemicals they may contain. Table VI lists approximately 40 complex substances that may act as metabolic antagonists for various named amino acids and their sources in different organisms.

TABLE I

SOME NITROGEN-CONTAINING GROUPS FOUND IN NATURALLY OCCURRING COMPOUNDS
IN HIGHER PLANTS

Structure	Name of group	Examples
R—NH$_2$	Amino	Glycine, γ-aminobutyrate
R—$\overset{\overset{\text{O}}{\|\|}}{\text{C}}$—NH$_2$	Amide	Asparagine, glutamine
R—$\overset{\overset{\text{O}}{\|\|}}{\text{C}}$—NHR	Peptide bond	Glutathione, urease
H$_2$N—$\overset{\overset{\text{O}}{\|\|}}{\text{C}}$O ~ PO$_3{}^{2-}$	Carbamoyl (carbamyl)	Carbamoyl phosphate, carbamoyl aspartate
R—NH$\overset{\overset{\text{O}}{\|\|}}{\text{C}}$—NH$_2$	Ureido	Citrulline, allantoin (ureides)
R—NH$\overset{\overset{\text{NH}}{\|\|}}{\text{C}}$—NH$_2$	Guanidino (guanido)	Arginine, monosubstituted guanidines
R—$\overset{\overset{\text{NH}}{\|\|}}{\text{C}}$—NH$_2$	Amidine	Indospicine
R—NH$\overset{\overset{\text{NH}}{\|\|}}{\text{C}}$—$\underset{\text{H}}{\text{N}}$ ~ PO$_3$H$_2$	N-Phosphoryl	N-Phosphorylarginine
R—$\overset{\overset{\text{O}}{\uparrow}}{\text{N}}$=N	N-Oxide	Cycasin
	(Cyclic amidine) Purine	Guanine, adenine, and synthetic and natural cytokinins
	Pyrimidine	Uracil, cytosine, thymine

TABLE I (Continued)

Structure	Name of group	Examples
R—C≡N	Cyano	β-Cyanoalanine, laetrile
	Pyrroline	Proline, hydroxyproline
$(CH_3)_3 \overset{+}{N}$—R	Betaine	Stachydrine
H_2N—$\overset{\overset{HN}{\parallel}}{C}$—NHC—NHR (with O above the C—NH)	Guanidylureido	Gigartinine
H_2N—$\overset{\overset{NH}{\parallel}}{C}$—NHO—R	Guanidinoxy	Canavanine
	Indole (indolyl)	Tryptophan, β-indoleacetic acid
	Imidazole	Histidine

TABLE II

PATTERNS OF DISTRIBUTION OF NONPROTEIN FREE AMINO ACIDS IN PLANTS[a]

Distribution	Examples
Universal	γ-Aminobutyrate, β-alanine, carbamoyl aspartate
Widespread	Pipecolic acid, α-aminoadipic acid, homoserine
Infrequent (specific)	Azetidine-2-carboxylic acid, β-pyrazol-1-ylalanine, N^2-(1-carboxyethyl)lysine
Infrequent (random)	γ-Methylglutamine, γ-substituted glutamic acids
Rare	Hypoglycine A, heptenoic acids, 2-[2-aminoimidazolin-4(5)-yl]-alanine

[a] (22).

TABLE III

Some Amino Acid Analogs, Homologs, and Substitution Products

Class of compound	Example	Homologs and substitution products		
Dicarboxylic amino acid	Aspartate	Glutamate	*threo*-γ-Hydroxy glutamate	δ-Hydroxy-α-aminoadipate
α-Keto acid	Oxaloacetate	α-Keto-glutarate	γ-Hydroxy-α-ketoglutarate	δ-Hydroxy-α-ketoadipate
Cyclic imino acid	Azetidine-2-carboxylate	Proline	Allo-4-hydroxy-proline	5-Hydroxypipe-colate
Diamino acid	α,γ-Diamino-butyrate	Ornithine	γ-Hydroxy-ornithine	δ-Hydroxylysine
Decarboxylation product	β-Alanine	γ-Amino-butyrate	γ-Hydroxy-aminobutyrate	α-Amino-δ-hydroxyvaleric acid
Amide	Asparagine	Glutamine	γ-Hydroxy-glutamine	δ-Hydroxy-α-amino-adipamide[a]
Carbamoyl or ureido product	γ-Ureido-α-aminobutyrate and carbamoyl aspartate	Citrulline	γ-Hydroxy-citrulline[a]	δ-Hydroxyhomo-citrulline[a]
ω-N-Acetyl derivative	N-Acetyl aspartate	N-Acetyl glutamate	N-Acetyl-δ-hydroxy-glutamate[a]	N-Acetyl-γ-hydroxy-adipate[a]
ω-Guanidino derivative	γ-Guanidine-α-amino-butyrate[a]	Arginine	γ-Hydroxy-arginine	γ-Hydroxy-homoarginine
Peptide	γ-Glutamyl aspartate	Glutathione	γ-Glutamyl amino acids	γ-Glutamyl peptide[a]

[a] Examples of compounds not yet known to occur in higher plants.

Obviously, the range of naturally occurring nitrogenous compounds continues to expand, and it reflects emphases that impinge upon the subject from without. The search for chemical carcinogens, especially nitrosamines present, or formed secondarily, in plants that may be used as foods for man or beast may add many such substances to the lists (140, 141, 161). The processing of large amounts of plant material and the scrutiny of residues that accumulate in great bulk, from the standpoint of their potential uses or abuses as pollutants, may again disclose substances (new or previously known) that would otherwise have escaped detection.

Areas in which such enquiries have been pursued are:

TABLE IV

SUGGESTED ROLE AND SIGNIFICANCE OF SOME NONPROTEIN NITROGEN COMPOUNDS[a]

Role or significance	Examples
Intermediates of synthesis	Carbamoyl aspartate (pyrimidines)
	Homoserine (threonine and methionine synthesis)
	O-Acetylserine (cysteine synthesis)
	β-Cyanoalanine (asparagine synthesis)
	Δ'-Pyrroline-5-carboxylic acid (proline synthesis)
Products from breakdown of proteins and nucleic acids	γ-Aminobutyric acid (from glutamic acid)
	β-Alanine (from uracil)
	β-Aminoisobutyric acid (from thymine)
Nitrogen-storage products	N-Ethylasparagine [as in cucurbits (Cucurbitaceae)]
	Azetidine-2-carboxylic acid [lily-of-the-valley (*Convallaria majalis*)]
	Guanidine and ureido compounds (forest trees)
	γ-Glutamyl peptides [onion (*Allium cepa*) and garlic (*A. sativum*)]
Nitrogen transport	Amides and ureides
	γ-Methyleneglutamine (across cell membranes)
	Homoserine [in peas (*Pisum sativum*)]
Metabolic or growth regulation	γ-Substituted glutamic acids (protein synthesis?)
	Azetidine-2-carboxylic acid (protein synthesis?)
	N-Formylmethionine (protein initiation)
	Indoleacetic acid (auxin activity)
	Di- and triisopentenylguanidines (cytokinins analogs)
	N-Substituted arginines (octopine and nopaline) (tumor growth)
Artifacts	Pyroglutamate[b] (ubiquitous)

[a] (22).

[b] Also known as pyrrolidone-2-carboxylic acid or 5-ketoproline. Pyroglutamate may yet prove to be a natural product; terminal glutamate residues in proteins may cyclize to give this structure [cf. Mazelius, M., and Pratt, H. M. (1976). *Plant Physiol.* **56,** 85–87].

1. The consequences of heavy uses of nitrogenous fertilizers (270)
2. Residues from large-scale processing of plant materials for food (5, 6, 28, 186, 239) fiber, or drugs (whether for pharmaceutical or social use) (e.g., 110)
3. The search for new compounds as nucleotide peptides, antibiotics, poisons, and toxins (105, 133, 152, 161, 193, 198)
4. The extension of the search for new nitrogenous compounds in plants to such hitherto unfamiliar environments as the sea (209), extraterrestial space (90), and even fossils (106, 128)

TABLE V

Possible Relationships between Nitrogen Compounds in Legumes
and the Presence or Absence of Seed Predators[a]

Legume species	Predator on seeds	Potentially toxic secondary chemicals in the host
Cassia grandis	Two Bruchidae species and at least two Lepidoptera species	No uncommon free amino acids
Entada polystachya	One Lepidoptera species larvae on green seeds	Unidentified uncommon amino acid
Gleditsia triacanthos	Infestation by bruchid species	5-Hydroxypipecolic acid
Gymnocladus dioicus	None	β-Hydroxy-γ-methylglutamic acid
Albizia julibrissin	None	2–3% albizziine; 2–3% S'-(β-carboxyethyl)cysteine
Enterolobium cyclocarpum	None	Albizziine
Dioclea megacarpa	Caryedes brasiliensis (Bruchidae)	5–10% Canavanine
Caesalpinia fonduc (Guilandina crista)	None	γ-Methylglutamic acid, γ-methyleneglutamic acid, γ-ethylideneglutamic acid
Hymenaea coubaril (in Costa Rica)	Rhinochenus spp. (Curculionidae) after pod resin lowered	No uncommon amino acids, no alkaloids
Hymenaea coubaril (in Puerto Rico)	None	No uncommon amino acids, no alkaloids
Griffonia simplicifolia	—	6–10% 5-Hydroxytryptophan

[a] Based on (196).

TABLE VI

Some Antagonists of Known Amino Acids[a]

Antagonist	Antagonizing amino acid	Producing organism
N-Acetyl-DON (duazomycin A) $NH_2\!=\!CHC\,CH_2CH_2CHCOOH$	Glutamine (inhibits GOGAT)[b]	Streptomyces
L-(N^5-Phosphono)methionine-(S)-sulfoximine	Glutamine	Streptomyces
Alanosine $O\!=\!NNCH_2CHCOOH$	Aspartic acid	Streptomyces
2-Amino-4,4-dichlorobutyric acid	Leucine	Streptomyces
2-Amino-3-dimethylaminopropionic acid	Leucine	Streptomyces
2-Amino-4-methylhex-5-enoic acid	Leucine	Streptomyces

TABLE VI (*Continued*)

Antagonist	Antagonizing amino acid	Producing organism
2-Amino-4-methylhex-4-enoic acid ⎫ β-Pyrazol-1-ylalanine ⎬	Phenylalanine	*Aesculus californica*
Azaserine	Glutamine, aromatic amino acids (inhibits GOGAT)	*Streptomyces*
Azetidine-2-carboxylic acid 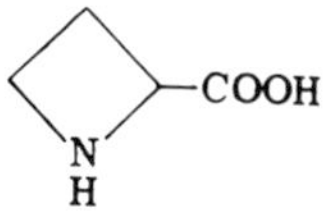	Proline	Liliaceae

Azaserine:
$$N_2{=}CHCOCH_2\,CHCOOH$$
$$\underset{O}{\|}\quad\underset{NH_2}{|}$$

Azetidine-2-carboxylic acid:

Azotomycin (duazomycin B)	Glutamine	*Streptomyces*

$$N_2{=}CHCOCH_2CH_2CHCOOH$$
$$|$$
$$NH$$
$$|$$
$$CO$$
$$|$$
$$N_2{=}CHCOCH_2CH_2CH$$
$$|$$
$$NH$$
$$|$$
$$HOOCCHCH_2CH_2CO$$
$$|$$
$$NH_2$$

Borrelidin 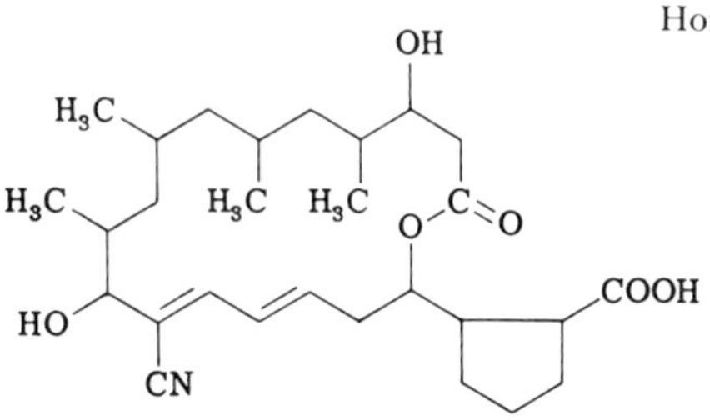	Threonine Homoserine	*Streptomyces*

Canavanine	Arginine	*Canavalia*

$$H_2NCNHOCH_2CH_2\,CHCOOH$$
$$\underset{NH}{\|}\qquad\underset{NH_2}{|}$$

O-Carbamoyl-D-serine	D-Alanine	*Streptomyces*

$$H_2NCOCH_2CHCOOH$$
$$\underset{O}{\|}\quad\underset{NH_2}{|}$$

(*continued*)

TABLE VI   (*Continued*)

Antagonist	Antagonizing amino acid	Producing organism
3-Cyclohexeneglycine	Homoserine, threonine, isoleucine	*Streptomyces*

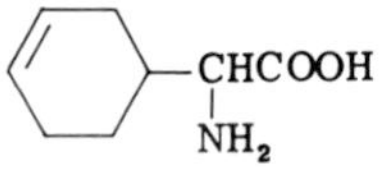		
D-Cycloserine	D-Alanine	*Streptomyces*
2,3-Diaminosuccinate	Aspartic acid	*Streptomyces*
6-Diazo-5-oxonorleucine (DON)	Glutamine	*Streptomyces*
L-2,5-Dihydrophenylalanine	Phenylalanine, tyrosine	*Streptomyces*
2-Amino-4-methylhex-4-enoic acid	Phenylalanine	*Aesculus*
Furanomycin	Isoleucine, valine	*Streptomyces*

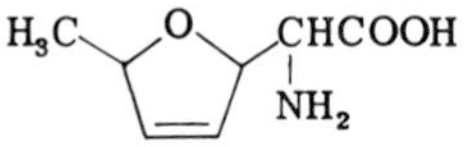		
Hadacidin	Aspartic acid	*Penicillium*

$$HC-N-CH_2COOH$$
$$\| \quad |$$
$$O \quad OH$$

Homoarginine	Arginine	*Lathyrus*
5-(O-Isoureido)-L-norvaline	Arginine	Bacteria
β-Hydroxy-L-leucine	Various amino acids	*Deutzia*
Ketomycin	Homoserine, threonine, isoleucine	*Streptomyces*

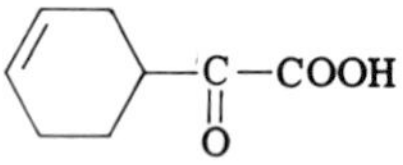		
α-(Methylenecyclopropyl)glycine	Leucine	Sapindaceae
Mimosine	Phenylalanine, tyrosine	*Mimosa*

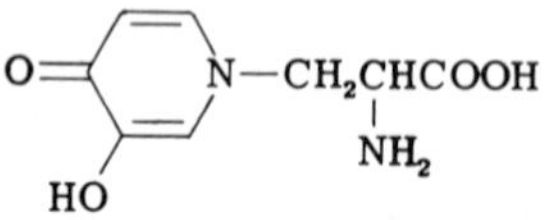		
Methionine sulfoxime	Glutamine; inhibits glutamine synthetase	—

a Other substances, particularly those in Appendix II, provide a wider range of alternate antagonists.

b See metabolic Charts 1 and 8 for GOGAT.

Under these categories many claims have been made. Large-scale processing of sugar beets (*Beta vulgaris*) has released residues in bulk (10^5 tons of residue yielded 50,000 kg of cationic nitrogenous compounds) (82). From this, 30 kg of azetidine-2-carboxylic acid, not previously found in the Chenopodiaceae, were obtained. This represented a natural level of 0.3 ppm, which was said to be below the sensitivity of routine chromatographic methods. Other compounds found that were familiar as in 1964 were pipecolic acid, baikiain, and some *N*-acylated diamino acids. Obviously, however, some caution is needed lest one attributes all such occurrences specifically to the crop plant in question or to its inevitable contamination with other flora and fauna.

But all these factors greatly complicate the recognition of normal nitrogenous constituents of plants. Therefore, a number of appendixes to this chapter, following on a useful pattern developed in the earlier one, are attached. These appendixes cope concisely with the number, identity, sources, and bibliographic citations of compounds that have accrued since 1964. Many other authors have addressed themselves to specific aspects of this whole problem as indicated in the references cited, namely, Bell (16, 17), Bidwell and Durzan (22), Davies (47), Drey (49), Fowden (79–82, 84, 85), Fowden *et al.* (88), Mauger and Witkop (150), Murakoshi and Hatanaka (170), Northcote (174), Oaks and Bidwell (175), Shrift (216), Synge (245), Thompson *et al.* (253), Umbarger (263, 264), Waley (275), and Weinstein (280).[2]

Appendix I (p. 200) presents the evidence for 23 substances that are derivatives of established protein amino acids in higher plants. Acceptance of the occurrence of amino acids actually in proteins of plant should follow rigorous criteria (94, 269). The compounds that are reported in this appendix do not meet all these criteria, and they should therefore be considered as substances "related to" the established protein amino acids but not necessarily as integral parts of the protein molecule as synthesized *de novo*, except where this is specifically substantiated. With special reference to microbial and animal proteins, Uy and Wold (265) believe that many protein amino acids (especially lysine and tyrosine, like proline) may be modified *in situ*. For these changes the term *posttranslational covalent modification* is used. If this principal is accepted, some 20 original protein amino acids may yield some 140 different derivatives. However, these are not established for plants, and their relevance for the release of the modified amino acids, as in breakdown,

[2]An effective summary of soluble nitrogen compounds is provided by G. R. Rosenthal in "Plant Non-Protein Amino and Imino Acids: Biological, Biochemical and Toxicological Properties." Academic Press, New York, 1982.

and their reincorporation, as in resynthesis and turnover, is quite unknown.

Appendix II (p. 204) lists new free amino and imino acids and their derivatives that occur free in plants and that have been recorded since 1964. The appendix as presented contains some 140 entries. Appendix II, as compiled, includes many substances already listed or referred to in Appendix I as possible constituents of protein; it also includes some alkaloids, cyanogenetic glucosides, simpler natural bases, purines and pyrimidines, as well as decarboxylation products of amino acids (i.e., amines). A large number of interesting nitrogenous compounds have been detected in urine, and though they may not yet be reported from plants, they may later be shown to have a botanical significance. For compounds not listed in Appendix II that occur free in micro-organisms, in fungi, or in animals, reference may be made to *Specialist Periodical Reports* [The Chemical Society (London)] for details.

In summary, the list of 140+ entries of Appendix II is a conservative compilation representing an even greater number of compounds that may occur free in nature; these may present a great diversity of chemical structures: (a) derivatives of monoaminomonocarboxylic acids (p. 204); (b) derivatives of unsaturated monoaminocarboxylic acids and their derivatives (p. 204); (c) derivatives of monoaminodicarboxylic acids and their amides (p. 206); (d) Diamino-, mono-, di-, and tricarboxylic acids and their derivatives (p. 208); (e) hydroxyamino acids and their derivatives (p. 210); (f) aromatic amino acids and their derivatives (p. 212); (g) heterocyclic compounds with amino or imino groups (p. 214); (h) *N*-methylated amino and imino acids (p. 227); and (i) sulfur- and selenium-containing amino acids and their derivatives (p. 229).

The primary purpose of Appendix III (p. 232) is to list the many compounds that are collectively substituents at the γ-position of glutamic acid, of the carbamoyl moiety, and of the amidine or guanidine moiety. The total number of entries in these three categories is of the order of 170. In Appendix IIId, the substituted guanidines have been implicated in some 70 instances in responses (as mutagens, as analgesics, as carcinogens, or as general antimetabolites) that endow them as a class, whether natural or synthetic, with particular broad biological significance. In summary, Appendix III consists of: (a) γ-glutamyl compounds (p. 232), (b) urea and carbamyl derivatives (p. 234), (c) free guanidine and amidino compounds (p. 241), and (d) selected properties of monosubstituted guanidines (p. 255).

Wherever, as in Appendixes I–III, one recognizes that α-amino groups exist in proximity to carboxyl groups, whether in free amino acids or the very large number of their derivatives by substitution, one may also expect that two families of chemical compounds may derive from them. These are amines by decarboxylations and keto acids by transaminations.

When this survey of nitrogen compounds occurring in plants was first compiled, appendixes devoted to these two categories of compounds were assembled. From the first observation of methylamine from glycine in 1875 to the present time, the decarboxylations of amino acids have yielded a very large number of amines, many of which arise from bacterial actions or stresses due to mineral deficiencies, diseases, etc. If such a survey now included all such examples from bacterial biochemistry, it would be greatly enlarged. Therefore, and because the involvement of these often volatile amines in further plant metabolism or physiological responses is meagerly documented, this large category of substances, their sources, identification, etc. are not presented here in full. Moreover, this omission is compensated because Smith (220–223) has reviewed these topics in detail.

However, because plants retain the carbon frameworks from the transamination reactions as keto acids and utilize these in ongoing metabolism or synthesis, Appendix IV, as a supplement to the earlier list of keto acids by Steward and Durzan (230), is given in detail.

Appendix IV (p. 261) deals with the keto acids of plants, their amino acid analogs, and their derivatives. Historically, the first recognized "ports of entry for nitrogen" into organic combinations were the keto acids of the tricarboxylic acid and glyoxylate cycles. In the earlier account, numerous keto acid analogs of amino acids were recognized. For example, the keto acid pyruvic aldol was known before its amino acid analog, γ-hydroxy-γ-methylglutamic acid. In the intervening years more keto acid–amino acid relationships have become evident as the more than 70 entries of Appendix IV show.

While much is covered in this chapter, of necessity much detail has been omitted because of space limitations. In particular, stereochemical studies on the metabolism of amino acids have been very well reviewed by Aberhart (2) and are not covered in this chapter.

III. The Biochemical Reactions of Nitrogen Metabolism in Plants

A. Nonprotein Nitrogen

The next task is to visualize the metabolic pathways by which carbon frameworks and nitrogen may pass from their various ports of entry into organic combination between nitrogen and carbon and to the many classes of compounds (simple and complex) that may be elaborated. To do this, 24 metabolic charts have been prepared, each of which presents a specific aspect of nitrogen metabolism through the use of detailed schemes accompanied by interpretive notes. The first series, Charts 1–19, concerns the metabolism of soluble nitrogen compounds. The second series, Charts 20–24 (p. 127) deal with protein synthesis.

CHART 1

PRIMARY NITROGEN ASSIMILATION: C_5 PORTS OF ENTRY[a,b]

Nitrogen port (C_5)	Nitrogen transfer to amino acids and amides	Some metabolic products of glutamate (cf. Chart 11)

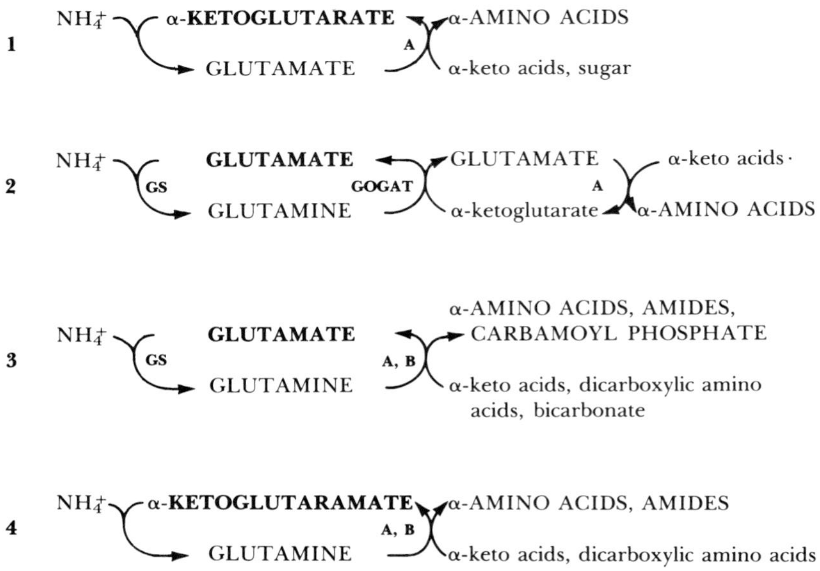

1
NH$_4^+$ → α-**KETOGLUTARATE** → α-AMINO ACIDS
GLUTAMATE ← α-keto acids, sugar (via A)

- Glutamic-γ-semialdehyde
- N-Acetylglutamate
- γ-Acetylglutamate
- γ-Glutamyl peptides
- Pyroglutamate (5-ketoproline, pyrrolidone-2-carboxylate)
- Proline

2
NH$_4^+$ (GS) → **GLUTAMATE** → GLUTAMATE → α-keto acids · α-AMINO ACIDS
GLUTAMINE ← α-ketoglutarate (via GOGAT, A)

- γ-Aminobutyrate
- Protein, glutathione
- Folate compounds
- Proline, histidine, arginine
- Lysine, δ-aminolevulinate, chlorophyll (Chart 14B)
- Carbamoyl phosphate

3
NH$_4^+$ (GS) → **GLUTAMATE** → α-AMINO ACIDS, AMIDES, CARBAMOYL PHOSPHATE
GLUTAMINE ← α-keto acids, dicarboxylic amino acids, bicarbonate (via A, B)

- Asparagine
- N^5-Ethylglutamine
- Purine-ring nitrogen (Chart 15)
- Pyrimidine amino nitrogen (Chart 15)
- Glucosamine
- Alanine, tryptophan
- Succinamic acid, succinimide

4
NH$_4^+$ → α-**KETOGLUTARAMATE** → α-AMINO ACIDS, AMIDES
GLUTAMINE ← α-keto acids, dicarboxylic amino acids (via A, B)

- Isosuccinimide-β-glycoside?
- N-Terminal pyroglutamate (protein)
- NAD-amide nitrogen

(continued)

CHART 1 (Continued)

^{a} Key: **1**, glutamate dehydrogenase: α-Ketoglutarate + NAD(P)H + NH$_3$ $\rightleftharpoons$ L-glutamate + NAD(P)$^+$ + H$_2$O; **2**, glutamine synthetase (GS): L-glutamate + NH$_3$ + ATP $\rightleftharpoons$ L-glutamine + ADP + P$_i$ + H$_2$O; and glutamate synthase (GOGAT): α-ketoglutarate + L-glutamine + NADPH (or NADH or ferredoxin red.) → 2-L-glutamate + NADP$^+$ (or NAD$^+$ or ferredoxin ox.); **3**, glutamine synthetase; **4**, α-ketoglutaramate dehydrogenase; **A**, transaminase; **B**, transamidase. Nitrogen acceptors are in **boldface** capitals; reaction products in lightface capitals.

^{b} Interpretive note for Chart 1: Reduced nitrogen (NH$_4^+$ or NH$_3$; see p. 115 for "enzyme–ammonia" intermediate) may enter in each of four possible ways by reactions mediated by specific enzymes (**1–4**) to produce a first-formed nitrogenous product. By a transaminase **A** or a postulated transamidase **B** nitrogen may then be transferred from the primary product to produce the range of compounds shown.

Enzyme system **1** of Chart 1 has had historical priority since the early 1930s (271). System **2** is designated by GS GOGAT {glutamine (amide): 2-oxoglutarate aminotransferase [oxidoreductase NADP]}. 2-Oxoglutarate is an alternative designation for α-ketoglutarate} after its discovery in bacteria GOGAT (see Chart 8) also occurs in plants, especially in leaves active in nitrate reduction (159). It is now accepted widely (160, 199) but may not function to the exclusion of enzyme System **1** (218). In fact, the glutamate dehydrogenase of System **1** may be preferred at high levels of NH$_4^+$ availability. Enzyme Systems **2** and **3** may operate at lower NH$_4^+$ levels. System **4** is distinguished by its acceptor α-ketoglutaramate, which produces glutamine. The extent of the contribution of this system, which has been detected in pine (*Pinus*) seedlings and soybean (*Glycine*) and pea (*Pisum*) seeds, to the overall assimilation of nitrogen is still unknown. Because germinating seeds, where System **4** is characteristically found, are largely concerned with remobilizing their nitrogenous reserves, it may have little relevance to the primary autotrophic intake of organic nitrogen.

A variation of glutamine synthetase, System **3**, has been found in plants and animals where bicarbonate accepts the amide nitrogen of glutamine to produce carbamoyl phosphate (cf. Chart 5). This alternative leads to the production of pyrimidines, citrulline, urea, and biotin.

Buchanan (31) has reviewed enzymes that catalyze the transfer of the amide group of glutamine to a substrate to form new C—N bonds. The deamidation of glutamyl and asparaginyl residues in proteins occurs readily and widely and is responsible for numerous *in vitro* artifacts (201).

CHART 2

PRIMARY NITROGEN ASSIMILATION: C_4 PORTS OF ENTRY[a,b]

Nitrogen port (C_4)	Nitrogen transfer to amino acids or amides		Some metabolic products
1 NH$_4^+$ ⤳ **OXALOACETATE** ⤴ ↘ ASPARTATE (A)	α-AMINO ACIDS ↗ ↘ α-keto acids, sugar		Of aspartate (cf. Chart 12) Aspartate-β-semialdehyde Carbamoylaspartate (cf. Chart 15) 5-Amino-4-imidazole *N*-succinocarboxamide (cf. Chart 15)
2 NH$_4^+$ ⤳ **ASPARTATE** ⤴ ↘ { AMP + PP$_i$ (microbes) or ADP + P$_i$ (plants) ASPARAGINE (ATP)	α-AMINO ACIDS or AMIDES ↗ **A or B** ↘ α-keto acids or dicarboxylic amino acids		Argininosuccinate Azetidine-2-carboxylate α,γ-Diaminobutyrate γ-Oxalylamino-α-aminobutyrate Methionine, threonine, alanine Homoserine, lysine γ-Glutamyl peptide 2-Hydroxysuccinamic acid
3 NH$_4^+$ ⤳ α-**KETOSUCCINAMATE** ⤴ ↘ ASPARAGINE	α-AMINO ACIDS or AMIDES ↗ **A or B** ↘ α-keto acids or dicarboxylic amino acids		Of asparagine Glutamine, aspartate Ammonia
4 NH$_4^+$ ⤳ **SUCCINIC SEMIALDEHYDE** ⤴ ↘ γ-AMINOBUTYRATE (A)	α-AMINO ACIDS ↗ ↘ α-keto acids		Of γ-aminobutyrate γ-Glutamyl peptide Glutamate

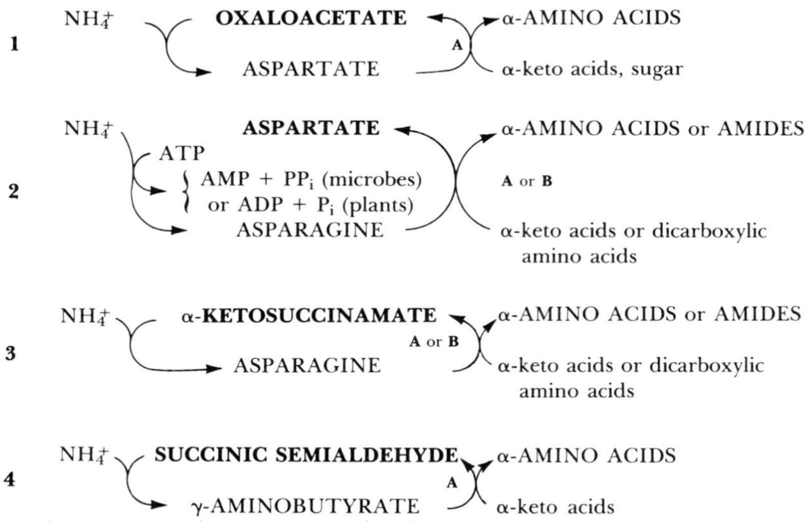

[a] Key: **1**, aspartate dehydrogense; **2**, asparagine synthetase; **3**, α-ketosuccinamate dehydrogenase; **4**, γ-aminobutyrate dehydrogenase; **A**, transaminase; **B**, transamidase. Nitrogen acceptors are in **boldface capitals**; reaction products are in lightface capitals.

[b] Interpretive note for Chart 2. Reduced nitrogen (NH$_4^+$ or NH$_3$) may here enter in four ways. Enzymes designated **1** to **4** all act to promote the primary entry of reduced nitrogen. Subsequently, nitrogen may be transferred by transaminases **A** or transamidases **B** to produce the range of compounds shown.

(*continued*)

CHART 2 (Continued)

Enzyme System 1 of Chart 2 has been a historical link between carbohydrate and nitrogen metabolism since the first knowledge of the Krebs cycle. The formation of free amino acids by the three types of reactions of the C_4 photosynthethic cycle have been described in Volume VII, Chapter 3, Fig. 14. System 2, negotiating the fixation of a second NH_2 group onto aspartate has been observed in potatoes and corn. Thereafter, the amide group of the resultant asparagine may be transferred to a keto acid acceptor, regenerating aspartate and so forming a range of amino acids. System 3 is analogous to System 4 of Chart 1 and produces asparagine, from which the amide group may be transferred to a keto acid to form a range of amino acids by transamination or transamidation while regenerating the acceptor. It has been observed in conifers and peas (*Pisum*) (96, 151). System 4 is analogous to System 1, but uses succinic-γ-semialdehyde as a substrate.

System 2 has proved difficult to establish in higher plants (22). Alternatively, glutamine has been postulated as an amide (NH_2) donor in lieu of NH_4^+ for the production of asparagine (e.g., 203, 204), as follows:

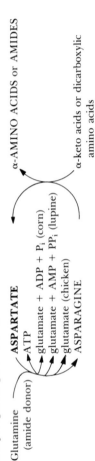

However, this enzyme system does not properly constitute a primary port of entry for nitrogen, but rather redistribution of glutamine nitrogen. It is shown here because of its close analogy to asparagine synthetase.

CHART 3

PRIMARY NITROGEN ASSIMILATION: C$_3$ PORTS OF ENTRY[a,b]

Nitrogen port (C$_3$)	Nitrogen transfer to amino acids and amides	Some metabolic products
1 NH$_4^+$ ⟋ **PYRUVATE** ↘ ALANINE	⟋ α-AMINO ACID **A** ↘ α-keto acid	Of alanine (cf. Chart 13) Numerous alanine β-substituted heterocyclic derivatives from O-acetylserine Ethylamine D-Alanine

2   C≡N$^-$ ⟋ **CYSTEINE**   ←   O-acetylserine
      ↘ β-CYANOALANINE → ASPARAGINE
           B
		Of β-cyanoalanine
		
		 γ-Glutamyl-β-cyano-alanine

[a] Key: **1,** alanine dehydrogenase; **2,** β-cyanoalanine synthetase; **A,** transaminases; **B,** β-cyanoalanine hydratase. Nitrogen acceptors in **boldface** capitals; reaction products in lightface capitals.

[b] Interpretive note for Chart 3. Reduced nitrogen may enter the C$_3$ port by two pathways: as NH$_4^+$ or NH$_3$ via pyruvate to alanine or as cyanide via cysteine to β-cyanoalanine and then to asparagine (41). Enzymes **1** and **2** catalyze the primary entry of the nitrogen, and subsequent reactions are mediated by transaminases **A** or a specific enzyme **B** that converts β-cyanoalanine to asparagine.

Enzyme System **1** has been recognized widely as a link between carbohydrate and nitrogen metabolism since the knowledge of glycolysis, the Krebs cycle, and transamination (15, 230, 271, and see Volume VII, Chapter 3). With the natural occurrence in plants of a wide range of alanine derivatives (Appendix II), the existence of analogous keto acids may be anticipated and their role as acceptors of nitrogen, as in System **1** here, may be presumed. System **2** is characteristic of plants that form cyanogenic glycosides, and its interest lies in the alternative route to asparagine so provided (39–41). The reaction has been found in lupine (*Lupinus*), sorghum (*Sorghum*), cotton (*Gossypium*), and cassava (*Manihot*).

CHART 4

Primary Nitrogen Assimilation: C_2 Port of Entry[a,b]

Nitrogen port (C_2)	Nitrogen transfer to amino acids	Some metabolic products of glycine (cf. Chart 14)

1   NH₄⁺ ⟍ GLYOXYLATE ⟍ α-AMINO ACIDS
          A
      ➤GLYCINE ⟍ α-keto acids
		Purines (cf. Chart 15)
		Sarcosine
		Glycine betaine
		N-Formyl- or N-hydroxyglycine
		Glutathione
		N-Methylphenylglycine
		m-Carboxyphenylglycine
		Porphyrins

[a] Key: **1,** glycine dehydrogenase; **A,** transaminases. Nitrogen acceptor in **boldface** capitals; reaction products in lightface capitals.

[b] Interpretive note for Chart 4. The importance of glyoxylic acid as a potential port of entry for nitrogen came relatively late with the discovery of its role in the glyoxylate cycle (cf. 14, 294, and Volume VII, Chapter 3, Fig. 9), which constitutes a bypass in the citric acid cycle (cf. metabolic Chart 9B). This system is unique among the ports of nitrogen entry described because it leads directly to glycine, which may then, by transamination reactions, give rise to various α-amino acids and, as shown in metabolic Charts 14A,B, to a great variety of other complex compounds, including chlorophyll, porphyrins, purines, peptides, etc.

CHART 5

PRIMARY NITROGEN ASSIMILATION: C_1 PORT OF ENTRY[a,b]

Nitrogen port (C_1)	Nitrogen transfer to amino acids, etc.	Some metabolic products of carbamoyl phosphate (cf. Chart 11)
NH_4^+ (or carbamate) (or glutamine + ATP) **1**	**BICARBONATE** 2 ATP 2 ADP + P_i glutamate + ADP CARBAMOYL **A** PHOSPHATE → CITRULLINE **B** → CARBAMOYL ASPARTATE	Arginine, citrulline Pyrimidines (from carbamoyl aspartate) Biotin Adenosine triphosphate N-Carbamoylputrescine Carbamoyl-β-alanine Ureidocytokinins (?) Thiamine from β-methyl aspartate

[a] Key: **1**, carbamoyl-phosphate synthetase; **A**, ornithine transcarbamylase; **B**, aspartate transcarbamylase. Nitrogen acceptor in **boldface** capitals; reaction products in lightface capitals.

[b] Interpretive note for Chart 5. Reduced nitrogen in the form of NH_4^+, NH_3, or (NH_2) may enter the C_1 port via bicarbonate to carbamoyl phosphate as mediated by carbamoyl phosphate synthetase (**1**) with ATP as the source of energy. From carbamoyl phosphate, transfer reactions mediated by the appropriate transcarbamylase **A** or **B** may lead to a variety of important compounds.

Although NH_4^+ participates with difficulty in enzyme System **1**, it may be replaced by glutamine as an effective donor of amide nitrogen (119, 120, 154, 179, and cf. System **2** of Chart 2). However, this does not constitute a primary port of entry for inorganic nitrogen. In certain microorganisms, formic acid will accept ammonia to produce formamide and water with formamide amidohydrolase (250). Whereas the chart shows the nitrogen products from carbamoyl phosphate to be carbamoylaspartate (e.g., 293) and citrulline (208), it should be noted that the phosphate moiety may pass to glucose as glucose 6-phosphate (147), and N-carbamoyl putrescine may be formed (126). Carbamoyl phosphate is unstable (4), and the related compound carboxyphosphate ($^-O_3C—PO_3{}^{2-}$, also known as carbonyl phosphate or carbonic phosphoric anhydride) has been proposed as a transient metabolic intermediate. It is formed by glutamine-dependent carbamoyl phosphate synthetase and has been proposed as an intermediate in reactions catalyzed by biotinyl carboxylases (288).

In plants lacking urease, such as *Chlorella*, allophanic acid ($NH_2CONHCOOH$) has been considered as an intermediate of urea amidolyase leading to the production of ammonia and carbon dioxide (metabolic Chart 18). In this case the ammonia comes from urea and may recombine with a single- or multiple-carbon acceptor to enter a range of nitrogenous compounds (109, 254, 282).

CHART 6

NITROGEN METABOLISM FROM C_2 AND C_3 PORTS[a,b]

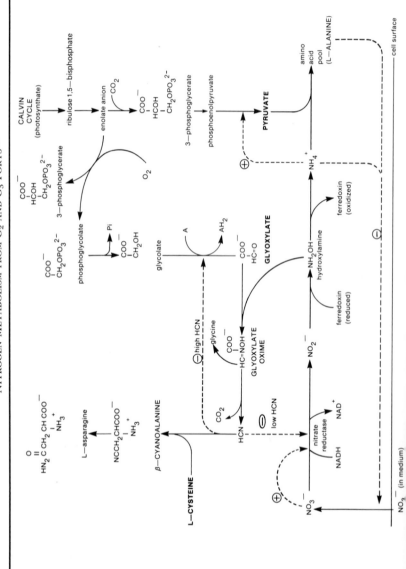

[a] Key: $\xrightarrow{\ominus}$, negative feedback control (inhibitory); $\xrightarrow{\oplus}$, positive feedforward control (stimulating). Nitrogen acceptors in **boldface** capitals; reaction products in lightface capitals.

[b] Interpretive note for Chart 6. This chart, modified from Solomonson and Spehar (224), traces nitrogen assimilation from nitrate and its reduction products (hydroxylamine and ammonia). Ammonium may enter at any of five ports of entry (cf. Charts 1–5) according to their number of carbons, but it is here shown as it may be accepted by C_3 products from photosynthesis leading to alanine via pyruvate. Hydroxylamine is shown combining with a C_2 acceptor (glyoxylate), which, via an oxime and reduction, may yield glycine. Alternatively, in plants that use cyanide, the glycine may arise from the oxime and the nitrogen be reincorporated at a C_3 port en route to L-asparagine. This metabolic system has implications for the regulation of nitrate assimilation (224).

As shown on this Chart, NH_4^+ is regarded as the principal inorganic substance actually assimilated into organic combination; there is, nevertheless, historical precedent for hydroxylamine via oximes (cf. 230). The degrees of freedom represented in Chart 6 raise obvious problems of metabolic regulation to determine the respective fates of hydroxylamine and ammonia.

Between the input of NO_3^- from the medium (as the nitrogen source) and the availability of C_3 compounds from photosynthesis (for the carbon), there lie several possibilities according to the gene-determined enzymes available.

Thus the supply of C_3 acceptors via photosynthesis links alanine production to green cells. Alternatively, the reduction of oxime to cyanide, leading to asparagine, links this pathway to cyanide-metabolizing plants and their specific enzymes (41). But the reduction of oxime to glycine may also be emphasized in glyoxylate-metabolizing plants as in photorespiration (metabolic Chart 8). Furthermore, alanine, NH_4^+, NO_3^-, or cyanide may control the overall pace of reaction pathways by their intervention at the specific enzymatic steps indicated.

CHART 7

NITROGEN METABOLISM FROM A C_3 PORT: PYRUVATE ALDOL[a,b]

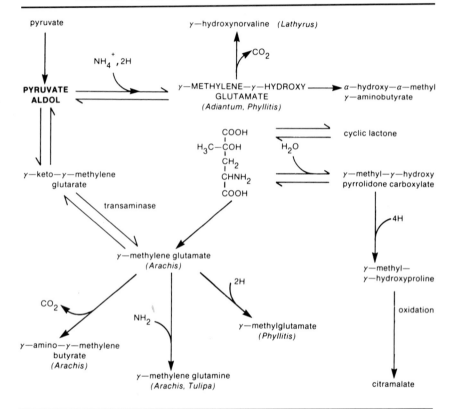

[a] Nitrogen acceptors in **boldface** capitals; reaction products in lightface capitals.

[b] Interpretive note for Chart 7. This chart traces nitrogen metabolism via NH_4^+ at an important C_3 port of entry (pyruvate; cf. Chart 3), but in this case in the form of its aldol, which yields γ-methyl-γ-hydroxyglutamate as the primary product. From this pivotal substance several pathways may flow as shown for the designated plants.

The amino acid product, γ-methyl-γ-hydroxyglutamic acid, was known in *Adiantum* before its keto acid acceptor (pyruvic aldol) was recognized, and from this substance the amino acid was eventually synthesized (cf. 230). Its relationships to γ-methylene derivatives as in liliaceous plants (especially *Tulipa*) and in *Arachis* and *Phyllitis* were first seen through the application of two-directional paper chromatography. The metabolic possibilities of γ-methyl-γ-hydroxyglutamic acid in these plants are evident. The general significance for other plants remains to be determined.

CHART 8

Nitrogen Metabolism from a C_5 Port: Glutamine and Glutamate Formation in Photosynthesizing Leaves[a,b]

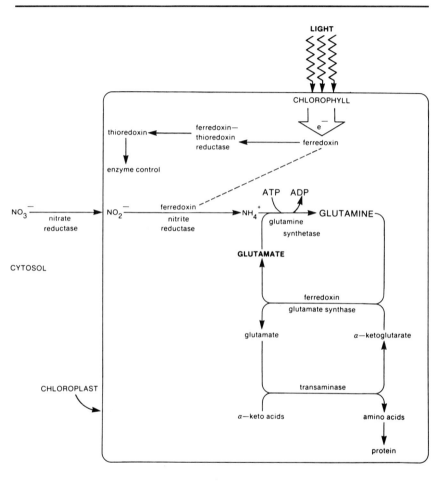

[a] Nitrogen acceptors in **boldface** capitals; reaction products in lightface capitals.

[b] Interpretive note for Chart 8. This chart, modified from Miflin and Lea (160), deals with metabolism of primary glutamine formed in chloroplasts from glutamate and ammonia derived from nitrate reduction in light. This reaction is mediated by glutamine synthetase (GS; see Chart 1) using energy derived from photosynthetic ATP. The amide nitrogen of glutamine produced by the GS reaction may be transferred to α-ketoglutarate (2-oxoglutarate) by the glutamate synthase reaction (GOGAT, for glutamine: 2-oxoglutarate aminotransferase) (for distinction between synthetase and synthase see p. 108). Details of the reactions are as follows:

(*continued*)

CHART 8 (*Continued*)

Glutamine synthetase (GS; EC 6.3.1.2)

$$\text{L-Glutamate} + NH_3 + ATP \xrightarrow{Mg^{2+}} \text{L-Glutamine} + ADP + P_i$$

Glutamate synthase (GOGAT; EC 1.4.1.13)

$$\text{L-Glutamine} + \alpha\text{-ketoglutarate} + NAD(P)H \xrightarrow[\text{microorganisms}]{\text{plants,}} 2\ \text{L-glutamate} + NAD(P)^+$$

or (EC 1.4.7.1)

$$\text{L-Glutamine} + \alpha\text{-ketoglutarate} \xrightarrow{\text{chloroplasts}} 2\ \text{L-glutamate}$$
$$+ 2\ \text{ferredoxin (reduced)} \qquad\qquad + 2\ \text{ferredoxin (oxidized)}$$

Unlike most glutamine aminotransferases, neither reaction can use ammonia as a substrate. Tyler (261) has suggested that enzymes able to use ammonia may be contaminated with glutamate dehydrogenase.

The sum of the reactions of the GS–GOGAT system is

$$\alpha\text{-Ketoglutarate} + NH_3 + ATP + NADPH \rightarrow \text{L-glutamate} + ADP + P_i + NADP^+$$

This reaction should be compared with glutamate dehydrogenase, which is energetically slightly less costly but very much less effective (see Chart 1):

$$\alpha\text{-Ketoglutarate} + NH_3 + NAD(P)H \rightarrow \text{L-glutamate} + NAD(P)^+ + H_2O$$

The GS–GOGAT system is powered by photosynthetically generated reducing power (as ferredoxin in chloroplasts) and ATP, or by the dehydrogenase system by NADPH + H$^+$ in nonphotosynthetic cells. The effectiveness by which light energy can be used to reduce and fix nitrogen goes far to explain the high levels of protein synthesis in illuminated green leaves. Furthermore, ferredoxin may be linked through ferredoxin–thioredoxin reduction to the thioredoxin control mechanism thought to be responsible for regulating chloroplast enzymes (32, and see Volume VII, Chapter 3, Section II,D).

CHART 9

FORMATION AND METABOLISM OF THE PROTEIN AMINO ACIDS: AMINO ACID FAMILIES[a]

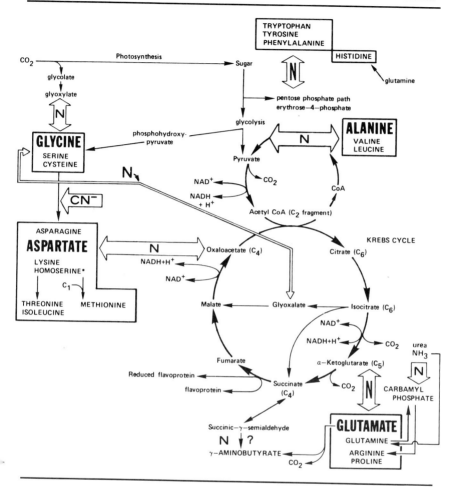

[a] Interpretive note for Chart 9. Nitrogen as ammonium or cyanide may enter into organic combination with various C_1 to C_5 carbon acceptors at eight points, as shown. The entry points for nitrogen are designated with open arrows marked N (for ammonium) or CN^-. Groups of amino acids derived from the same carbon framework are called "families." Each family is placed within a box. The "family head" from which other members of the family are derived is printed in **boldface** capitals.

The formation of the 20 protein amino acids and γ-aminobutyrate from carbon derived from photosynthesis and respiration proceeds through nitrogen ports of entry via reactions involving nitrogen transfer, oxidoreduction, hydrolases, lyases, isomerases, and synthetases. Aspartate, glutamine, and glycine provide nitrogen for the formation of purines and pyrimidines. The following should be noted: the entry of a C_2 moiety as acetyl-CoA into the Krebs cycle; the oxidations by H removal; H accepted by NAD^+, which then passes over the electron-transport system eventually to oxygen to form water; the removal of CO_2 by decarboxylations; intermediates of the Krebs cycle showing the number of carbons per molecule and the possible bypass via glyoxalate (glyoxylate cycle). The basic steps in the biosynthesis of the protein amino acids in the various families of amino acids are in many ways similar to those found in microorganisms (e.g., 263, 264).

CHART 10

SYNTHESIS AND METABOLISM OF AROMATIC AMINO ACIDS AND HISTIDINE

(A) Phenylalanine, tyrosine, and tryptophan[a,b]

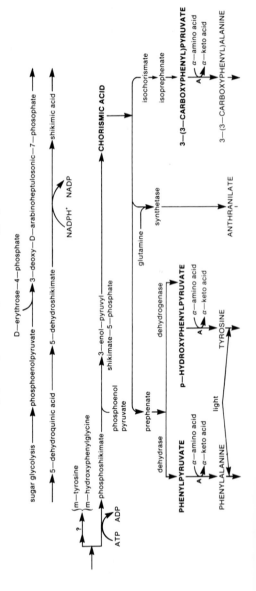

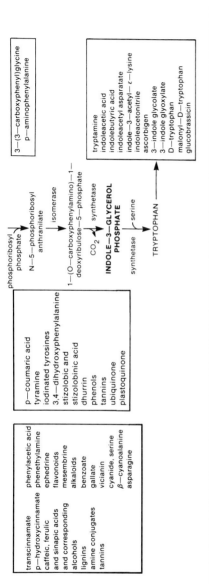

3—(3—carboxyphenyl)glycine
p—aminophenylalanine

```
phosphoribosyl phosphate
        ↓
N—5—phosphoribosyl anthranilate
        ↓ isomerase
1—(O—carboxyphenylamino)—1—deoxyribulose—5—phosphate
        ↓ CO₂  synthetase
INDOLE—3—GLYCEROL PHOSPHATE
   synthetase ↓ serine
TRYPTOPHAN ⟶
```

tryptamine
indoleacetic acid
indolebutyric acid
indoleacetyl asparatate
indole—3—acetyl—ε—lysine
indoleacetonitrile
ascorbigen
3—indole glycolate
3—indole glyoxylate
D—tryptophan
malonyl—D—tryptophan
glucobrassicin

transcinnamate   phenylacetic acid
p—hydroxycinnamate   phenethylamine
caffeic, ferulic   ephedrine
and sinapic acids   flavonoids
and corresponding   mesembrine
alcohols   alkaloids
lignins   benzoate
amine conjugates   gallate
tannins   vicianin
   cyanide, serine
   β—cyanoalanine
   asparagine

p—coumaric acid
tyramine
iodinated tyrosines
3,4—dihydroxyphenylalanine
stizolobic acid
stizolobinic acid
dhurrin
phenols
tannins
ubiquinone
plastoquinone

a Key: **A**, transaminase; nitrogen acceptors in **boldface** capitals; reaction products in lightface capitals. Compounds in box have a common precursor.

b Interpretive note for Chart 10A. This chart gives an idea of the large number of metabolites formed from phenylalanine, tyrosine, and tryptophan. The aromatic and heterocyclic amino acids originate from glucose via heptulose and shikimic acid (46, 99). The ammonia lyases of phenylalanine and tyrosine give rise to *trans*-cinnamic acid and *p*-coumaric acid, respectively. Phenylalanine and tyrosine, the *p*-hydroxy derivative of phenylalanine, are the precursors of many important metabolites leading to coenzymes (ubiquinone), tannins, lignins, and inhibitors commonly found in seed coats (226). With few exceptions all biological benzene derivatives are synthesized through these two amino acids. Thus the side chain is the important component, and the amino acid component is concerned with placing the cyclic side chain appropriately in protein molecules. Phenylalanine, tyrosine, and 3-(3-carboxy-phenyl)alanine are formed by transamination from various α-amino acids to the substituted pyruvic acid as the nitrogen acceptor. Tryptophan or β-indolealanine is synthesized via a parallel pathway (see 283–285). The entry of nitrogen into tryptophan is mediated by synthetases. The first takes glutamine and chorismic acid to produce anthranilic acid. The second takes serine and indole-3-glycerol phosphate to form L-tryptophan. Each of these aromatic amino acids may give rise, as shown on the chart, to a number of secondary products (146). The control of enzymatic reactions associated with phenylpropanoid metabolism by substrate supply has been examined (148), and enzymatic controls in the biosynthesis of lignins and flavonoids have been reviewed (98). Many indole derivatives are important as auxins or growth regulators (231). The biological activities of indoleacetylamino acids and their use as auxins have been evaluated by Hangarter *et al.* (101).

(*continued*)

CHART 10 (*Continued*)

(B) Histidine[c,d]

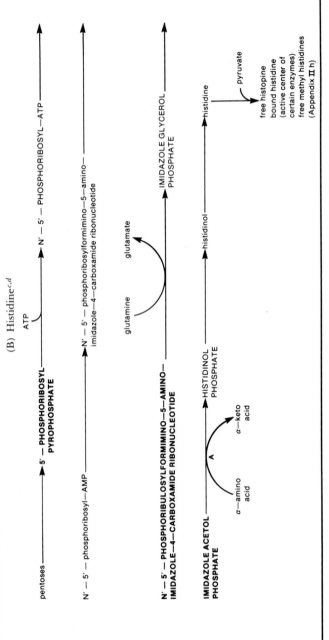

pentoses —————→ **5′ — PHOSPHORIBOSYL** —————→ N′ — 5′ — **PHOSPHORIBOSYL—ATP** —————→
                    PYROPHOSPHATE
                              ↑
                             ATP

N′ — 5′ — phosphoribosyl—AMP —————→ N′ — 5′ — phosphoribosylformimino—5—amino— —————→
                                    imidazole—4—carboxamide ribonucleotide

                                              glutamine   glutamate

N′ — 5′ — **PHOSPHORIBULOSYLFORMIMINO—5—AMINO—** —————→ **IMIDAZOLE GLYCEROL**
IMIDAZOLE—4—CARBOXAMIDE RIBONUCLEOTIDE   **PHOSPHATE**

IMIDAZOLE ACETOL —————→ **HISTIDINOL** —————→ histidinol —————→ histidine
PHOSPHATE   **PHOSPHATE**

        A
  α—amino   α—keto
  acid   acid
                                                              pyruvate

                                                    free histopine
                                                    bound histidine
                                                    (active center of
                                                    certain enzymes)
                                                    free methyl histidines
                                                    (Appendix II h)

[c] Key: **A,** transaminases. Nitrogen acceptors are in **boldface** capitals; reaction products are in lightface capitals.

[d] Interpretive note for Chart 10B. Chart 10B shows that the essential amino acid histidine originates from carbon of the pentose cycle and the nitrogen of glutamine. Histidine may be regarded as imidazolylalanine, and as such it is the benzene analog of phenylalanine. In the biosynthesis of histidine, nitrogens are added in two reactions. In the first, nitrogen of ATP combines with 5′-phosphoribosyl pyrophosphate. A second nitrogen derived from the amide of glutamine completes the formation of the imidazole group. The complete biosynthesis of histidine requires 10 enzymatic steps. When bound in protein, histidine residues serve as the active site of many enzymes.

Histidinol phosphate is formed by transamination from an amino acid nitrogen donor to imidazole acetol phosphate. Histidine may also be converted into histopine (i.e., free histidine with its α-amino group attached to pyruvate). Histopine is formed in certain crown-gall tumors through

86

CHART 11

GLUTAMATE FAMILY OF AMINO ACIDS: COUPLING BETWEEN THE TRICARBOXYLIC ACID AND ORNITHINE CYCLES[a,b]

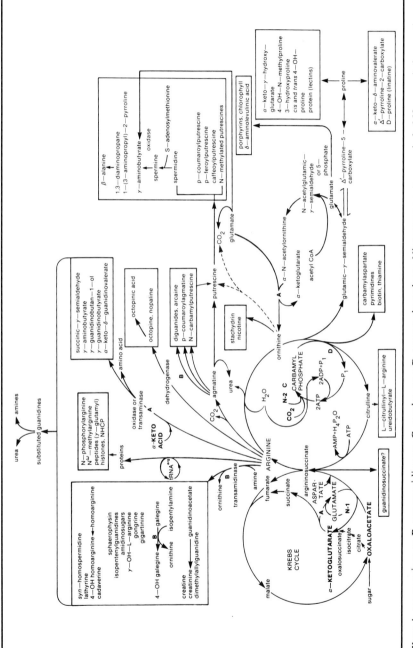

[a] Key: **A**, transaminases; **B**, transamidinases; **C**, synthetases; **D**, transcarbamylases. Nitrogen acceptors are in **boldface** capitals; reaction products are in lightface capitals. N-1, Glutamine, asparagine, or NH₃; N-2, carbamate, glutamine, or NH₃; NHCP, nonhistone chromosomal protein.

(continued)

CHART 11 (*Continued*)

[b] Interpretive note for Chart 11. Metabolic cycles with different ports of entry for inorganic nitrogen are linked with one another to produce the wide array of nitrogenous compounds found in the glutamate family of amino acids. Compounds with common origin are shown in boxes, and references to their natural occurrence may be found in Appendix IIIa–c.

Although some doubts have been expressed about the operation of the ornithine cycle in plants (22, 171, 197), the key enzymes and many of the intermediates have been found (208). The cycle is well established in animals and microorganisms (195, 273). The net reaction for the ornithine cycle is

$$2NH_3 + CO_2 + 3ATP + 2H_2O \rightleftharpoons urea + AMP + 2ADP + 2P_i + PP_i$$

Some evidence supports the occurrence of partial reactions of the cycle mediated by arginase rather than the entire cycle. This creates a dilemma because free urea is not commonly found in plants. The scarcity of urea may be attributed to its high permeability, the occurrence of urease, and the rapid incorporation of urea nitrogen into many compounds, including DNA, RNA, and protein (58, 59).

The chart also shows probable linkages between the Krebs cycle and the ornithine cycle. The main point of interaction is argininosuccinate, which arises from citrulline on the one hand and from aspartate and glutamate on the other; both pathways have been demonstrated in plant mitochondria. Where two cycles connect, aspartate is a nitrogen acceptor from glutamate and a donor to citrulline.

The linkage between these two cycles provides many points of departure. From aspartate and glutamate via arginine a great variety of arginine derivatives are produced. Many of these transformations involve enzymes designated transamidinases (they transfer the "amidine" group to an amine acceptor). Alternatively, many plants have the enzyme arginase, which removes the nitrogen of arginine as urea and ornithine, which in turn provides, through ornithine, a further point of departure to cyclic imino derivatives (e.g., proline). Any residual ornithine not metabolized as described here needs to pick up a carbamoyl moiety (from a C_1 acceptor system as in Chart 5) to regenerate citrulline and arginine.

The formation of arginine via carbamoyl phosphate may remove the nitrogen not used in pyrimidine biosynthesis through carbamoyl aspartate. Arginine formation removes the nitrogen from the amino acid pool normally used for transamination. Arginine has a central role in the formation of a wide range of monosubstituted guanidines with important physiological properties (Appendix IIId), and the formation of arginine-rich histones and chromosomal proteins with amino acids that are open to posttranslational modification during growth and development. At least 1000 of the 1500 enzymes now known to act on negatively charged substrates or to require anionic cofactors employ arginyl residues as the complementary, positively charged recognition site on the enzyme (135). The structure of several members of the glutamate family of amino acids can be modified after incorporation into protein (e.g., the hydroxylation of proline). Posttranslational modifications of arginine (methylation and phosphorylation) and of glutamic acid (pyroglutamic terminal moiety) occur (180, 265). Modifications of arginine have implications in molecular evolution (121, 277). The metabolism of monosubstituted guanidines has been variously reviewed (22, 197, 266).

CHART 12

Aspartate Family of Amino Acids

(A) Aspartate, lysine, threonine, and isoleucine[a,b]

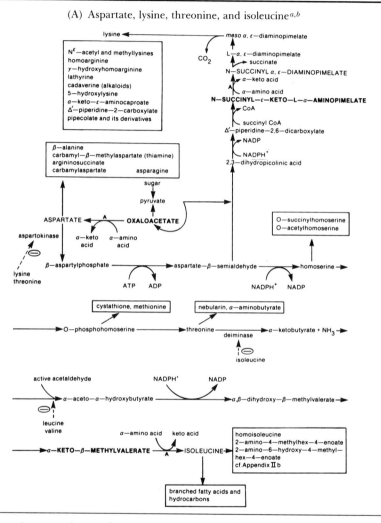

[a] Key: **A**, transaminases; ⊖, feedback inhibitors. Nitrogen acceptors are in **boldface** capitals; reaction products are in lightface capitals.

[b] Interpretive note for Chart 12A. Aspartate, lysine, threonine, and isoleucine are formed through the C_4 port of entry with oxaloacetate as the nitrogen acceptor. The formation of aspartate requires transamination because oxaloacetate, unlike α-ketoglutarate, does not react directly with ammonia. Aspartate and its semiamide, asparagine, are important as nitrogen-storage compounds and as nitrogen donors for the synthesis of other amino acids. Aspartic acid is also a precursor of purines and pyrimidines (Chart 15A,B).

(*continued*)

CHART 12 (*Continued*)

In the reactions stemming from aspartate there are two distinct branches, one leading to lysine and the other, via homoserine, to methionine, threonine, and isoleucine. Lysine, the diamino counterpart of the dicarboxylic amino acids in the aspartate family, is synthesized via α,ε-diaminopimelic acid. An alternative pathway to lysine involves α-aminoadipic acid, but no organism is known to utilize both pathways (38).

Threonine may be regarded as β-methylhydroxyalanine and can be converted to serine, glycine, and isoleucine. Threonine is formed by isomerization of homoserine through esterification of the hydroxyl group by phosphate and its subsequent elimination. Although serine may be synthesized from threonine, their major pathways of biosynthesis and metabolism are distinct. The synthesis and metabolism of homoserine and its esterification have been variously described (91, 164).

The metabolic pathways of the aspartate family are subject to feedback controls. The first point of control is at aspartokinase, which is regulated by lysine and threonine (30). The first enzyme in the threonine–methionine branch, homoserine dehydrogenase, is also under similar control (287). Other aspects of enzymic control in this family of compounds have been reviewed (158).

CHART 12 (*Continued*)

(B) Methionine and related compounds[c,d]

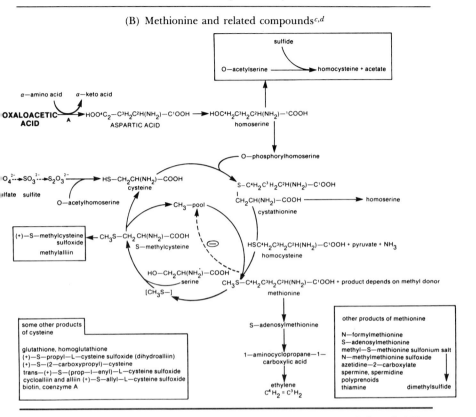

[c] Key: **A**, transaminase; ⊖, feedback inhibition. Nitrogen acceptors are in large **boldface** capitals; reaction products are in large lightface capitals.

[d] Interpretive note for Chart 12B. Methionine and related compounds are formed from aspartate via the C_4 port of entry. Sulfur compounds are listed in Appendix IIi. Many are found in *Allium* spp. (95). Methionine has been viewed as an intermediate that gives rise to the production of ethylene via 1-aminocyclopropane-1-carboxylic acid (3). The carbons of aspartate and the intermediates of this pathway are numbered to show their relationships and the derivation of ethylene. 1-(Malonylamino)cyclopropane-1-carboxylic acid has been identified in germinating peanut (*Arachis hypogaea*) seeds (111a). This compound does not serve as a source of ethylene during germination. A pool of compounds function as a donor of methyl groups. The CH_3 pool, through N^5-methyltetrahydrofolic acid, contributes to the biosynthesis of methionine. The precursor for the methyl donor is N^5,N^{10}-methylenetetrahydrofolic acid, which originates in part from serine. Methionine exerts a feedback inhibitory control in some plants on the formation of N^5,N^{10}-methylenetetrahydrofolic acid by inhibiting glycine decarboxylase (157, 158). The effect of gibberellic acid and methionine on the C_1 metabolism of carrot (*Daucus carota* var. *sativus*) discs and on the net production of metabolically important folate compounds has been described (75). Enzymes that use tetrahydrofolic acid derivatives as requisite cofactors for catalyzing transformations involving C_1 transfer and hydroxylation have been reviewed (19).

CHART 13

ALANINE FAMILY OF AMINO ACIDS[a,b]

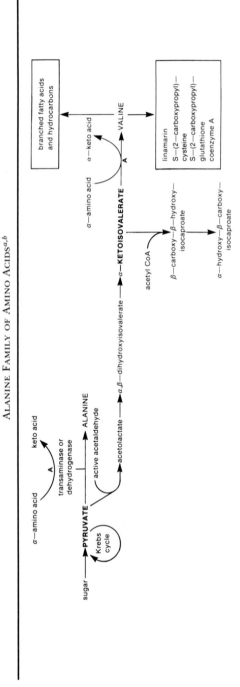

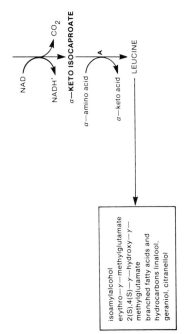

NAD

NADH⁺

CO_2

α—**KETO ISOCAPROATE**

α—amino acid

α—keto acid

LEUCINE

isoamylalcohol
erythro—γ—methylglutamate
2(S),4(S)—γ—hydroxy—γ—
  methylglutamate
branched fatty acids and
hydrocarbons linalool,
geraniol, citranellol

---

a Key: **A**, transaminase. Nitrogen acceptors in **boldface** capitals; reaction products in lightface capitals.

b Interpretive note for Chart 13. The formation of alanine occurs by transamination or by a dehydrogenase reaction with pyruvate as the C_3 acceptor. Other amino acids in this family are formed by transamination from appropriate α-keto acids that are derived from pyruvate. The α-keto acid analog of valine, α-ketoisovalerate, is a branch point because it is also the substrate of the first step in leucine biosynthesis. A large number of other amino acids that are not formed by this pathway or directly from alanine may nevertheless be considered as substituted alanines. These are listed in Appendix II.

CHART 14

Glycine–Serine Family of Amino Acids

(A) Serine, glycine, and cysteine[a,b]

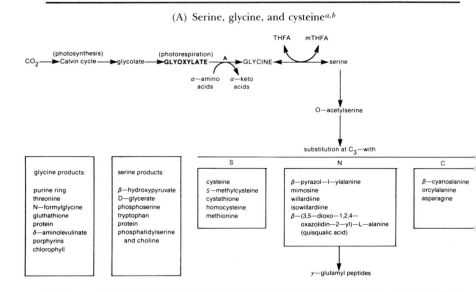

[a] Key: **A,** transaminase; THFA, 5,6,7,8-tetrahydrofolic acid (tetrahydropteroylglutamic acid); MTHFA, 5,10-methylenetetrahydrofolic acid. Nitrogen acceptor in **boldface** capitals; reaction product in lightface capitals.

[b] Interpretive note for Chart 14A. The carbon required for this family comes from the glycolate pathway (cf. Volume VII, Chapter 3, Fig. 9). Serine is formed by the transfer of a methyl carbon derived from glycine to a second molecule of glycine. After the formation of the key intermediate, O-acetylserine, a number of substitution reactions can occur at C_3. The substitutions may involve moieties containing sulfur, nitrogen, or carbon, each of which forms a subclass of metabolic derivatives found in plants (Appendix II). These may occur in germinating seeds in the form of γ-glutamyl peptides (Appendix IIIa). Two compounds derived from O-acetylserine, not listed on the chart, are ascorbalamic acid and β-(2-β-D-glucopyranosyl-3-isoxazolin-5-on-4-yl)alanine (Appendix IIg). The pathway from serine to cysteine consists of an acetyl-CoA-dependent activation of serine to O-acetylserine and a reaction with sulfide. The sulfide is derived from sulfate by a complex mechanism that may involve thioredoxin (259, 264).

CHART 14 (*Continued*)

(B) Succinate–glycine cycle: synthesis of porphyrins[c]

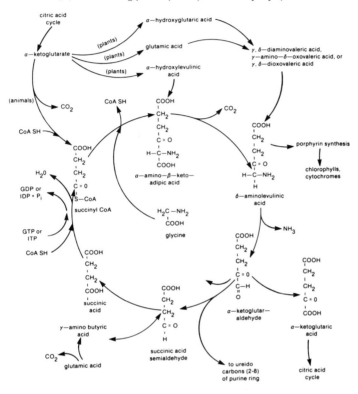

[c] Interpretive note for Chart 14B. In animals and bacteria, glycine combines with succinyl-CoA to form δ-aminolevulinic acid (ALA), the porphyrin precursor. The nitrogen and α carbon of glycine become, respectively, the porphyrin nitrogens and the methylene bridges of the tetrapyrrole molecule (12). Carbon from the succinate–glycine cycle can also be diverted to carbons 2 and 8 of the purine ring (Chart 15B). In plants, ALA is the first confirmed intermediate in tetrapyrrole synthesis for chlorophyll (13). In some leaf tissues and algae ALA is known to be formed from glutamate or α-ketoglutarate through one or more of several possible intermediates, as shown, rather than by glycine–succinyl-CoA condensation. Although the synthesis of ALA takes place in chloroplasts, the enzymes are coded by nuclear genes and formed in the cytoplasm. CoA, Coenzyme A; CoASH, reduced coenzyme A.

CHART 15

BIOSYNTHESIS OF PYRIMIDINE AND PURINE NUCLEOTIDES

(A) Pyrimidines[a,b]

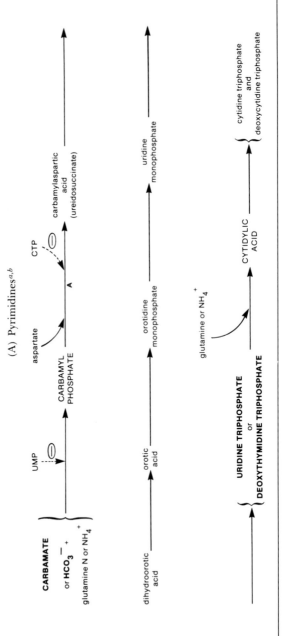

[a] Key: **A**, aspartate transcarbamylase; ⊖, feedback inhibition. Nitrogen acceptors in **boldface** capitals; reaction products in lightface capitals.
[b] Interpretive note for Chart 15A. The pyrimidines and their imidazole-condensed structural derivatives, the purines, are the most common and important biological compounds with two nitrogen atoms per heterocyclic ring.

De novo nucleic acid synthesis occurs through pyrimidine metabolism with orotic acid as a key intermediate and without the occurrence of free pyrimidine bases (15, 102, 205–207, 296). Pyrimidines released during breakdown of nucleic acids may be salvaged for nucleic acid biosynthesis (Chart 16). Control of the rates of metabolic reactions of pyridine nucleotides has been reviewed by Krebs (131). Orotic acid, uracil, and serine may serve as precursors for the cyclic guanidine, lathyrine (Appendix IIIc).

(B) Purines[c,d]

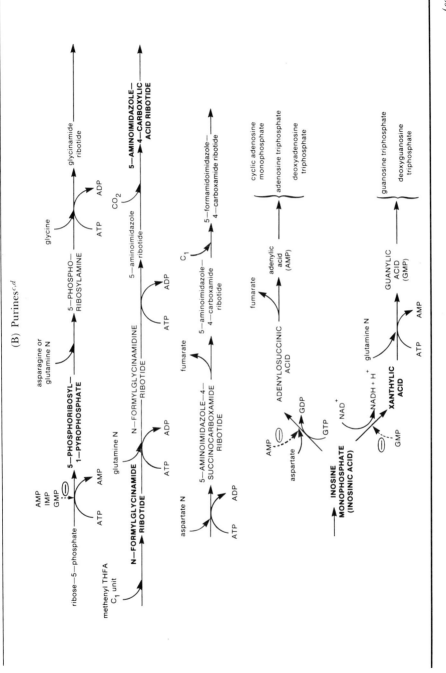

(continued)

CHART 15 (*Continued*)

c Key: THFA, 5,6,7,8-tetrahydrofolic acid; AMP, IMP, GMP, adenosine, inosine, and guanosine 5′-monophosphate; ⊖, feedback inhibition.

d Interpretive note for Chart 15B. Nitrogen-transfer reactions from glutamine, aspartate, glycine, and possibly asparagine and ammonia contribute to the biosynthesis of purine nucleotides containing adenine, guanine, and their derivatives (15, 102, 296). Adenosine triphosphate is one of the most important metabolites in the cell, especially for the control of energy metabolism (e.g., 10). The biosynthetic pathways for pyrimidines and purines have quite different starting points. The earliest precursor of the purines is already conjugated with ribose phosphate, whereas in pyrimidine synthesis the ring is closed before this conjugation occurs. A feature of purine nucleotide biosynthesis is that ATP is need for the formation of GMP, whereas reciprocally, GTP is a necessary coenzyme for the formation of AMP.

CHART 16

SALVAGE MECHANISMS FOR PURINES AND PYRIMIDINES[a,b]

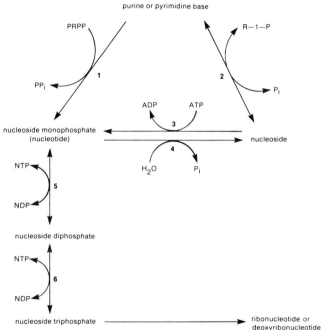

Enzyme reactions

1. Phosphoribosyltransferase (ribose only)

$$\text{Base} + \text{PRPP} \rightarrow \text{nucleotide} + \text{PP}_i$$

2. Nucleoside phosphorylases (riboside or deoxyriboside)

$$\text{Base} + \text{ribose-1-phospate} \rightleftharpoons \text{nucleoside} + \text{P}_i$$

3. Nucleoside kinases (riboside or deoxyriboside)

$$\text{Nucleoside} + \text{ATP} \overset{K^+}{\rightleftharpoons} \text{nucleotide} + \text{ADP}$$

4. 5′-Nucleotidases

$$\text{Nucleotide} + \text{H}_2\text{O} \rightarrow \text{nucleoside} + \text{P}_i$$

5. Nucleoside monophosphate kinases (riboside or deoxyriboside)

$$\text{N}'\text{MP} + \text{N}''\text{TP} \overset{K^+}{\rightleftharpoons} \text{N}'\text{DP} + \text{N}''\text{DP}$$

6. Nucleoside diphosphate kinases (riboside or deoxyriboside)

$$\text{N}'\text{DP} + \text{N}''\text{TP} \overset{K^+}{\rightleftharpoons} \text{N}'\text{TP} + \text{N}''\text{DP}$$

[a] Key: base, purine or pyrimidine base; MP, DP, TP, mono-, di-, or triphosphate; N′, N″, nucleosides; nucleoside, base–ribose (or deoxyribose); nucleotide, nucleoside–phosphate; PRPP, 5-phosphoribosyl pyrophosphate.

[b] Interpretive note for Chart 16. When purine or pyrimidine bases are released from nucleotides or nucleosides they do not necessarily undergo further breakdown but may be "salvaged" for resynthesis of nucleotides (67, 102, 191). The pathways of salvage and interconversion of nucleotides and nucleosides may involve one or more of the enzymes 1–6 as shown on the chart.

CHART 17

PYRIMIDINE BREAKDOWN: FORMATION OF β-AMINO ACIDS[a]

(A) Thymine degradation

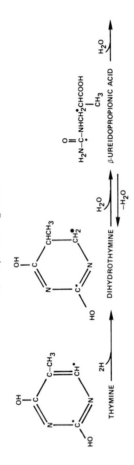

(B) Cytosine and uracil degradation

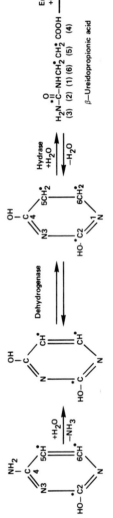

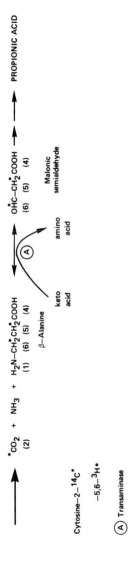

$^*CO_2$ + NH_3 + $H_2N-CH_2\overset{*}{C}H_2CH_2COOH$ $\longrightarrow$ $O\overset{*}{H}C-CH_2\overset{*}{C}H_2COOH$ $\longrightarrow$ PROPIONIC ACID
(2)   (1) (6) (5) (4)   (6) (5) (4)

β-Alanine   Malonic semialdehyde

Cytosine–2–$^{14}C^*$

$-5,6-{}^3H\cdot$

Ⓐ Transaminase

keto acid

amino acid

a Interpretive note for Chart 17A,B. The degradation of pyrimidines restores their nitrogen to the cellular pool of nitrogen available for all purposes. β-Alanine and β-aminoisobutyric acid are commonly found in small amounts in higher plants (Appendix II of 230). They originate largely from the catabolism of pyrimidines. The occurrence of these pathways has been confirmed in germinating pine (*Pinus banksiana*) seedlings by supplying radioactive pyrimidines labeled either with ^{14}C in the 2 position (C^* in the chart) or ^3H in the 6 or 5–6 positions ($H\cdot$ in the chart). All products so designated have been detected, primarily in the cytoplasm of cells (191). When salvage of the pyrimidine moiety occurs, low levels of radioactivity appear in the ribonucleic acids but not in DNA.

β- or γ-Amino acids may arise via other pathways. A β-alanine transaminase occurs in rape seedlings (258), and the formation of β-alanine and γ-aminobutyric acid from polyamines (e.g., spermine and spermidine) have been reported in maize (*Zea mays*) seedlings (248, 249).

CHART 18

PURINE BREAKDOWN: DEGRADATION OF UREIDES[a,b]

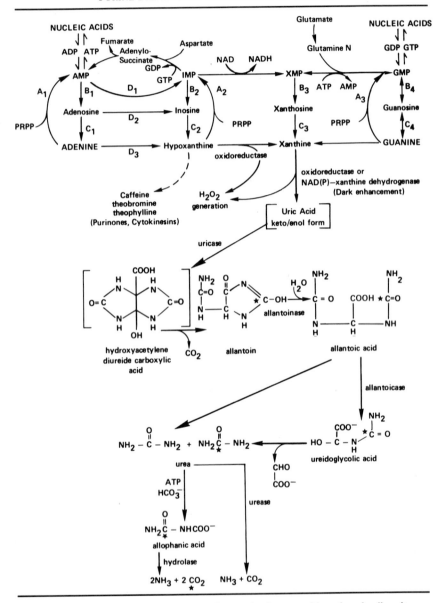

[a] Key: A_1, adenine phosphoribosyltransferase; A_2, hypoxanthine phosphoribosyltransferase; A_3, guanine phosphoribosyltransferase; B_1–B_4, nucleotidases or phosphatases; C_1–C_4, nucleotide phosphorylases or nucleoside hydrolases; D_1–D_3, deaminases; *C, ^{14}C; PRPP, 5-phosphoribosyl pyrophosphate. Transient intermediates or keto–enol forms in square brackets.

102

CHART 18 (*Continued*)

[b] Interpretive note for Chart 18. Adenine and guanine derivatives released from nucleic acids and nucleotides may be oxidized to ureides and subsequently to urea (166, 167). A further source of urea is the cleavage of monosubstituted guanidines from arginine metabolism, as in Chart 11, with the release of amines. In the plants most studied, purines are maintained in the form of ureides and guanidines until they are subsequently degraded through urea (which does not accumulate) to ammonia and carbon dioxide (197).

Although hypoxanthine and xanthine are common intermediates in a breakdown of purines from nucleic acids in plants, insects may produce uric acid as a final excretory product. Although uric acid as such is not regarded as a plant product, it may well (as the chart shows) be a transient intermediary from xanthine to allantoin and allantoic acid, which are frequently found (Appendix IIIB). Hypoxanthine may give rise to some growth regulators (27) and to familiar alkaloids, for example, theophylline (1,3-dimethylxanthine), theobromine (3,7-dimethylxanthine), and caffeine (1,3,7-trimethylxanthine) in tea [*Camellia* (*Thea*) *sinensis*] leaves, coffee (*Coffea*) beans, cocoa (*Theobroma cacao*), etc. (243, 244).

Although nucleic acids in nucleoproteins are basic components of the hereditary material, substantial metabolism and turnover of these compounds and their derivatives clearly takes place that is not directly associated with the genetic code. Interrelations with other aspects of nitrogen metabolism may occur. The production of allophanic acid is also considered in Chart 5.

CHART 19

RELATIONSHIPS OF SOLUBLE NITROGEN COMPOUNDS
TO PROTEIN AND NUCLEOPROTEIN METABOLISM[a]

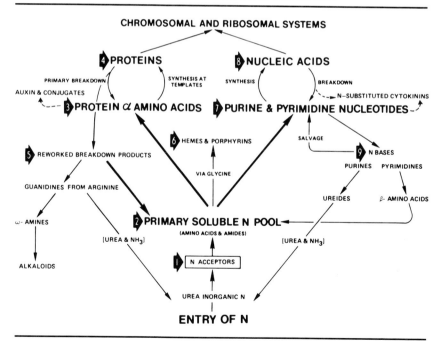

[a] Interpretive note and key for Chart 19. This chart summarizes the relationships of all the nitrogenous substances found in plant cells as they relate to the synthesis of proteins

103

(*continued*)

CHART 19 (*Continued*)

and nucleoproteins. It does not deal with the mechanisms by which RNA, DNA, and ribosomes are involved in protein synthesis; these appear later in Charts 20–24. It does, however, recognize in its upper section the incorporation of proteins and nucleic acids into the structure of chromosomes, ribosomes, and subcellular organelles.

The details are summarized here by reference to the numbered arrows 1–9 on the chart:

1. These are acceptors from carbohydrate metabolism with 1–5 carbon atoms (see Charts 1–5).

2. This is the pool of soluble nitrogen compounds, which includes free amino acids and amides, the primary products of the nitrogen acceptors as specified in Charts 1–5 and some end products of degradative metabolism.

3. The 20 protein amino acids are products resulting from nitrogen transfers from compounds in the primary soluble-nitrogen pool (specified in Charts 1–5 and 9) for the principal amino acid families that contribute to protein synthesis and that arise from its breakdown.

4. Proteins are involved in reversible cycles of synthesis and breakdown or turnover; the nitrogen from breakdown products contributes to resynthesis or may be reworked as shown later.

5. The reworked breakdown products contribute much nitrogen in the form of amides (asparagine and glutamine) and soluble free amino acids (protein and nonprotein) to the reusable soluble-nitrogen pool (cf. metabolic Charts 10–16; Appendix II–IV). Nitrogen is also directed to other not reusable end products, including alkaloids from amines and other major secondary products of metabolism (e.g., guanidines, including γ-glutamyl peptides not specifically shown on the charts, etc.). Whereas the nitrogen of protein breakdown is returned as nitrogen-rich storage products, much of the carbon is respired away via the citric acid cycle, which also replenishes the carbon and energy needed when these nitrogen-rich reserves are reworked in subsequent syntheses (see also Chapter 3).

6. Carbon as succinate from the citric acid cycle with nitrogen from glycine leads to a wide range of hemes and porphyrins (cf. Chart 14B).

7. Nitrogen as carbamoyl phosphate, glycine, aspartate, and glutamine with carbon from ribose (Chart 15A,B) produces the purine and pyrimidine ribonucleotides for nucleic acid synthesis.

8. In the synthesis of nucleic acids some ribonucleotides are reduced and modified to the deoxyribonucleotides required for DNA in the nucleus or other organelles. Also, some of the nucleotide bases in nucleic acids are further modified *in situ* during the fabrication of the messenger RNA that serves as the template for protein and enzyme synthesis. The dilemma of protein synthesis is that it occurs at surfaces (e.g., templates) that provide information (via mRNA) and receive amino acids (via tRNA) for the condensation at a ribosomal surface which itself is a ribosomal protein. Nucleic acids are also involved in other major directions as in the nucleoproteins, which in the form of chromatin, with its genes, provide the physical basis of inheritance.

9. The primary products of nucleic acid breakdown (the nucleotides) yield by further breakdown their respective nitrogen bases. These have several possible fates. The pyrimidine bases yield β-amino acids, the nitrogen of which may be donated to the reusable soluble pool (Chart 17), and the purine bases via stored ureides may contribute nitrogen to the soluble pool via urea as a transient (Chart 18). Finally, some of the bases may be salvaged for resynthesis of nucleic acids (Chart 16).

1. Ports of Entry for Nitrogen: C_5 to C_1[3]

The metabolic charts are presented first from various "ports of entry" for nitrogen (Charts 1–5). The entry of nitrogen at the various C_5 ports is shown on Chart 1 with its interpretive note, together with the principal pathways from the various acceptors (whether C_5 keto or amino acids, or even a ketoamino acid) via identified enzyme systems to their resultant end products. Thereafter, and following a similar style, Chart 2 shows similar schemes for C_4 ports of entry (which may again be a keto, amino, or a ketoamino acid). Another variant on C_4 ports of entry includes succinic semialdehyde, as on Chart 2. Metabolic Chart 3 relates to C_3 ports of entry (both keto and amino acid). Chart 4 relates to a C_2 port of entry, and Chart 5 shows a C_1 port of entry.

Clearly, the complexity of the biochemistry from the ports of entry for nitrogen (as designated in Charts 1–5), has increased progressively [cf. Chibnall (36), Virtanen (271), Steward and Durzan (230), Beevers (15)].

These charts (and as amplified by their interpretive notes) should not suggest that all five ports of entry may operate in any one plant or at any particular time. Although several variants of carbon acceptors (i.e., C_1 to C_5) are possible, all of them may not function simultaneously. Also, if a given acceptor system may lead to many end products (e.g., as from glutamate in the C_5 system), this does not imply that all the possibilities will be fulfilled. Furthermore still other ports of entry may yet be found. For example, in animals, ammonia can react with ATP to produce phosphoramide and ADP with an ammonia kinase (10). This system is still unknown in plants.

Therefore, the challenge is to know and understand how plants discriminate among these various possibilities and so endow a particular system with the specificity by which it is characterized. These degrees of freedom will also need to be considered with reference to reaction sites in cells or to parts of the plant body and to changes that may be brought about by environments and nutrition.

Also, concepts outlined in Charts 1–5 apply in different degrees to organisms that have autotrophic or heterotrophic nutrition in greater or lesser degree. Some blue-green algae, with their photosynthetic and nitrogen-fixing abilities have cells that may be as completely autotrophic as any known. But even the recognized autotrophy of mature higher green plants is more a property of the organism than of its individual cells.

---

[3]In keeping with current biochemical conventions, the notation C_5 denotes a 5-carbon molecule, whereas C-5 denotes the fifth carbon in the molecule.

Thus the ports of entry for nitrogen may even vary inasmuch as the synthesizing cells produce their carbon compounds directly, as in leaves, or receive them via translocation, as in roots, or even exogenously, as in heterotrophic cell and tissue cultures. In fact, through the "alternation of generations" and the reproductive functions of seed plants, there is a vast developmental area in which, from spore to embryo to germinated seedling, the organism is not autotrophic but heterotrophically dependent upon a parent sporophyte. Indeed, it is the evolved nutritional efficiency of sporophytes dependent upon their organization with roots, leaves, and vascular systems that enables them to support the seed habit.

Therefore, with this range of autotrophic and heterotrophic nutrition, there are many choices to be made concerning the ports of nitrogen entry that are feasible or appropriate with their consequential impact upon carbon and nitrogen metabolism (Charts 1–5).

2. Reactions that Elaborate Amino Acids and Their Derivatives

The C—N bond is established initially through reactions shown in metabolic Charts 1–5, and some further elaborations into other classes of compounds are shown in Charts 6–8. Six general classes of reactions elaborate amino acids and their nitrogenous products in the reactions shown in Charts 9–19.

First, nitrogen can be transferred from one compound to another in a system described as a donor:acceptor group transferase. The donor is often a cofactor (coenzyme) charged with the group to be transferred. Reactions, called transaminations, that transfer the amino group have the general equation

$$\underset{R^1-CHR^2}{\overset{NH_2}{|}} + \underset{R^3-C-R^4}{\overset{O}{\|}} \rightleftharpoons \underset{R^1-C-R^2}{\overset{O}{\|}} + \underset{R^3-CHR^4}{\overset{NH_2}{|}}$$

The transamination reaction can be considered as an oxidative deamination of the donor primary amino acid or amine linked with the reductive amination of the acceptor α-keto acid (Tables VII and VIII). Deuterium and tritium oxide have been used as tools for the study of the formation of amino acids labeled at the C-2 during transamination reactions (59, 65, 267). Introduction of label into other specific positions in the amino acid leads to conclusions about other reactions outlined later. Although transaminase reactions might be regarded as oxidoreduction, the unique and distinctive factor is the transfer of the amino group.

Transaminations involving keto- and aminoinositols and glutamine

2. NITROGEN METABOLISM 107

TABLE VII

EXAMPLES OF AMINO-NITROGEN TRANSFER REACTIONS[a]

Reaction	Cofactor requirement
1. Glycine aminotransferase Glycine + α-ketoglutarate ⇌ glyoxylate + L-glutamate	Pyridoxal–phosphate–protein; glycolate in chloroplasts is converted to glyoxylate in microbodies
2. Alanine aminotransferase L-Alanine + α-ketoglutarate ⇌ pyruvate + L-glutamate	Pyridoxal–phosphate–protein
3. Alanine ketoacid aminotransferase L-Alanine + α-ketoacid ⇌ pyruvate + L-amino acid	Pyridoxal–phosphate–protein
4. Aspartate aminotransferase L-Aspartate + α-ketoglutarate ⇌ oxaloacetate + L-glutamate	Pyridoxal–phosphate–protein
5. Glutamate synthase L-Glutamine + α-ketoglutarate ⇌ 2L-glutamate	NAD(P)H or ferredoxin cofactor

[a] See (31, 45, 154). For example, aminotransferases (α-amino acid donor) (α-keto acid acceptor). Other transfer reactions involving oximes, amides, and amidines also occur. For these reactions see IUPAC–IUB Commission (1979). "Enzyme Nomenclature." Academic Press, New York.

have been implicated in the production of some antibiotics (35). In plants, aspartate-α-ketoglutarate transaminase (EC 2.6.1.1, aspartate aminotransferase) participates in the formation of new amino acids in the "electron shuttle" (104), in intercellular transport of metabolites during C_4 photosynthesis (103), and in the glycolate pathway in cellular compartments (255).[4] Isoenzymes of transminases are specifically located in subcellular organelles and compartments (142).

Second, amino compounds can be modified by oxidoreductases or dehydrogenases. The substrate that is oxidized is regarded as the hydrogen donor. The term oxidase is used only where molecular oxygen is the acceptor. L-Amino acid oxidases have been reported but are rare in plants. Evidence exists of thylakoid-bound L-amino acid oxidases in chloroplasts (144). Oxidoreductase activity varies with the group in the hydrogen donor, which undergoes oxidation, and with the type of acceptor involved. In reactions involving the coenzyme nicotinamide, this cofactor is generally regarded as the acceptor even if the reaction is not readily demonstrable.

[4]A discussion of the nitrogen metabolism associated with C_4 photosynthesis and the glycolate pathway will be found in Volume VII, Chapter 3.

TABLE VIII

SIMPLE ALIPHATIC PRIMARY MONOAMINES FOUND IN PLANTS,
WITH SOME OF THEIR KNOWN *IN VITRO* AND HYPOTHETICAL PRECURSORS[a]

Amine	Structure	Formation by Decarboxylation of	Formation by Transamination
Methylamine	H_3C-NH_2	Glycine[b]	Formaldehyde
Ethylamine	$H_3C-CH_2-NH_2$	Alanine	Acetaldehyde
n-Propylamine	$H_3C-(CH_2)_2-NH_2$	2- or 4-Aminobutyric acid	*n*-Propanal
n-Butylamine	$H_3C-(CH_2)_3-NH_2$	Norvaline	*n*-Butanal
n-Amylamine	$H_3C-(CH_2)_4-NH_2$	Norleucine	*n*-Pentanal
n-Hexylamine	$H_3C-(CH_2)_5-NH_2$	1-Aminoheptanoic acid	*n*-Hexanal
n-Heptylamine	$H_3C-(CH_2)_6-NH_2$	—	*n*-Heptanal
n-Octylamine	$H_3C-(CH_2)_7-NH_2$	—	*n*-Octanal
Isopropylamine	CH_3 $\vert$ $H_3C-CHNH_2$	2-Aminoisobutyric acid[b]	Acetone[b]
Isobutylamine	CH_3 $\vert$ $H_3C-CHCH_2-NH_2$	Valine	Isobutanal
Isoamylamine	CH_3 $\vert$ $H_3C-CH(CH_2)_2-NH_2$	Leucine	Isopentanal
2-Methylbutylamine	CH_3 $\vert$ $CH_3CH_2CHCH_2-NH_2$	Isoleucine	2-Methylbutana[l]
2-Thioethylamine	$HS-(CH_2)_2-NH_2$	Cysteine	—
3-Methylthiopropylamine	$H_3C-S-(CH_2)_3-NH_2$	Methionine	—
3-Thiopropylamine	$HS-(CH_2)_3-NH_2$	Homocysteine	—
Ethanolamine	$HO-(CH_2)_2-NH_2$	Serine	Glycolaldehyde[b]
2-Propanolamine	$CH_3CHCH_2-NH_2$ $\vert$ OH	Threonine[b]	—

[a] From Smith (222).
[b] Hypothetical precursor only.

Third, enzymes termed hydrolases can achieve the hydrolytic cleavage of C—N or other bonds (e.g., C—C and C—O), as in the action of urease (34, 118). These hydrolases may act on peptides, amides, esters, and other bonds. The reaction involves the hydrolytic removal of a group from the substrate and its transfer to a suitable acceptor (e.g., water). Fourth, lyases are enzymes that eliminate C—N, C—C, C—O, and other groups to leave double bonds, or by adding groups to double bonds [e.g., argininosuccinate lyase (195, 213, 214)]. Terms such as decarboxylase and dehydratase are sometimes used, and where the reverse reaction is much more important or the only one demonstrable the term synthase (not synthetase) is used in the name [e.g., glutamate synthase (Chart 1, enzyme system 2)]. The lyases may catalyze the elimination of a β-or γ-substituent from an α-amino acid, followed by its replacement by some other group. In the overall but not partial replacement reaction, unsaturated end products are not formed. The reaction is often a two-step one in which enzyme-bound or α-, β-, or β,γ-unsaturated amino acids are formed.

Fifth, a molecule can be modified geometrically or structurally by isomerism brought about by enzymes [e.g., racemases, epimerases, cis–trans isomerases, etc. (47)]. In such cases the change in substrate involves oxidoreduction, but because hydrogen donor and acceptor are in the same molecule and no oxidized product appears, the enzymes are not classified as oxidoreductases.

Sixth, when two molecules are joined with the hydrolysis of a pyrophosphate bond in ATP or similar triphosphate, the enzymes involved are classified as ligases or synthetases (10). The bonds formed may then be high-energy bonds, as in the action of glutamine synthetase (Chart 1) or enzymes that acylate RNA (Chart 22).

Control mechanisms for enzyme pathways therefore may be linked to each of the six classes of enzyme-mediated reactions just described.

Finally, the origin of amino acids from proteins, be they storage, structural, or enzymatic, should not be overlooked. The earlier history of the role of proteins was reviewed by Chibnall (36), Fowden (83), Greenstein and Winitz (94), McKee (151), Steward and Durzan (230), and Vickery (269).

Enzymes that act on peptide bonds such as the proteases have not been extensively studied, except for a few such as papain, and the topic of protein turnover has been given special and separate attention from this point of view in Chapter 3.

3. Biosynthesis of Families of Amino Acids and Their Derivatives

Charts 9–19 therefore present a comprehensive view of the later nitrogen metabolism of plants and, in this context, of various postulated reactions and sequences. These reactions involve group transfers, as of amino nitrogen, and eventually give rise to free ammonia or urea and amines from guanidines and ureides.

Anabolic reactions from the first products of nitrogen metabolism are shown in Charts 6–8. Chart 6 refers to the C_2 and C_3 ports of entry, as they link up with glyoxylate via the tricarboxylic acid cycle or glycerophosphate from the Calvin cycle. Chart 7 shows schemes that relate the C_3 port of entry to the pivotal substance γ-methyl-γ-hydroxyglutamic acid. Chart 8 visualizes the entry of nitrogen from nitrate into amino acids of leaves via a C_5 port of entry. The scheme in Chart 9 portrays the interlocking of metabolic cycles involved in photosynthesis and respiration as they yield the pivotal carbon frameworks that serve as the ports of entry for nitrogen and as focal points for the elaboration of families of amino acids commonly found in protein and some other complex nitrogenous derivatives. Chart 10A envisages the lines of synthesis from the glycolysis of sugar that lead eventually to phenylalanine, tyrosine, and anthranilic acid and the numerous substances derived from them and thence to tryptophan. Chart 10B deals with histidine biosynthesis. Chart 11 deals with the formation of the glutamate family of amino acids and the ramifications of metabolism that stem from the Krebs cycle via α-ketoglutarate and oxaloacetate as they couple with the ornithine cycle to produce arginine and the vast array of amines and guanidino and ureido compounds that occur (see Appendix III).

Chart 12A returns to the Krebs cycle and its C_4 port of entry to trace the ramifications that stem from oxaloacetic acid through the aspartate family to compounds that include threonine, isoleucine, and lysine and their products. Chart 12B shows the input of reduced sulfur compounds such as cysteine and amino acids derived from aspartate during the formation of methionine and even leading to ethylene. Chart 13 depicts the formation of alanine, valine, and leucine from a presumptive C_3 acceptor of nitrogen and pyruvate. Chart 14A deals with the glycine family and shows interconversions with O-acetylserine to produce a wide range of amino acids by substitution with groups containing nitrogen, carbon, and sulfur. Chart 14B shows the relationships between glycine and the Krebs cycle and the reactions leading to the formation of porphyrin precursors.

Charts 15A and 15B start with such simple compounds as aspartic acid, glutamine, and even urea and ammonia (carbamoyl phosphate) to trace the routes to pyrimidine and purine nucleotides. Chart 16 describes salvage mechanisms for the nucleotides, whereas Charts 17 and 18 outline the degradation of pyrimidine and purine bases to β-amino acids and ureides. Chart 19 in a broad scheme shows how complex substances (proteins and nucleic acids) are formed from the simpler compounds and give rise again through turnover to various categories of substances that may be reused or may contribute to the diversity of the soluble nitrogenous compounds in plants.

4. Nitrogen Metabolism: Overview

Nitrogen metabolism has different salient features in various parts of the plant body. Roots are organs in which nitrate absorption and reduction aided by organic substrates ultimately derived from shoots is of prime importance. Even so, the metabolism of nitrogen in root tips, where cells multiply, will have features distinct from those of cells where absorption from the soil solutions is the prime function. Because the reduction of nitrate and its supply to other parts of the plant body are discussed in Chapter 4, these aspects are not emphasized here. Furthermore, nitrogen fixation and the symbiotic relations between nitrogen-fixing organisms and their hosts are discussed in Chapters 1 and 4. Nevertheless, attention should be focused here on the reduction of nitrate, utilizing energy and metabolites from photosynthesis primarily in leaves, as the starting point for the numerous reactions outlined in Charts 6–8.

But in the whole economy of nitrogen in the plant body the synthesis of protein is a feature of dividing cells wherever they occur (i.e., in the apices of shoot and root and any other sites of meristematic activity). Nevertheless, green leaves are still to be regarded as prime centers for protein formation. Also, via the essential breakdown of leaf protein and the translocation of nitrogen to other organs where resynthesis and storage occur (i.e., organs of storage and/or perennation), the nitrogen economy of plants needs to be seen as an integrated pattern involving both soluble-nitrogen compounds and synthesized protein as well as its eventual turnover.

The discussion, therefore, now turns first to sites and reactions whereby the soluble-nitrogen metabolism may be initiated and regulated and, later, to sites where soluble-nitrogen may be converted to protein and then regenerated by its breakdown or turnover.

5. Nitrate Assimilation in Leaves

a. Nitrate Reduction. The reduction of nitrate and other oxidized forms of inorganic nitrogen usually ends with ammonia (107, 108).[5] The rate-controlling and most-regulated step in nitrate assimilation seems to be the conversion of nitrate to nitrite by nitrate reductase. Metabolic Chart 6 shows the scheme of Solomonson and Spehar (224) for the regulation of nitrate assimilation; in this scheme carbon dioxide fixation and nitrate assimilation are coordinately controlled by the intracellular ratio of O_2 to CO_2.

Even so, it is clear that nitrate assimilation involves complex interrelationships with products of photosynthesis that may ultimately deter-

---

[5]Two references by Hewitt (107) and Hewitt *et al.* (108) conveniently update much of the prior knowledge of nitrogen assimilation as it was presented in various chapters of Volume III of this treatise in the context of the inorganic nutrition of plants.

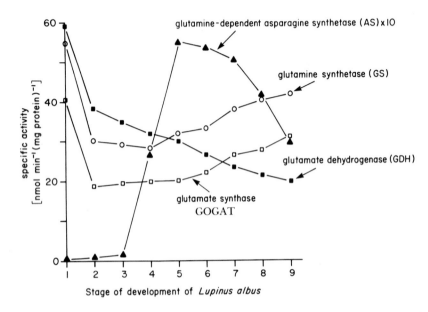

Fig. 1. Levels of various enzymes involved in ammonia assimilation in maturing and germinating *Lupinus albus* cotyledons. Stages of development are: 1, maturing seeds that had reached full size but that were still green, prior to the drying-out period; 2, dry mature seeds; 3, seeds soaked 24 hr in water, no sign of radicle emergence; 4, 3-day-old seedlings, radicle 1–2 cm; 5, 5-day-old seedlings, radicle 2–3 cm, hypocotyl 1 cm; 6, 7-day-old seedlings, radicle 4 cm, hypocotyl 2 cm; 7, 9-day-old seedlings, radicle 6 cm, hypocotyl 5 cm; 8, 12-day-old seedlings, radicle 8–10 cm, hypocotyl 8 cm; 9, 15-day-old seedlings, radicle 10 cm, hypocotyl 10 cm. The seedlings were grown in the dark at 25°C. Redrawn from Lea and Fowden (138).

mine the carbon acceptor for the reduced nitrate. (These relationships implicate carbon compounds, cofactors, and energy and cytoplasmic reaction systems that involve glycolysis and photorespiration.) Furthermore, the reduction of nitrate involves an internal electron-transport path to and from specific electron donors and acceptors such as the pyridine nucleotides and the reductase enzyme itself (Chart 8). In this electron-transport pathway, NADH is considered to be the electron donor for the reduction of nitrate to nitrite. It has been observed that NAD^+ is a cytoplasmic cofactor and that nitrate reductase may be located in the cytoplasm rather than in chloroplasts. Even so, the enzyme is light dependent and the required energy is derived ultimately from photosynthesis (see Chapter 3).

Nitrate rarely accumulates and is rapidly reduced to ammonium by nitrite reductase and reduced ferredoxin. The oxidized ferredoxin is returned to the reaction system by ferredoxin-NADP reductase. This reaction is almost certainly located in chloroplasts. The details and the mechanism of reactions for enzymes in nitrate assimilation have been summarized by Beevers (15).

Once ammonia has been formed, its further assimilation involves a number of enzyme systems (Charts 1–5). An example of the levels of various enzymes involved in ammonia assimilation in cotyledons of *Lupinus albus* is shown in Fig. 1. More recently, protoplasts from expanding leaves of *Pisum sativum* have been used for the isolation of cell organelles in the study of intracellular distribution of enzymes of nitrate assimilation (278). Nitrate reductase and glutamate synthase were located wholly in chloroplasts. Glutamine synthetase was distributed between chloroplasts and the cytoplasm, with 60% in the former.

b. Ammonia Assimilation in Chloroplasts. Once nitrate is reduced to ammonia, its nitrogen is now thought to be accepted by the glutamine synthetase–glutamate synthase (GS–GOGAT) system (Charts 1 and 8). Prior to this idea, glutamate dehydrogenase was thought to be the main route of entry for ammonia and nitrogen. In some cases this may still be so, especially in the presence of high levels of ammonia (160, 218). The difficulty, however, was that the dehydrogenase enzyme, concerned with the combination of ammonia and α-ketoglutarate, had too low an affinity for its substrate, ammonia. An earlier alternative regarded glutamine synthetase as the acceptor and glutamine as the product. The difficulty here was that the entering nitrogen would be fixed in an amide group instead of the amino group most needed for transaminations. The reasons behind the current postulates of the GS–GOGAT system are as follows.

If glutamate dehydrogenase were the primary enzyme, then labeled ammonia should be incorporated first into the amino group of glutamate and then into the amide group of glutamine. Under the GS–GOGAT system labeled ammonia is combined first with glutamate by GS and becomes the amide group of glutamine. Subsequently, the GOGAT enzyme transfers the amide nitrogen of glutamine to the alpha carbon of an acceptor (α-ketoglutarate) to give glutamate. (Essentially this requires an amide-aminotransferase, which is what GOGAT achieves.) In this way, one molecule of glutamate generates one molecule of glutamine and then two molecules of glutamate, of which one is available to repeat the entry of more ammonia. This circuitous system is, nevertheless, more efficient than the direct reductive amination of the α-keto acid. Therefore, if the GS–GOGAT system operates, the nitrogen of $^{15}NO_3{}^-$, [^{15}N]amide glutamine and [^{15}N]aminoglutamate should all be equally accessible for the synthesis of all new protein α-amino acids. Indeed, this seems to be so (160, 199, 218). These metabolic steps have been supported by studies in which selective enzyme inhibitors (e.g., methionine sulfoximine for GS and azaserine for GOGAT) discriminate between these enzymes and the reactions they catalyze (160). However, the postulated role of glutamine in the GS–GOGAT system only calls for catalytic amounts of glutamine, whereas in many cells and under many conditions, glutamine is a major component of the soluble nonprotein nitrogen. Although highly localized distributions of the operative enzymes in the cells could outweigh this disparity, actual evidence has not been forthcoming. On the contrary, the high levels of glutamine, typical of actively synthesizing cells, have been traditionally regarded as contributing to the pace of directed synthesis (230).

But what of the well-known contrasted roles of asparagine and glutamine? These differences might be ascribed to very great specificity of the amide-aminotransferase enzymes of the GS–GOGAT system so that access of the amide of asparagine to the system is denied. On this point the evidence is fragmentary (160, p. 311).

Some other salient questions may be raised. For example, have the formulations of ammonia assimilation, as described previously, shed light on the paramount role of green leaves in the light as organs of protein synthesis, in which the nitrogen has appeared to be canalized through glutamine? Or in fact do they shed light on the more general idea that "active pools" of nitrogen en route to protein pass over glutamine and not asparagine? Also, whereas syntheses in green leaves that pass over glutamine are paramount in the light, how does the converse occur in the dark, with asparagine as a more frequent end product?

Answers to these questions may be as follows. First, the efficiency by which leaves make protein from the nitrogen of glutamine may be largely explained by the participation of photosynthetic ATP in the GS–GOGAT reaction. But even so, to place the amide nitrogen into the α-amino position for protein synthesis needs reduction, and this is apparently negotiated by ferredoxin (160). Second, when protein synthesis occurs in organs not directly stimulated by light, then the GS–GOGAT system would need to be activated by ATP from other sources, such as the speeded operation of the Krebs cycle by active, aerobic metabolism. Third, whereas the GS–GOGAT formulations have been and are useful in explaining nitrogen metabolism via glutamine in leaves in the light, the converse, where protein breakdown occurs in the dark, with asparagine as a more prominent end product, is not as easily comprehended under the GOGAT postulates. Fourth, these interpretations and the publications on which they are based all rely on the implicit assumptions that each enzyme, its substrates, and the products are free of contaminants [e.g., transglutaminases (78)] or artifacts and are identical with their counterparts *in situ*. This is, however, a very difficult area of experimentation even for the most experienced investigators. It raises such problems as the maintenance of asepsis and bacteria-free conditions during extraction and assay; the purity and/or the chemical stability of some components of assay systems, even of glutamine itself, and of the keto acid intermediates; the design of assay systems so that they shall not *in vitro* be physiologically unreal or oversimplified compared with conditions *in vivo*; and, finally, the degree of senescence of leaves from which the enzyme systems were obtained. Furthermore, where enzymes react with ammonia, there is in every case according to Dixon *et al.* (48) a transitional metal ion more or less tightly bound to the enzyme and the frequently postulated "enzyme–ammonia" intermediate in reactions catalyzed by these enzymes may, therefore, involve ammonia complexed to the bound metal ion. All this being so, and recognizing the complexity of interpreting reactions of even a very limited range of compounds, the question arises whether organized cells *in situ* operate more simply and directly than the reactions in test tubes imply.

c. Reactions Regulated by Light in Chloroplasts (Chart 8). The fixation of energy for plants and their biochemistry revolves around chloroplasts in leaves in the light. The harnessing of light energy to biosynthesis, via systems that transport protons and electrons from water to a range of reduced products and energy transduction involving ATPase complexes, are now familiar in the reduction of carbon dioxide to carbohydrates (see Volume VII, Chapters 2 and 3) (215). But in the context of this chapter, the events in leaves in the light also need to be integrated with the conversions of the fixation products of nitrogen (see previous discussion) to their ultimate destinations in the wide array of essential and secondary nitrogenous products (see metabolic Chart 19 and the Appendixes). The extent to which this involves enzymes, special proteins and regulation by cofactors is the concern of this section. There is here, however, an analogy between the group transfers involving nitrogen moieties with the better known flow of electrons, of hydrogen, of phosphate, and of energy in leaves.

The first focus should be on the catalytic mechanisms in chloroplasts that negotiate light-activated reductions. Among these ferredoxin and the more recently appreciated thioredoxin are crucial. Regulated by these special proteins are sets of enzymes that strategically determine the

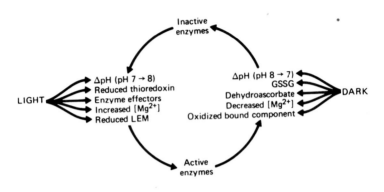

FIG. 2. Controls in light and darkness mediated by enzymes and the ferredoxin–thioredoxin systems. LEM signifies membrane-bound reductants that are light-effect mediators. Enzymes regulated by the thioredoxin system are: fructose-1,6-bisphosphatase (gluconeogenesis), NADP glyceraldehyde-3-phosphate dehydrogenase, phosphoribulokinase (photosynthesis), sedoheptulose-1,7-bisphosphatase, NADP malate dehydrogenase, phenylalanine ammonia-lyase, and ribonucleotide reductase (cell cycle). After Buchanan *et al.* (32).

course of syntheses and metabolism in light and darkness. Figure 2 shows some salient enzymes that are regulated by ferredoxin and thioredoxin in light and darkness.

The ferredoxin–thioredoxin systems of chloroplasts constitute electron-carrier proteins that function in a variety of ways. They convert ribonucleotides to deoxyribonucleotides for DNA synthesis (32, 251). Thioredoxin regulates enzymes involved in photosynthesis and in amino acid metabolism, particularly phenylalanine ammonia-lyase. Therefore, thioredoxin is considered to be a link between light and enzymes that use energy-rich products formed by light (e.g., ATP and NADPH). In this way, thioredoxin regulates enzymes that promote biosynthesis in the light and specifically may affect the light-activated products of the compounds listed in Chart 10 as derived from phenylalanine or tyrosine. A corollary is that analogs of phenylalanine [e.g., α-amino-4-methylhex-4-enoic acid (89)] may inhibit the lyase and also the amino acid activation, which is the first step toward protein synthesis (see Chart 22 and its interpretive note).

Finally, it should be pointed out that corn (*Zea mays*) leaves, even while photosynthesizing actively, are capable of releasing ammonia during senescence (74).

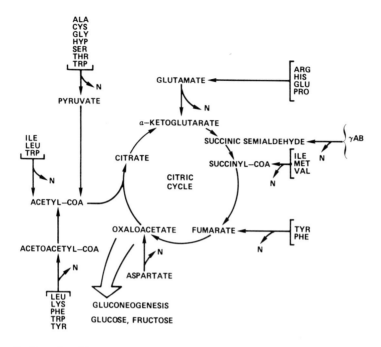

FIG. 3. Organic acid intermediates formed from the carbon skeletons of amino acids. The amino acids commonly found in protein are released from protein by proteolysis. The amino nitrogen is recycled via the amides glutamine and asparagine, and the carbon can be recycled through gluconeogenesis to glucose and fructose. ALA, alanine; ARG, arginine; CYS, cysteine; GLU, glutamate; GLY, glycine, HIS, histidine; HYP, hydroxyproline; ILE, iso-leucine; LEU, leucine; LYS, lysine; MET, methionine; PHE, phenylalanine; PRO, proline; SER, serine; THR, threonine; TRP, tryptophan; TYR, tyrosine; VAL, valine; γAB, γ-aminobuty-rate.

6. Gluconeogenesis and Reactions that Release Ammonia and Urea

As noted earlier, free ammonia and urea are difficult to detect in plants. Nevertheless, reactions occur that release these substances, as in gluconeogenesis (i.e., the rebuilding of carbohydrates from deaminated residues of amino acids), shown in Fig. 3. The analog of gluconeogenesis had been postulated earlier in the history of nitrogen metabolism for the linkage between a carbon cycle and a nitrogen cycle in plants (see 230). Gluconeogenesis occurs in germinating seeds that contain considerable fat and lipid reserves [e.g. castor bean (*Ricinus communis*), most gym-nosperms). Organic acid residues from deaminated amino acids are do-nated to the Krebs cycle at various points (Fig. 3). Davies (46) has exam-

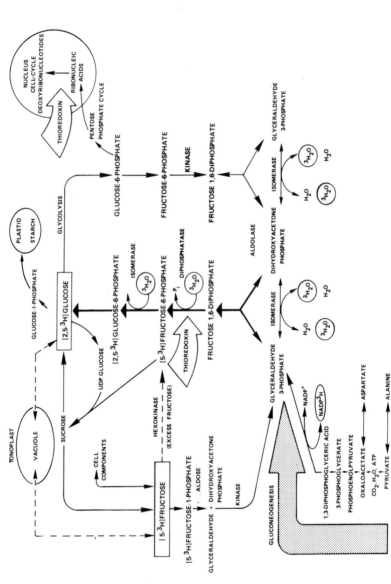

Fig. 4. Gluconeogenesis of alanine and aspartic acid and the metabolic fate of tritiated water during this process. The entry of tritium into covalent combinations is facilitated by thioredoxin at fructose 1,6-bisphosphate and during the conversion of ribonucleotides to deoxyribonucleotides. An important role for fructose 2,6-bisphosphate in the

ined the role of phosphoenolpyruvate carboxykinase in gluconeogenesis.

The fate of two pivotal amino acids, alanine and aspartate, in gluconeogenesis is illustrated in Fig. 4. In this system tritium from radioactive water can be inserted covalently into specific positions of the sugar being formed from the amino acids.

Reactions that release ammonia and urea are shown in Tables IX and X, respectively. Historically, the enzymes that have been most important in these reactions are urease (240) and phenylalanine ammonia-lyase for the release of free ammonia, but with some involvement of amide-cleaving enzymes when, as in senescent organs, sufficient ammonium acceptor is not always available. The enzyme that has been historically most important in the release of urea and an accompanying amine is arginase, which is active in many germinating seeds (e.g., 97). Urea is also formed during the degradation of purines (166 and Chart 18). Intermediates of purine metabolism such as the ureides have attracted investigators be-

TABLE IX

Examples of Reactions that Produce Urea and Amines from Guanidines with Special Reference to Microorganisms and Higher Plants[a]

Guanidine	Amine	Source	Reference
L-Arginine	L-Ornithine	Lupinus luteus, Trifolium, Angelica sp.	1909. A. Kiesel. Z. Physiol. Chem. **60**, 460.
			1910. A. Kiesel. Z. Physiol. Chem. **67**, 241.
			1922. A. Kiesel. Z. Physiol. Chem. **118**, 267.
		Pinus pinaster seeds	1959. Y. Guitton. C. R. Hebd. Seances Acad. Sci. Ser. D **245**, 590.
γ-Guanidino-butyric acid	γ-Aminobutyric acid	Panus tigrinus	1967. J. Miersch and H. Reinbothe. Phytochemistry **6**, 485.
		Picea glauca buds	1969. D. J. Durzan. Can. J. Biochem. **47**, 771.
		Pseudomonas putida	1972. C. S. Chou and V. W. Rodwell. J. Biol. Chem. **247**, 4486.
		Penicillium roquefortii	1972. G. Brunel-Capelle et al. Physiol. Veg. **10**, 559.

(continued)

TABLE IX (*Continued*)

Guanidine	Amine	Source	Reference
Octopine[b] (N-carboxyethyl-arginine)	Octopinic acid (N²-D-1-carboxy-ethyl)-L-ornithine	Crown galls	1965. A. Ménagé and G. Morel. *C. R. Hebd. Seances Acad. Sci. Ser. D* **261**, 2001.
	N²(1,3-Dicarboxy-propyl)ornithine	Crown galls	1977. J. L. Firmin and R. G. Fenwick. *Phytochemistry* **16**, 761.
Guanidino-succinic acid	L-Aspartic acid	*Pseudomonas chlororaphis*	1972. S. Milstein and P. Goldman. *J. Biol. Chem.* **247**, 6280.
Guanidinoacetic acid	Glycine	*Penicillium roquefortii*; also arginase and γ-guanidinobutyrate-urea hydrolase	1967. A. Brunel *et al. C. R. Hebd. Seances Acad. Sci. Ser. D* **264**, 2777.
		Cyclohydrase	1968. G. Brunel-Capelle and A. Brunel. *C. R. Hebd. Seances Acad. Sci. Ser. D* **266**, 590.
Agmatine	Putrescine	Bacteria	1956. K. Miyaki and H. Momiyama. *J. Biochem.* **27**, 765.
	N-Carbamoyl-putrescine	Potassium-deficient barley (*Hordeum vulgare*)	1964. T. A. Smith and J. L. Garraway. *Phytochemistry* **3**, 23.
Homoarginine	Homoagmatine	*Lathyrus sativus*	1973. S. Ramakrishna and P. R. Adiga. *Phytochemistry* **12**, 2691.
Canavanine	Canaline	*Canavalia* seeds	1940. M. Damodaran and K. G. A. Narayanan. *Biochem. J.* **34**, 1449.
		Caragana spinosa	1967. R. Toepfer *et al. Biochem. Physiol. Pflanz.* **161**, 231.
	Ureidohomoserine	*Streptomyces*	1957. H. Kihara and E. E. Snell. *J. Biol. Chem.* **226**, 485.

[a] It is conceivable that other naturally occurring amines thought to be decarboxylation products of amino acids may also be derived by amidino or urea hydrolase action (e.g., lysine from homoarginine in *Lathyrus* sp.).

[b] Nopaline (lysine derivative) and histopine (histidine derivative) are also known (Appendix II).

TABLE X

Some Reactions that Yield Free Ammonia[a]

Enzyme	Reaction
	Hydrolyzing reactions
Urease	Urea + $H_2O \to CO_2$ + $2NH_3$
Phenylalanine ammonia-lyase	Phenylalanine + $H_2O \to$ *trans*-cinnamate + NH_3
Tyrosine ammonia-lyase	Tyrosine + $H_2O \to$ *trans-p*-coumarate + NH_3
Histidine ammonia-lyase	Histidine + $H_2O \to$ *trans*-urocanate + NH_3
Adenine deaminase	Adenine + $H_2O \to$ hypoxanthine + NH_3
Guanine deaminase (guanase)	Guanine + $H_2O \to$ xanthine + NH_3
Cytosine deaminase	Cytosine + $H_2O \to$ uracil + NH_3
Adenosine deaminase (nonspecific adenosine deaminase)	Adenosine + $H_2O \to$ inosine + NH_3; adenosine compound + $H_2O \to$ inosine compound + NH_3
Cytidine deaminase	Cytidine + $H_2O \to$ uridine + NH_3
5′-Adenylic acid deaminase	5′-Adenylic acid + $H_2O \to$ inosinic acid + NH_3
Adenosine diphosphate (ADP) deaminase	ADP + $H_2O \to$ IDP + NH_3
Pterin deaminase	Pterin + $H_2O \to$ deaminopterin + NH_3
β-Ureidopropionase	β-Ureidopropionate $\to$ β-alanine + CO_2 + NH_3
Ureidosuccinase	L-Ureidosuccinate $\to$ L-aspartic acid + CO_2 + NH_3
	Oxidative formation of ammonia
L-Amino-acid oxidase	L-Amino acid + O_2 + $H_2O \to$ corresponding α-keto acid + H_2O_2 + NH_3
D-Amino-acid oxidase	D-Amino acid + O_2 + $H_2O \to$ corresponding α-keto acid + H_2O_2 + NH_3
Glycine oxidase	Glycine + $\frac{1}{2}O_2 \to$ glyoxylate + NH_3
	Desulfurase and dehydrolase
Homoserine dehydratase	L-Homoserine $\to$ α-ketobutyrate + NH_3
L-Serine dehydratase	L-Serine $\to$ pyruvic acid + NH_3
L-Threonine dehydratase	L-Threonine $\to$ α-ketobutyrate + NH_3
	Amidases
Glutaminase	Glutamine $\to$ glutamic acid + NH_3
Asparaginase	Asparagine $\to$ aspartic acid + NH_3
Amino acid amidase	Amide $\to$ amino acid + NH_3
Aminoimidazolase	4-Aminoimidazole $\to$ glycine derivative + NH_3
AMP amidase	AMP—$NH_2 \to$ AMP + NH_3
	Dehydrogenases
L-Alanine dehydrogenase	L-Alanine + NAD^+ + $H_2O \to$ pyruvate + NADH + H^+ + NH_3
Glutamate dehydrogenase	L-Glutamate + $NAD(P)^+$ $H_2O \to$ α-ketoglutarate + NAD(P)H

(continued)

TABLE X (*Continued*)

Enzyme	Reaction
L-Cysteine sulfinate dehydrogenase	L-Cysteine sulfinate + $NAD^+ \rightarrow$ β-sulfinyl pyruvate + NH_3 + NADH + H^+
	Amine oxidases
Amine oxidase (mono-amine oxidase)	Monoamine + O_2 + $H_2O \rightarrow$ corresponding aldehyde + NH_3 + H_2O_2
Amine oxidase (copper-containing diamine oxidase)	Diamine + O_2 + $H_2O \rightarrow$ corresponding aldehyde + NH_3 + H_2O_2
	Reductive formation of ammonia
Hydroxylamine reductase	NH_2OH + cytochrome$_{red}$ $\rightarrow$ NH_3 + cytochrome$_{ox}$ NH_2OH + NADH $\rightarrow$ NH_3 + NAD^+ + H_2O

a See IUPAC–IUB Commission (1979). "Enzyme Nomenclature." Academic Press, New York.

cause of their production of urea, their common occurrence in the saps of many plants and trees, and more recently because of their structural relationships to cytokinins (27).

7. Summation

The various figures, tables, appendixes, and metabolic charts, as already referred to, in their entirety relate to a great array of nonprotein nitrogen compounds that may be present in plants. They present conceivable metabolic pathways for their synthesis, turnover, or complete degradation. To appreciate the diversity and range of the nitrogen compounds of plants one may scan the Appendixes I–IV. To see the possible reaction patterns by which this diversity may arise one may scan the metabolic Charts 1–19. Necessary and popular as this general approach is, and despite its tribute to over 50 yr of modern biochemical work on intermediary metabolism, one should be realistic about its ultimate meaning. It is one thing to set down pathways and schemes that are biochemically feasible, even with known *in vitro* enzyme systems, group-transfer reactions, specific coenzymes of oxidation–reduction, etc. that may exist especially in microorganisms; it is quite another to show in *specific* cases that these systems actually operate *in vivo*—this is especially so if one should make the demand that the overall picture be established for a particular organism.

Ultimately, therefore, plant physiologists or crop scientists are faced with the need to understand specific plants in their entirety and with the integration of nitrogen metabolism and other aspects of nutrition and metabolism and their responses during development and to environments. The facing of these challenges, in drawing on the knowledge gained in the classical period (as documented in Volume IVA) and supplemented by some later comprehensive investigations of specific plants, is the objective of Section IV of this chapter. For the linear array of amino acids combined in proteins we have an established dogma that the basis is the genetic code, inherent in the DNA of chromosomes. The metabolic charts depict "linear arrays" of sequential enzymatic steps constituting the families of metabolites and their reaction "trees." These lineages must also be inherited; but what and where is the physical basis for these inherited patterns in the organization of cells and organisms? In microorganisms the regulation of amino acid biosynthesis occurs at two levels. These are the regulation of enzyme formation and of enzyme activity (14, 263, 264). In higher plants, however, further provision needs to be made, in ways not yet wholly clear, for the "plugging in" at appropriate times and places of great blocks of information to steer the metabolism along the routes that it clearly takes.

To examine this problem further it is well to see it first in relation to the metabolism of proteins (i.e., their synthesis, breakdown, and turnover), as in the following section.

B. PROTEINS: SYNTHESIS AND METABOLISM

1. Introduction

Great changes have taken place in the interpretation of the synthesis and metabolism of proteins since the accounts given in (Volume IVA, pp. 548–573). New advances are based on new knowledge of the chemistry and structure of specific proteins (219, 296); of the means of transmitting genetic information via the interlocking roles of nucleic acids, nucleotides, and proteins (100, 113, 172, 296); of the relationships between structure and enzymatic functions (7, 23, 117); and of the roles of proteins in membranes (33, 137). Because much of this new knowledge is covered from the standpoint of biochemistry in Chapter 3 of this volume, the aspects to be covered here will be those needed to understand the synthesis and turnover of proteins in relation to plant metabolism.

Before presenting details of protein synthesis as summarized in the second series of metabolic charts, the general picture may be stated as follows. The genetic code, already familiar in 1965 (Volume IVA, pp.

570 ff.), relates the sequence of bases in the DNA, or in the mRNA transcript, to the sequence of amino acids in proteins (122, 286). The 20 amino acids used in protein synthesis are each coded by groups of three bases called codons, but another three codons (UAA, UAG, and UGA) represent signals for termination points in polypeptide synthesis. (Because for most amino acids there is more than one code word, the code is described as degenerate. Different codons that specify the same amino acid usually differ only in the last base of the triplet.)

Once the DNA sequence is transcribed by RNA polymerase to form mRNA, the sequence of codons in mRNA is "read" by tRNA molecules that bring the appropriate amino acid to the template surface where proteins are to be synthesized. The first step in the mechanism of protein synthesis has been called "translation" because the information in the language of the four bases (A, G, C, and T) paired with each other is copied in complementary strands of nucleic acids and eventually specifies the nature and sequence of the amino acids in protein. In eukaryotes with more stable mRNAs there is control over gene expression at the translational level (176).

Proteins are synthesized in the amino to carboxyl direction by the sequential addition of amino acids to the carboxyl end of the growing polypeptide chain. The acids are activated in the form of aminoacyl-tRNAs in which the carboxyl group of an amino acid is joined to the 3' terminus of a tRNA. The recognition and linking of an amino acid to its corresponding tRNA is catalyzed by an aminoacyl-tRNA synthetase reaction that is driven by energy supplied as ATP. For each of the 20 protein amino acids there is at least one kind of tRNA and activating enzyme. It is the specificity of the enzyme toward the amino acid tRNA complex that prevents nonprotein amino acids from being incorporated into protein.

There are three stages in polypeptide synthesis, namely, the initiation, the elongation, and the termination of the polypeptide chain. Initiation results from the binding of the initiator tRNA to the starting codon in the mRNA. The initiator tRNA occupies a specific site on the ribosome surface (P site), which is one of two binding sites on the ribosome for tRNA. The binding of another aminoacyl-tRNA to the second site on the ribosome (A site) is the signal that starts the elongation of the polypeptide chain. The first peptide bond is formed between the amino group of the incoming aminoacyl-tRNA and the carboxyl group of N-formylmethionine, a nonprotein amino acid, carried by the initiator tRNA. The resultant dipeptide–tRNA complex is first bound with the initiator tRNA complex at the P site, and the first amino acid of the polypeptide as its tRNA complex is bound at the A site. In the next step, which is translocation of the peptide, the P site is vacated, and the complex, hitherto at the A site, moves to the P site, while the initiator leaves the ribosomal surface. The A site is then free to be occupied by another amino acid–tRNA complex.

As so described, translation has proved to be a very complex process and it is now conceived to involve a large number of ancillary macromolecules to build even a single polypeptide. Also, after the amino acid is inserted in its proper protein sequence, there are more than 140 specific posttranslational modifications that can be made with the 20 protein amino acids, of which only a few are recognized as readily reversible. Examples of covalent protein modifications are (a) phosphorylation–dephosphorylation, (b) acetylation–deacetylation, (c) adenylation–deadenylation, (d) uridylation–deuridylation, (e) methylation–demethylation, and (f) SS–SH interconversions. Transient enzyme–substrate complexes that are part of reaction mechanisms (132) represent another type of protein modification.

In some bacteria, amino acids are activated for peptide synthesis by the formation of enzyme-bound thiolesters. The amino acid so activated is attached to a sulfhydryl group of one of two enzymes rather than to the 3′ terminal group of a tRNA. Peptide synthesis is then initiated by the interaction of the two enzymes. The amino acid sequence of the peptide is determined by the spatial arrangement and specificity of the enzymes, one of which contains a thiol residue that is thought to carry the growing peptide from one site to another on the same enzyme.

Plant cells synthesize oligopeptides such as glutathione by a different mechanism. In all such cases, the energy is derived from ATP. Glutathione is a ubiquitous tripeptide that contains one γ-peptide linkage rather than the α-linkage usually found in peptides (153). Glutathione is formed in two steps:

$$\text{L-Glutamic acid + L-cysteine + ATP} \xrightarrow[\text{Mg}^{2+},\ \text{K}^+]{\text{synthetase}} \text{γ-glutamylcysteine + ADP + P}_i \qquad (1)$$

$$\begin{array}{l}\text{γ-Glutamylcysteine} \xrightarrow{\text{glutathione}} \text{γ-glutamylcysteinylglycine (glutathione)}\\ \text{+ glycine + ATP} \quad \overset{}{\text{synthetase}} \qquad\qquad\qquad \text{+ ADP + P}_i\end{array}$$
$$(2)$$

The specificity of the first enzyme seems to vary in some plants and yields a variety of γ-glutamyl peptides (Appendix IIIa) the functions of which are still unclear.

A large literature on hydroxyproline-rich glycopeptides has accumulated (e.g., 215). The original attention to these substances in plants (in contrast to their prevalence as collagen, etc. in animals) came somewhat unexpectedly from aseptically cultured carrot (*Daucus carota* var. *sativus*) explants (see Volume IVA, Chapter 4, Section V,B), which contained large amounts of insoluble hydroxyproline even though this amino acid was not conspicuous, free or combined, in the carrot tissue from which the cultures were derived. They were later isolated from the ambient medium of tobacco (*Nicotiana*) cultures. These compounds are therefore prominently associated with cultured cells or explants, and discussion of them will be deferred to the volume that will deal specifically with the problems of growth and the isolated culture of cells and tissue explants. This will permit these compounds to be discussed more in relation to cells, their walls, their cytoplasms, and their inclusions rather than identifying them with nitrogen metabolism per se. Their limited range of amino acids combined with sugars and their often lower molecular weight than proteins places them "off the main course" of nitrogen and protein metabolism, which is the main concern of this chapter. This, however, does not preclude some special significance of proline–hydroxyproline relationships in the economy of nitrogen in plants, which will be taken up in the volumes concerned with growth.

The cyclopeptide alkaloids are another expanding group of polyamide plant bases composed of amino acid residues in common and highly modified forms (see 181). Very little is known about their biosynthesis.

2. Protein Synthesis: A Synopsis

The incredibly complex molecular events involved in protein synthesis to arrange the amino of even one protein in linear array (as, e.g., that given in Chapter 1, Fig. 2) lend themselves to, even require, schematic or pictorial representations by which they may be visualized. The concepts current in 1965 were illustrated in this way in Volume IVA (Figs. 52, 53A,B, and 54 and in the accompanying text). Later, in 1971, the earlier schemes were adapted to newer knowledge (231, see Figs. 8.3, 8.4, and 8.6). But, since then, the molecular biology literature (as in issues of *Science, Nature, Scientific American*,[6] etc.) is replete with conceptual interpretations of the ways that genes, enzymes, energy, molecules, and surfaces work coherently to make a protein. The portrayal of these events has become a veritable interpretive biochemical art form. Such

---

[6]See Section IV in P. C. Hanawalt (ed.) (1980), collected papers entitled "Molecules to Living Cells." Freeman, San Francisco.

schemes convey an impression of the sequential biochemical steps albeit they do not, and cannot, do justice to the physical complexity of the organizations within which they are compressed and regulated. What follows therefore is but a brief synopsis of the whole system as it may operate *in vivo*.

Current knowledge of the molecular biology of protein synthesis is largely derived from microorganisms and certain hemoglobin systems (e.g., 43, 44, 117). However, the bacterial model, although useful, does not apply strictly to higher plants. Therefore, Charts 20–24, with explantory or interpretive notes, present for higher plants a view of events that intervene between the synthesis of amino acids *de novo* and their ultimate binding in polypeptides and proteins. Chart 20 sets the stage for polypeptide synthesis, as in a replicating plant cell, and shows how the information for protein synthesis relates to the structure of chromatin and to the events of the cell cycle. Chart 21 describes the transcription and processing of the genetic information in DNA into the messenger RNA that is to be translated into the protein-synthesizing ribosomal surface. Charts 22A and 22B illustrate RNA molecules, their synthesis, turnover, and role in the transfer of amino acids to the protein-synthesizing surface. Chart 23 deals with ribosomes and their role in polypeptide synthesis. Chart 24 gives the genetic code and shows a ribosomal cycle in protein synthesis together with a current model for secretion of proteins through membranes. If these charts and their notes are read in sequence, an overall current view of protein synthesis should emerge. Clearly, however, the information comes from many sources, derived from many experimental systems, and the purpose of this form of presentation is to achieve an integrated account potentially applicable to higher plants.

METABOLIC CHARTS 20–24

20. Protein synthesis: introduction to polypeptide formation during cell replication
21. Protein synthesis: transcription
22. Protein synthesis: formation, turnover, and function of tRNA
    A. Structure of a typical tRNA
    B. Synthesis, turnover, and function of tRNA
23. Protein synthesis: peptide formation on ribsomes
24. Protein synthesis: summary (the genetic code and the formation and secretion of proteins)

CHART 20

PROTEIN SYNTHESIS: INTRODUCTION TO POLYPEPTIDE FORMATION
DURING CELL REPLICATION[a,b]

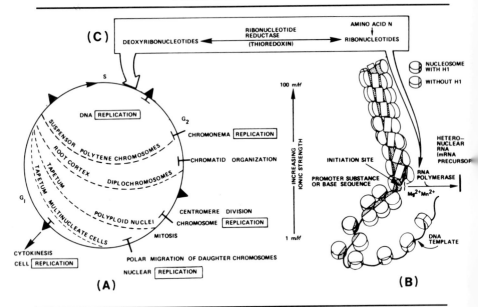

[a] (A) The mitotic cell cycle in higher plants. This part of the chart relates the behavior of DNA to the replication of cells. Based on Brown and Dyer (29). (B) A single strand of chromatin, showing how and where messenger RNA is formed. Redrawn from Thoma *et al.* (252). (C) The flow of nitrogen via ribonucleotide precursors into deoxyribonucleotides for DNA and ribonucleotides for RNA.

[b] Interpretive note for Chart 20. (A) Summarizes the normal mitotic cell cycle, with four modified or curtailed cycles of special cell types (dashed lines). Of the six essential stages in the cycle associated with cell replication, one (the S stage) is concerned with the replication of all the DNA in the nuclear genome. Virtually all double-stranded DNA exists naturally in a negatively supertwisted form (42a, cf. 295). The super coiling is mediated by a DNA gyrase requiring ATP, and in bacteria negative supercoiling is essential for replication. Eventually, replicated DNA appears in the chromatin of the nucleus of the daughter cell, as in B (cf. 136, 177). G_1 and G_2 are intervening periods of interphase where there is no DNA accumulation. Curtailed cell cycles produce a variety of cell types with an increased nuclear content.

CHART 20 (*Continued*)

(B) Shows how the information replicated in the dividing cell is "packaged" for insertion into the nucleus of the daughter cell. This part of the chart shows the structure of a very small segment of a chromosome in which part of the chromatin thread is shown extended as would happen at low ionic strengths when the nucleoprotein coat, in the form of nucleosomes containing at least four and usually five histones (H1, H2A, H2B, H3, H4), is "loosened" (114, 130, 252). H1 is probably bound to the section of DNA linking the subunits. Where a segment of DNA containing a gene is exposed, RNA polymerase may bind with DNA and begin the process of transcription.

Transcription describes the events leading to polypeptide synthesis that require the synthesis of several categories of RNA. At one site the messenger RNA (mRNA) may be manufactured, at other sites transfer (tRNA) and ribosomal (rRNA) RNA may be made. Numerous nonhistone chromosomal proteins (NHCP) are now found to be associated with chromatin (e.g., 187-189, 276), some of which appear to respond to insecticides (190). At sites where transcription is initiated or in progress, they make the site accessible to RNA polymerase with its cofactors and facilitate the process of transcription (51). At the site (→) where transcription is initiated, RNA polymerase takes incoming ribonucleotides (Chart 15A,B) and manufactures the appropriate RNA (mRNA, tRNA, or rRNA).

The present chart illustrates how information needed for protein synthesis is transmitted to the nuclei of daughter cells, embodied and protected in their chromosomes, and how, at specific gene sites, the information they contain is transcribed in the synthesis of ribonucleic acids for later use in the protein translation sequence. These early events make demands for nitrogen in the form of ribonucleotides, as shown in part (C), which are modified to deoxyribonucleotides for DNA replication (251) or used as ribonucleotides, which transcribe the information in the DNA into RNA. The following charts show how this information is conveyed to and used at the protein-synthesis templates on ribosomes in the cytoplasm. The synthesis of RNA on nucleolar genes has been visualized by electron microscopy (162, 163), as has the substructure of chromatin depicted in Chart 20 (178). This subject has been reviewed (77).

Inevitably, the rather rigidly prescribed biochemical patterns as presented in the metabolic charts and schemata in this chapter will need to become more flexible to accomodate the regulatory controls of metabolism during development and in response to nutrition and environments, as demanded in Sections IV and V. Trends in these directions that are already evident have portents for the future.

Even as this text goes to print the 30-year-old work by B. McClintock has been recognized by a Nobel Prize. This work prompted her to postulate that some hereditary units of corn could become unexpectedly mobile and when transposed to their new situations could alter the behavior of the cells. The roles of viruses that bring "foreign" genetic information into cells, that properly relate to other volumes, are now familiar. But more recent evidence indicates that very much smaller protein-free "naked" molecules of RNA, called viroids, when introduced into plant cells may replicate and intervene to alter metabolism of the host cells (for summary see *American Scientist* **71**, 481–489).

CHART 21

Protein Synthesis: Transcription[a,b]

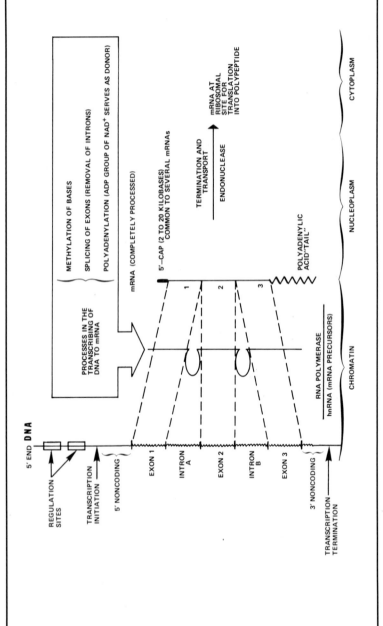

a Modified from Crick (43).

b Interpretive note for Chart 21. Hitherto, and in bacterial systems, the genes were supposed to be in continuous linear arrays on the DNA strand. Each gene specified one enzyme. Now, however, and in higher organisms, there may be intervening regions of DNA ("introns") that do not carry the amino acid code (43). Intervening sequences also occur in the mitochondrial genome (20). The useful information is in regions referred to as "exons," which specify sequences of amino acids, as in a protein or enzyme. When synthesis of a protein or enzyme is called for, the early events of transcription are promoted by a substance that "loosens" the DNA–histone complex at sites where RNA polymerase can evoke the synthesis of transcribed DNA (i.e., RNA). For protein synthesis this transcript is called *heteronuclear RNA* (hnRNA), and it eventually becomes the messenger RNA (mRNA) that is transported into the cytoplasm (52, 247). In the processing of the first DNA transcript to condense it into mRNA (containing the amino acid code or exons), the introns are eliminated, and during this process methylation of bases, resplicing of the exons, and completion of the mRNA by combination with polyadenylic acid occurs (1). Once the mRNA is transcribed and processed, it is ready to be transported into the cytoplasm.

Although the schemes have derived recently from bacterial and mammalian systems, it should be recalled that B. McClintock postulated, years ago, regions on the chromosomes of maize (*Zea mays*) that were equivalent to the introns shown in this chart. Now these mechanisms are commonly reported in other organisms (173). Furthermore, most plants rely upon systems of genes.

Gene systems of plants have been categorized by Grant (93). The systems are based on the nature, size, and position of the genetic elements in the cell. For example, plants may have

1. Mendelian genes (classical gene; major gene; modifier gene, in part; multiple factor, cistron; structural gene) (many species contain genes with intervening nucleotide spacer sequences of unknown function)
2. Polygenes [modifier gene, in part; redundant DNA; repetitive DNA (e.g., 145)]
3. Controlling elements (heterochromatin block, in part)
4. Cytoplasmic genes (plasmagene, organelle gene)

Gene systems often interact with each other in the following ways:

1. Two or more Mendelian genes controlling successive steps in a metabolic or developmental sequence, forming a "serial gene system"
2. Two or more Mendelian genes with similar and usually cumulative actions, forming a "multiple gene system"
3. Numerous polygenes, forming a "polygenic system"
4. Combination of a Mendelian or major gene with modifiers, forming a "modifier system"
5. A combination of controlling genes with positive chromosomal genes, forming an "oppositional gene system"
6. Combinations of chromosomal and cytoplasmic genes to specify proteins in organelles (e.g., ribulose-1,5-bisphosphate carboxylase in chloroplasts)

The inheritance of amino acid content in cereals has been reviewed (172), and the evolutionary changes in genome, chromosome size, and DNA content outlined (192, 225).

CHART 22

PROTEIN SYNTHESIS: FORMATION, TURNOVER, AND FUNCTION OF tRNA

(A) Structure of a typical tRNA[a,b]

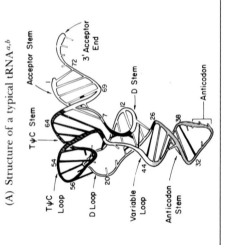

[a] From Rich (199a). Copyright 1976 by the American Association for the Advancement of Science.

[b] Interpretive note for Chart 22A. The "cloverleaf" structure of yeast tRNA[Phe] is shown with the ribose–phosphate backbone depicted as a coiled tube. The numbers refer to the nucleotide residues in the sequence; the first base sequence of a tRNA was determined by Holley *et al.* (111). Hydrogen-bonding interactions between bases are shown as cross-rings. Tertiary interactions between bases are solid black. Bases that are not involved in hydrogen bonding to each other are shown as shortened rods attached to the backbone (193a). The molecule has a terminal attachment site (3' acceptor end) to which the amino acid is specifically combined. The configuration of the rest of the tRNA molecule gives it the specificity just to recognize the proper protein amino acid and also to allow its combination as brought about by an aminoacyl-tRNA synthetase. Finally, the tRNA molecule has the accessory structures to direct the amino acid to its proper place on the protein-synthesizing surface. The acceptor strand contains modified bases that have cytokinin activity (167, 231).

132

(B) Synthesis, Turnover, and Function of tRNA[c][d]

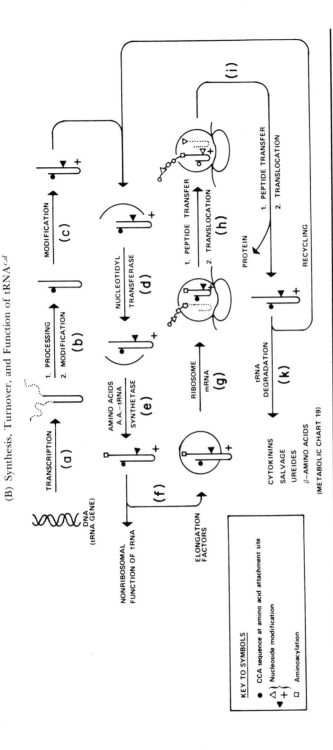

[c] After (125, 193a).

[d] Interpretive note for Chart 22B. Initially, as at (a), RNA polymerase transcribes a tRNA gene to produce a tRNA precursor (125). This contains extra nucleotides on the 5' as well as the 3' end of the transcribed tRNA. Sometimes the precursor contains more than one tRNA. Some nucleotides may be modified in their base content at the precursor stage.

(continued)

133

CHART 22 (*Continued*)

Second (b), processing enzymes recognize tRNA precursors and strip the tRNA of extraneous nucleotides.

Third (also at b), some processed tRNAs lack the C-C-A base sequence at the 3' end. tRNA nucleotidyltransferase (d) recognize these incomplete tRNAs and add on the C-C-A base sequence.

Fourth (c), nucleosides at various positions on the tRNA are altered by many different tRNA-modifying enzymes, which in most cases discriminate among the specific tRNAs. In some tRNAs as much as 16% of the bases are modified.

Fifth (e), the resultant tRNA is aminoacylated by the appropriate aminoacyl-tRNA synthetase through a specific recognition process. In the cytoplasm of plants, the initiator tRNA for protein synthesis is not formylated although the *N*-formylmethionine in tRNA is found in bacteria, in chloroplasts, and in mitochondria of plants (143; also see Charts 23 and 24). Methionyl-tRNA is involved in the insertion of methionine into protein. In barley (*Hordeum*) plants, *N*-feruloylglycine has been suggested as a "starter" for protein synthesis (Appendix I). The charging of the tRNA with an amino acid involves two steps:

$$\text{Amino acid} + \text{ATP} \underset{}{\overset{\text{Mg}^{2+}}{\rightleftharpoons}} \text{aminoacyl-AMP} + \text{PP}_i \qquad (1)$$

$$\text{Aminoacyl-AMP} + \text{tRNA} \longrightarrow \text{aminoacyl-tRNA} + \text{AMP} \qquad (2)$$

Aminoacyl-tRNA synthetases have altered active sites to prevent the incorporation of analogs of amino acids into protein (79, 81–83, 86, 87). This specificity [as at (e)] is one of the most critical points in maintaining the fidelity of protein synthesis and in discriminating between protein and nonprotein amino acids.

Sixth (f), an "elongation factor" forms a complex with tRNA in the presence of GTP, so that the complex may deliver aminoacyl-tRNA to the "A" site of the ribosome.

Seventh (g), the tRNA–amino acid complex then recognizes the appropriate site on the ribosomal surface where the processes of peptide formation then occur (h).

Eighth, tRNA is relesed from the ribosome and recycled. Alternatively, tRNA may be degraded by various nucleases into mononucleotides, which then rejoin the nucleotide pool (k).

In plants it has been postulated that the turnover of tRNA releases modified nitrogen–bases that have cytokinin activity, but this concept is not as yet completely established (167, 231). Labeled cytokinins fed to plant cells can be isolated from all types of RNA molecules (167).

Thus a tRNA molecule performs throughout its "life cycle" a variety of functions for which its structure gives it specificity. The general features and recognition of tRNAs by aminoacyl-tRNA synthetases have been reviewed (211, 296).

CHART 23

PROTEIN SYNTHESIS: PEPTIDE FORMATION ON RIBOSOMES[a]

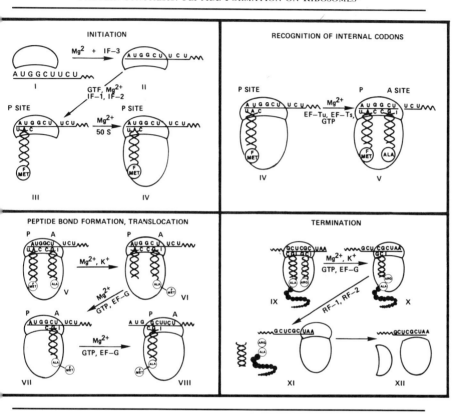

[a] Interpretive note for Chart 23. The semilunar cap (I) of the ribosome represents the free 30-S subunit. Initiation of protein synthesis involves the attachment of mRNA to a 30-S subunit (I) to form a complex (II); this process requires Mg^{2+} as well as an initiation factor IF-3 (cf. 296).

In the bacterial model, an attachment at the P site of fMet–tRNA occurs in response to the initiation codon AUG on the mRNA to form a complex (III). The addition of the 50-S ribosomal unit is aided by Mg^{2+} and produces a complex (IV). The chart shows alanine as the next amino acid coded for on the mRNA. This will be recognized at the so-called A site. The enzymatic recognition to establish the placement of the tRNA–alanine complex at the A site involves several cofactors designated as EF-Tu, EF-Ts, and GTP (V). Peptide-bond formation between methionine and alanine, located at P and A sites, respectively, occurs on the ribosome and is aided by K^+ (VI). After peptide-bond formation the tRNA moiety, being relieved of its amino acid, vacates the P site, and the dipeptide formed is still bound at the A site (VII). Subsequently, however, the whole complex of mRNA with the dipeptide attached, hitherto at the A site, moves to the P site (VIII). Having formed the peptide bond in this way, the A site has been vacated to receive another tRNA–amino acid complex (IX). This process continues until the peptide is completed (X), when a termination codon

(continued)

CHART 23 (*Continued*)

(UAA) on the mRNA and a specific release factor negotiate the release of the completed polypeptide from the ribosome. The mRNA is now released from the ribosome (XII), which dissociates into its 30- and 50-S subunits (XII).

Although a complete model on these lines has not been demonstrated for higher plants, it is interesting and suggestive that wheat germ, because of its availability and quantity, has been a source of the ancillary factors used in artificial systems reconstituted from animal and bacterial components. The synthesis and processing of maize storage proteins has been reported in toad eggs (137). This suggests some degree of compatibility between plant and animal protein-synthesizing systems. The initiation, elongation, and termination of genetic messages have been reviewed (200).

CHART 24

PROTEIN SYNTHESIS: SUMMARY
(THE GENETIC CODE AND THE FORMATION AND SECRETION OF PROTEIN)[a]

(A)

Second letter

First letter		U	C	A	G	
U		UUU UUC Phe / UUA UUG Leu	UCU UCC UCA UCG Ser	UAU UAC Tyr / UAA UAG Ter	UGU UGC Cys / UGA Ter / UGG Trp	U C A G
C		CUU CUC CUA CUG Leu	CCU CCC CCA CCG Pro	CAU CAC His / CAA CAG Gin	CGU CGC CGA CGG Arg	U C A G
A		AUU AUC AUA Ile / AUG Met	ACU Acc ACA ACG Thr	AAU AAC Asn / AAA AAG Lys	AGU AGC Ser / AGA AGG Arg	U C A G
G		GUU GUC GUA GUG Val	GCU GCC GCA GCG Ala	GAU GAC Asp / GAA GAG Glu	GUU GGC GGA GGG Gly	U C A G

(B)

CHART 24 (*Continued*)

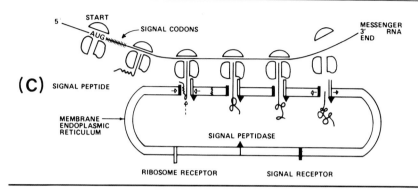

 Interpretive note for Chart 24. (A) The genetic code, showing the three-base code words representing each of 20 individual protein amino acids in the DNA and as transmitted via mRNA to the ribosomal surface. Ter, terminate.

(B) Shows how bacterial ribosomes may be assembled from 30 and 50 S subunits, incorporating the tRNAs (as in Chart 22) to manufacture a specific polypeptide (I–VIII in Chart 23), and how polysomes comprise a multiple system capable of producing sets of several polypeptides. Having completed their functions, polysomes and ribosomes break down to their subunits and repeat the process (134, 184, 289). The model for plants as related to the biosynthesis of legume storage proteins has been reviewed (24, 296).

(C) Describes a hypothetical model (149) for situations in which proteins formed by ribosomes adjacent to membranes of the endoplasmic reticulum are thought to be secreted across them. This requires that the mRNA coding for a given protein also have a special segment of bases, called signal codons, specifying in the newly synthesized peptide a binding property to sites in the membrane. In addition, the ribosomes may be bound to ribosome receptors in the membrane. When the newly synthesized protein binds its signal sequence to the membrane receptor and the ribosome binds to its receptor, a pore in the membrane opens, allowing the protein to pass into the interior space of the endoplasmic reticulum. Proteins with signal sequences of amino acids have been found in chloroplasts. The arabinoglycoproteins or β-lectins at the plasmalemma may be another category of proteins that fit this model (cf. 296). The synthesis and processing of zein on protein-body membranes has been described (33). The process of proteolysis has been reviewed (112, 296, and see Chapter 3).

The accuracy of the protein-synthesizing mechanism is very high, but errors occasionally do occur. For microorganisms, the level of error for protein synthesis from mRNA is one wrong amino acid for every 10^4 correct ones. For DNA replication, errors in specification of bases are no more than 1 in 3×10^4.

3. Protein Synthesis in Organelles and at Different Sites[7]

Finally, it should now be recognized that organelles which are distinctively endowed with their own DNA (i.e., chloroplasts and mitochondria) are separate centers for protein synthesis. For example, the

---

[7]A complementary account of the biochemistry of protein synthesis, with particular reference to RuBPcase, will be found in Chapter 3, Section IV, A. (Eds.)

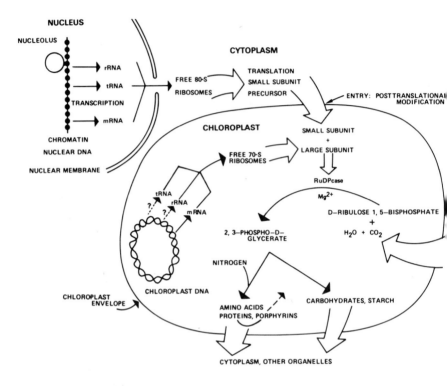

Fig. 5. Events in the synthesis of ribulosebisophosphate carboxylase. Based on Bradbeer *et al.* (26) and Ellis (73).

synthesis of the enzyme ribulosebisphosphate carboxylase (RuBPcase) is specified by two genetic systems—one in nuclei, the other in chloroplasts (26, 73). Nevertheless, the steps of the protein synthesis at the two sites are deemed to be consistent with views outlined in the metabolic Charts 20–24.

RuBPcase consists of large and small subunits of greatly differing molecular weights. The large subunit is encoded by the chloroplast genome and is synthesized inside this organelle (Fig. 5). The small subunit is encoded by the nuclear genome and is synthesized as a precursor by cytoplasmic ribosomes. This precursor crosses the chloroplast envelope with the removal of part of the amino acid sequence, and it then combines with the large subunit in the stroma to form the "holoenzyme." Furthermore, the nucleic acids of chloroplasts are quite different from those in the nucleus. The principal ribosomal RNAs of chloroplasts are smaller. Chloroplast ribosomes do not have the 5.85-S rRNA found in the cytosol and *N*-formylmethionyl-tRNA rather than the methionyl-tRNA that initiates protein synthesis. Chloroplasts, however, do contain a 5-S rRNA and a 4.5-S rRNA. These RNAs are common to ribosomes of flowering plants and do not contain modified nucleotides. The nucleotide sequences and properties of these rRNAs have been described (25, 72). The nucleotide sequence used by ribosomal DNA in coding for 16-S rRNA in the DNA of ribosomes from *Zea mays* chloroplasts resembles that of the RNA of *Escherichia coli* 16-S rRNA; this is compatible with the prokaryotic origin of chloroplast ribosomes (212).

The mitochondrial genome, as judged from work largely on yeast, codes for a few of the proteins of the respiratory chain. It contains genes for tRNA and rRNA molecules of the mitochondrial translation system.

But, after all this has been said and schemes drawn and interpreted, it is here that the subject can be most frustrating. Weber *et al.* (279) have attempted to examine how the regulation of key enzymes may relate to the reprograming of gene expression. If one properly regards the proteins (their synthesis, turnover, and structural roles in plants) as the ultimate ends to which the unique nitrogen metabolism of plants is directed, we are still left with biological problems that are easier to pose than to resolve. These relate to the superiority of the organized systems over the best that the unorganized ones can do. Such problems are:

1. Why is noncellular protein synthesis still so far below what cells, particularly growing cells *in situ* in higher plants, can achieve?

2. Why are the anabolic processes (e.g., of photosynthesis and protein synthesis) so much more difficult to emulate in cell-free systems than their catabolic counterparts?

3. If green leaves in the light are paramount as organs of protein synthesis, which also occurs in growing regions where cells multiply and in organs of perennation where proteins are stored, do all these syntheses proceed in similar ways and subject to similar controls?

4. Because protein breakdown and the reuse of breakdown products occur so frequently, how does the doctrine of "turnover," as between organs and also intracellularly, now stand as an integral aspect of the vital machinery?

5. It is now recognized that RuBPcase is the ubiquitous plant protein. This being so, how far does the diversity of other enzymes account for the great and ever-expanding range of electrophoretically detectable proteins that seems to accompany development?

6. Is there some more deep-seated biological lesson to be learned for the differences between animals [which so jealously safeguard the identity of their proteins (e.g., 7)] and plants, which cannot in the same way react to reject "foreign" proteins? Is there a need to recognize the importance of the reversible covalent modification of protein through

methylation, phosphorylation, and dephosphorylation as in glycogen metabolism or calmodulin in the animal body (127, 132, 180, 256, 257)?

It must be admitted that easy answers to these and other questions do not emerge from the schemes summarized and interpreted earlier. But can new light and understanding be drawn from the current status of molecular genetics and the recombinant DNA technology (217, 296)?

The most interesting aspects of molecular genetics that may impinge on problems of proteins and nitrogen metabolism in relation to development and environments are the now-evident signs of cytoplasmic DNA in organelles. Therefore, one might conceive that variants in cells or organs during development might be induced by cytoplasmic determinants such as plasmids, "foreign" DNA, or any changes in the DNA of cytoplasmic organelles. These ideas gain some credence from the following examples. The ability of *Agrobacterium tumefaciens* to induce neoplasms in tobacco, with the consequential changes in metabolism, is now attributable to the invasion of the host cells by plasmid DNA from the bacteria (27, 50, 210). Plant-tumor reversal is then associated with the loss of foreign DNA (290). Although cases of male sterility are related to nuclear genes, examples are now known in which cytoplasmic DNA influences their expression (262). These examples (and others from parasites and symbiosis) usually, however, affect the morphology first and the metabolism later.

The powerful techniques based on recombinant DNA technology illustrate what may be done to influence the responses of simple organisms like bacteria and viruses (e.g., 117). These have involved the introduction into the genome of the organism of "foreign" DNA, whether synthesized by man or enzymatically "cut and spliced" from the DNA of an organism (123, 124). The results and potentialities for the synthesis thereby of substances are unquestionable and dramatic. But even so, are there any signs or prospects that, in their normal development and responses to environments, the metabolic behavior of the organs of higher plants is responsive to regulation by similar but spontaneous changes in the DNA content of the cells in question? The prime argument that this is not so comes from the evidence that cells of mature organs, with characteristic form and composition, when isolated and cultured, exhibit their essential totipotency and revert to the normal responses of zygotes.

Therefore, nature, which has obviously discovered means of acquiring diversity of morphology and metabolism during development while preserving the constancy of inheritance through DNA, seems to have achieved these ends without recourse to devices of the recombinant DNA type. The well-known and obvious examples of genetic crossing-over and the segregation of the products of recombination that occur at meioses are not precedents here, for once they have occurred they commit the organism metabolically and morphologically, throughout its life cycle.

4. Protein Turnover and Cellular Metabolism

Most concepts of proteins regard them as constituent substances that only have structural, storage, or enzymatic roles. Any subsequent breakdown that may occur is then associated with senescence or translocation and resynthesis. There is, however, the alternative, implicit in "turnover," that the active protein complement is in a dynamic state of breakdown and resynthesis in metabolizing cells so that it becomes an integral part of the dynamic machinery. The analysis of agricultural productivity by Penning de Vries (183) takes the naive view that the elimination of protein turnover would increase the efficiency of crop production. This view ignores the roles of protein turnover in the pacing and maintenance of metabolic activities, of such energy-related functions as salt accumulation and respiration and in the integration of carbon and nitrogen metabolism in the organism. The passage of carbon from protein and its ultimate fate in respiration or resynthesis is an integral part of the course of plant metabolism. In this overall system, and in the nitrogen economy of plants, nitrogen is conserved through turnover whereas carbon may be dissipated. This is in sharp contrast to the situation in animals, where turnover is more concerned with the retention and recycling of carbon and the wastage of nitrogen.

A variety of techniques for detecting and measuring protein turnover are summarized in Chapter 3. The final outcome, however, is that protein turnover is a well-established reality, with differences only of intensity occurring between various systems, such differences being reflections of the metabolism with which protein turnover is associated in different situations. For example, more rapidly growing cultured cells are distinguished by a more rapid pace of turnover than quiescent or slowly growing cells, and the rates of the turnover are directly related to the rates of respiration (21). The situation in leaves is somewhat more complex. In developing leaves [e.g., of wheat (*Triticum*) or tobacco (*Nicotiana*)] the rate of protein turnover parallels the rate of leaf expansion (105a). However, in cotyledons (e.g., of *Antirrhinum majus*) that are developing into photosynthetic organs, protein turnover parallels their biochemical development even though their morphological development may be virtually complete (105b). There are sharp contrasts in the metabolism of leaves as between conditions in the light and in the dark. That protein synthesis is emphasized in the light, whereas breakdown is emphasized in darkness, is consistent with similar contrasts in turnover between these two environmental situations.

An important feature of the earlier work on protein turnover is that the amino acids derived from protein breakdown in the turnover cycle are not simply reused for the synthesis of new proteins, but instead their nitrogen is reworked into the soluble pool and their carbon is used, to a greater or lesser extent, as a substrate for respiration (21). The pace of cellular metabolism in general was thought to be linked to and perhaps controlled by the pace of protein turnover. Studies of the nitrogen metabolism associated with photorespiration show that the relationship between protein turnover and cellular metabolism may, in this context, be even more specific (44a).

The key point is that the C_2 cycle of photorespiration, an obligatory part of photosynthetic carbon fixation, requires intense nitrogen metabolism.[8] Details of the metabolism of both carbon and nitrogen in the photorespiratory C_2 cycle are given in Volume VII, Chapter 3, Section III and Figs. 9 and 10 of this treatise. Briefly, under normal circumstances, during the net fixation of three molecules of carbon dioxide, two molecules of glyoxylate are formed. These enter the C_2 cycle and are converted to glycine. One molecule of glycine is decarboxylated and deaminated, and its remaining methylene carbon is transferred to the other molecule of glycine to make serine. Serine is deaminated, and the remaining three-carbon acid is eventually converted to phosphoglyceric acid and reenters the carbon reduction cycle. The formation of the two molecules of glycine requires two amino groups. One comes by transamination from serine, and the other comes from the cellular pool of ammonia. The decarboxylation–deamination of one of the glycine molecules serves to maintain the pool. However, to be available for renewed glycine synthesis, ammonia must first be refixed by the GS–GOGAT system. Thus the rate of ammonia fixation must be very rapid during active photosynthesis, attaining a rate of at least one-third, on a molecular basis, of the rate of net carbon dioxide fixation. If any ammonia escapes the cellular pool and is excreted, the rate may be even higher.

Cullimore and Sims (44a) have shown that *Chlamydomonas* cells excrete ammonia during photorespiration, but their data indicate that the ammonia so released is derived ultimately from protein turnover. Furthermore, they show that ammonia recycling is relatively unimportant in this

---

[8]It is interesting to recall that ideas of the connections between photosynthesis and protein synthesis in the light and breakdown in the dark seemed to be required (297) even before modern knowledge of either photosynthesis or photorespiration had been developed (see Volume IVA, p. 586).

alga, which is not seriously affected when GS is blocked by inhibitors. It may be concluded that the source of much of the ammonia needed to form one of the two molecules of glycine is derived by transamination from amino acids released by protein turnover. This also suggests the further possibility that part of the apparent photorespiratory output of carbon as well as of ammonia may, in fact, be derived from the oxidation of amino acids released by protein turnover.

The importance of GS–GOGAT, and hence of ammonia recycling, to higher plants was emphasized by Somerville and Ogren (224a). They showed that mutants of *Arabidopsis* lacking this enzyme system could only survive in an atmosphere without oxygen, in which the oxidative steps of photorespiration cannot take place. The difference in this respect between *Arabidopsis* and *Chlamydomonas* undoubtedly relates to the continuous availability of ammonia to the latter in its nutrient medium.

The central role of photorespiratory nitrogen metabolism, and the very large flux of ammonia that must be handled by the ammonia-refixing enzymes, suggest that this may be an important point at which the metabolism of photosynthetic cells is regulated. For example, photorespiration is known to be affected by stress, as is protein turnover (42). It is probable that some of the "light respiration" of bean leaves (i.e., dark respiratory processes that continue in light, as distinct from photorespiration) observed by Mangat *et al.* (147a) results from protein turnover, as occurs in *Chlamydomonas*. Light respiration is controlled by the level of photosynthetically produced ATP, which in turn depends on the rate of photosynthesis and photorespiration. Thus the overall pace of these aspects of cellular metabolism is tightly integrated and mediated by reactions of ammonia release and refixation in protein turnover and photorespiration. This gives credence to earlier discussions of the relationship between protein turnover and cellular metabolism, from the ideas of Gregory and Sen to the more specific experiments on cells and tissue cultures from carrot and other species (21), and emphasizes the concept of protein *metabolism* as distinct from the individual events of protein synthesis and breakdown.

A point that should not be overlooked is that useful energy transfers (for example, for salt uptake) may flow from conformational changes in protein structures,[9] or from changes in their orientation in membranes; these may be mediated by their partial breakdown and resynthesis. All this emphasizes the dynamic role of proteins as essential components in the working of the molecular machinery of cells, supplementing, in this respect, roles merely as catalysts or as structural entities. In this context,

[9]See Volume VII, Chapter 2, Section II,D.

therefore, protein turnover is seen to be as an essential feature of metabolism.

5. *Resumé: Controls of Amino Acid and Protein Metabolism*

At the outset one should recognize that each organism is uniquely endowed with genetic characteristics that set biochemical limits to its preformance. But within these limits there is a great range in the role of nitrogen as it is affected by factors that intervene during development or that may be imposed by variable, but controllable, features of the environment or nutrition.

In the long course of development from dividing cells (zygotes or totipotent somatic cells) through a life cycle completed at senescence and/or meiosis, there are many points of control at which the course of nitrogen metabolism can be visualized as vulnerable to decisive shifts or change (e.g., 268). In many such situations exogenous factors (e.g., photoperiodicity, diurnal or seasonal temperature effects, etc.) may be recognized as obvious "triggers" that switch metabolism from one course to another and often the means by which the external stimulus, as it were, releases an endogenous trigger linked to physiologically active biochemical substances or systems. In other situations, equally obviously, biochemical shifts may follow upon developmental events, events creating changed morphological milieux with far reaching biochemical consequences. Thus the control mechanisms are of different kinds and operate at different levels. All of the controls, however, must be accommodated within the scope of the total information contained in a given genome.

First, the selection from the available information via induction or repression of genes is governed by the morphological situation of the cells by mechanisms that are not perfectly understood, though they are obviously specific to the different organs and tissues.

Second, the level at which selective control operates is that of the enzymatic, because enzymes are the products of gene action. Enzymatic reactions are in turn subject to many controls. Among these are the controls exercized by small molecules, notably through positive or negative feedback (i.e., activation or inhibition, respectively, of an enzyme by a compound that is a product of its reaction or of the metabolic sequence of which it is a part). A typical example is found in the synthesis of methionine as shown in Fig. 6. Control is often also exerted by small molecules that are not components of the metabolic sequence, as for example the effect of lysine on the synthesis of methionine also shown in

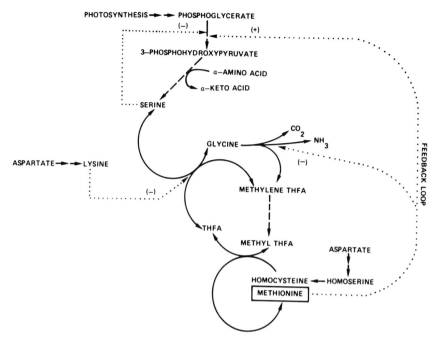

FIG. 6. Reactions in the biosynthesis of methionine showing single step (→), multistep (−−→), and feedback control (····→) reactions and the metabolic interlock of several reactions. THFA, Tetrahydrofolic acid. After Miflin (157).

Fig. 6. Such controls are often allosteric; that is they act at sites on the enzyme other than that which binds its substrate (the active site). These controls may serve two distinct purposes. First, they may regulate the pace of a given metabolic sequence (e.g., the direct regulation of the first step in methionine synthesis by serine and methionine, as shown in Fig. 6). Second, they may also coordinate the pace of different, but associated, reaction sequences (e.g., the control of methionine synthesis by the product of the aspartate–lysine sequence, also shown in Fig. 6).

A third level of control is exerted by chemical regulators such as hormones and growth regulators that may themselves be products of metabolic sequences. In the literature of auxins, cytokinins, gibberellins, etc., the examples are innumerable.

But finally, the overriding controls of amino acid and protein metabolism are often environmental or nutritional, modulated by such factors as photoperiodism or diurnal and seasonal fluctuations in temperature, which make their impact probably via the action of growth regulatory substances upon development and morphogenesis.

C. Integration of Metabolism, Heredity, and Development

Figure 7 shows two cyclical schemes. One (cycle B) represents the input of exogenous nutrients (including sources of nitrogen) and suggests the range of cellular organizations and biochemical mechanisms that utilize inherited information to fabricate the complexity of substance and vital machinery as cells metabolize and multiply. The other (cycle A) portrays the course of plant development through ontogeny from single cells, whether they are zygotes or totipotent cells. Here also heredity determines the course of this morphogenetic cycle. But the events portrayed as for cells at B may be modulated by the various morphological milieux under which given cells occur as at A. Thus the two interlocking cyclical systems of Fig. 7 are, in fact, "two faces of the same coin" in the currency of regulatory control. Thus the course of nitrogen metabolism is variously subject to controls that may operate at very different levels.

The "reductionist" approach—which has yielded the rich biochemical returns of molecular biology—interprets the events of nitrogen metabolism at the cellular and subcellular levels. But to interpret how these events are integrated into an organismal plan a further holistic approach is needed in which cellular machinery becomes subject to overriding developmental and environmental controls.

The life cycle of an angiosperm begins at fertilization with the establishment of its unique characteristics in the diploid zygote, which embodies all the information to carry the plant through its sporophyte development until at meiosis and reduction division the inheritance is segregated into the brief gametophytic development of the male and female parts of the life cycle. It is an article of faith, however, that metabolically (as at cycle B) as well as morphologically (as at cycle A) the entire course of development, structure, and form uniquely depends on the inheritance carried by DNA. Equally, however, all the events of metabolism and biosynthesis of anabolism and catabolism that make the organism work are made feasible, or are limited, by the same complement of inherited information that is contained in the single-celled zygote and the totipotent somatic cells that can carry that information through development. Cells, whether in growing regions or developed organs, are the smallest units that, at the outset, can be regarded as receiving the full complement of required inherited information. This may be expressed as organs develop from growing regions in very different ways, in an

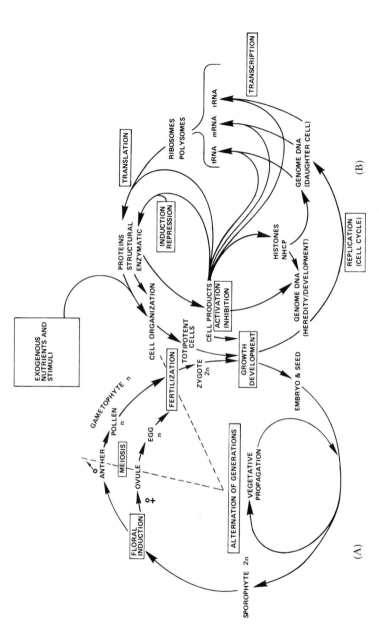

Fig. 7. Schematic life cycle integrating cellular metabolism and heredity with development, structure and form. (A) Life cycle: sporophyte and gametophyte. (B) Nuclear–cytoplasmic cycle of cell determination. NHCP, Nonhistone chromosomal protein; tRNA, transfer RNA; mRNA, messenger RNA; rRNA, ribosomal RNA; *n*, haploid; 2*n*, diploid.

economy based on "divisions of labor" as between organs, tissues, cells, and even organelles. The problems of integration, however, face the challenge that the flow of energy and matter through cycle A (in terms of form) and cycle B (in terms of molecules, enzymes, and metabolism) is equally guided by the same inheritance. But the spread in size, form, and function encompassed in the comparison of events in A and B, respectively, is matched by the great spread in the time span of the events to be controlled and correlated.

Gene-controlled chemical reactions may operate on an active scale of microseconds. The "half-lives" of enzymes may range from hours to weeks [possibly dependent on their arginine content (121, 277)]. Cell multiplications in active growing regions of angiosperms may be of the order of a day or so, but crucial steps in the cell cycle (like DNA synthesis) may vary over a time scale of minutes (274, 292). Individual long-lived cells in a perennial plant body have been estimated to retain their viability for upward of 100 years, as indeed may some dormant seeds, and the longevity of the longest-lived trees is of the order of thousands of years. Also, the estimated energy costs in synthesizing specific amino acids (10) may vary over a great range (from 18 equivalents of ATP to make serine from 3-phosphoglycerate to 51 for lysine from pyruvate or aspartate to 78 to synthesize tryptophan).

To encompass and integrate this range of complexity in concepts recognizing that, when fully understood, plants in their environments do not defy physical laws is an interpretive task too formidable to be undertaken here. But, at least, one might describe it as viewing plants, with their inherited information, their morphogenetic potentials, and their metabolic and biochemical diversities as "integrated energy–entropy machines" that make highly improbable events feasible in their environments.

Thus there remain problems of ultimate causation. It may often be satisfying to point to stages in cell multiplication on the one hand, or features of ontogeny on the other (Fig. 7), at which the course of nitrogen biochemistry is regulated. It is, nevertheless, chastening to recognize that so often it is an aspect of development, whether endogenously or exogenously triggered, that places the cells in the morphological setting in which they are able to respond and metabolize as they do.

All this being so, the evidence to be summarized in Section IV is marshalled to show the range of responses in nitrogen metabolism that are brought about, without irreversible change in the genome or ultimate biochemical potentiality, solely by developmental sequences and environmental stimuli.

IV. Nitrogen Metabolism during Growth and Development

A. INTRODUCTION

To this point the current status of nitrogen metabolism has been treated in terms of (i) the nitrogenous compounds, their feasible biochemical reactions and their catalyzing enzyme systems; (ii) their linkages into metabolic pathways and cyclical systems; and (iii) current concepts of protein synthesis. However, a full understanding requires that all this knowledge be adaptable to the unique characteristics of particular plants. Generalities are so often formulated in terms of *the* plant that they apply, as such, to *no* plant in particular. Therefore, this section attempts to see nitrogen metabolism comprehensively, in terms of specific plants and situations that their growth and development present, selected on the basis of our own experience.

This approach raises formidable problems of regulation that are even more difficult to resolve now than when Schulze and Prianischnikov or Vickery and Chibnall investigated such plants as lupine (*Lupinus*) tobacco (*Nicotiana*) comprehensively. Paradoxically, it is the very accumulation of knowledge in the interim that now presents the dilemma. This newer knowledge extends from the whole array of biochemical reactions that need to be coordinated (Section III) to the heterogeneity of subcellular organization, with its organelles, membranes, and particles (Volume VII, Chapter 1) as the systems within which the reactions occur. Regulation must also include the known diversity between cells, tissues, and organs, on the one hand, and the responses of plants to environments, on the other. Whereas all this detail may seem baffling, there is some simplification when it is viewed *in toto*. Plants are subject to major controlling factors (endogenous and exogenous) that seem to "plug in" large blocks of information simultaneously, or to turn "master switches" so that the essential "decisions" are made at higher levels than simply "enzyme by enzyme" or "gene by gene." The diagram of Fig. 7 links biochemical events at the cellular and subcellular levels to morphogenetic events in a life cycle. The higher levels of determination operate where and when organs are initiated, as seen in growing regions; it is in the development of organs so determined that the biochemical potentialities of the cells are so variously exploited.

This section, therefore, treats nitrogen metabolism descriptively, in broad, even holistic, terms as a necessary prelude to understanding where and how the links between inherited information and its full expression in *both* morphogenesis and metabolism are made.

In our earlier chapter (230) the intimate involvement of nitrogen metabolism with other metabolic processes and vital functions were either reviewed or anticipated. The metabolic processes included protein synthesis and turnover; respiration and photosynthesis; the essential features of growth, such as cell multiplication and enlargement; and the accumulations of solutes and of water. All this occurs concomitantly with the creation of molecular architecture as plants respond to mineral nutrients and to environmental factors that modulate morphogenesis, etc. (227; 230, pp. 511 ff.) These themes have expanded in the interim; their further discussion will now concern

1. Nitrogen metabolism and development in terms of different organs, all of which have the same genetic constitution as they develop from a common zygote—this is, therefore, a study in diversity
2. Nitrogen metabolism in terms of various environmental factors that, within the potentialities of a given genome, trigger aspects of morphogenesis
3. Nitrogen metabolism in the context of cells that are growing and developing as seen through the responses of actively proliferating aseptic cultures, containing cells that are totipotent, in contrast to quiescent cells in mature tissues or organs of the plant body

Although these topics are developed for specific systems with which we are most familiar, they can be more generally introduced as follows.

Cells, which are engaged in self-duplication with the multiplication of all their cellular organelles, whether in apical meristems or in unspecialized proliferating parenchyma (e.g., from carrot root phloem), synthesize the nitrogenous compounds responsible for the plant's molecular architecture and morphology; they do this with the minimum of stored, nonprotein, soluble reserves (such as amino acids and amides, etc.). Later, as cell division subsides, when such cells enlarge and vacuolate, they tend to accumulate solutes, which decrease the internal activity of water as they develop turgor. The first such solutes to accumulate in cells proliferating in a full heterotrophic medium are organic (sugars, organic acids, nonprotein-nitrogen compounds), and they reflect both the relative nitrogen and carbohydrate supplies in the medium and the presence of exogenous growth factors. The composition of cells still proliferating contrasts very sharply with the resting cells of intact storage organs. But once embarked on renewed growth the proliferating cells interact with (i) light and darkness and (ii) photoperiod and temperature (227, 231). However,their responses to levels of essential inorganic nutrients (especially of the trace elements and of sources of nitrogen like nitrate and ammonium) also reflect, at the cellular level, the range of responses in nitrogenous composition that also occur in whole plants (see references cited in 227).

But when initials form primordia that develop into organs (leaves, internodes, buds, etc.), their morphological commitment carries with it a striking degree of metabolic commitment. This commitment is expressed in great differences (despite their unchanged genomes) in nitrogenous composition between leaf, stem, and root, between organs (e.g., leaves or bulb scales) at different levels on an axis, or between tubers, aerial shoots, etc. Such changes, easier to observe than to explain causally, were made evident in mint (*Mentha*), potato (*Solanum tuberosum*), or tulip (*Tulipa*), and, of course, in many other plants (see 230, pp. 480 ff.)

When major morphogenetic stimuli cause shifts in shoot apices from vegetative to reproductive growth, the repercussions on nitrogen metabolism are as great, if less obvious, as those in terms of form. These shifts may result primarily from innate potentialities triggered only by endogenous stimuli activated at the appropriate stage in ontogeny, as in the banana plant, or by similar circumstances also triggered by appropriate temperature regimes, as in tulip bulbs that are large enough to flower. Some plants (e.g., mint or potato), bear shoots that respond to lengths of day, interacting with temperature (especially night temperature) to produce vegetative organs of perennation (stolons in mint; stolons and tubers in potato) that illustrate also the effects of environments on both form and nitrogen metabolism. In these situations the expression of the innate, genetic, metabolic capacity of the presumptively totipotent cells of the plants in question is responsive first to inherent ontogenetic controls even as these are often activated by highly specific environmental controls. However the crucial events are initiated, their outcome may be anticipated by visible responses in the form, even the composition, of the apical growing points of the shoots in question. Therefore, in the final analysis, the descriptive biochemistry may describe *what* happens (or has happened)—it may not answer *why* it comes about.

Similar problems of metabolic regulation arise in almost any situation in which vegetative buds begin to enter dormancy [as in maple (*Acer*)] in late summer, remain dormant through winter, but have their rest "broken" by a sufficient exposure to cold, with consequent changes in their metabolism. Some conifers may grow at latitudes such that they may be exposed to extremes of low temperature while they preserve their leaves and maintain dormancy of their buds, whereas others, under near-tropical conditions, avoid the most extreme of these hazards.

These general observations now need to be examined in the light of specific examples. One might appropriately begin with zygotes and young embryos in ovules and trace the nitrogen metabolism of particular angiosperms throughout their development. Though logical, this approach faces at the outset the limitations of size and accessibility of zygotes and very young embryos and the complexity of the changing heterotrophic nutrition they first receive. Attention will therefore be addressed first to a particular plant, namely, the potato. However, after exploring the range of effects observable in the growth of vegetatively propagated buds[10] (or it might be of embryos of seeds) one can then return to some problems of earlier development in so far as they may be reflected in the responses of aseptically cultured cells and tissues, induced to grow again and, especially so, when as in carrot the cultured cells can be shown to be totipotent and capable of developing into somatic embryos and plants (228, 231, 232).

B. Nitrogen Metabolism and Morphogenesis

1. Responses of Potato (Solanum tuberosum) to Environments

The responses of potato plants, clonally propagated from buds and grown under full nutrient conditions in fully programmed growth chambers, constitute a convenient "case history" of nitrogen metabolism and morphogenesis as regulated by environments. This account draws upon theses of Moreno (165) and Roca (202), based on work with one of us (F.C.S.), and on a later summary (233) that stressed

1. How diversified the morphological responses of potatoes are to endogenous stimuli during their development and to the controlled environments under which they have been grown (i.e., under long and short days interacting with higher and lower day and/or night temperatures). These effects are to be seen in the general habit of the plants as grown and in the organization of the growing regions where they are initiated.

2. That the environments changing the total crops and forms of leaves, the branching habits of shoots, the initiation of flowers, and the

[10]This develops the theme of *integration and organization* that concluded in 1965 (Volume VIC), as in Chapter 12, pp. 401–412.

development of stolons and tubers also modify the nitrogenous composition of leaves and tubers, respectively.

3. That the nitrogen metabolism of potato plants differs, with respect to the balance between the total soluble nonprotein nitrogen (SN) compounds and the bulk protein nitrogen (PN), in different sites in the plant body, notably in leaves versus tubers. This relationship also changes when thin slices of tubers have been aerated and reactivated (see Volume II, Chapter 4, pp. 352 ff.). Large differences also occur in the relative composition of the complement of soluble nitrogen compounds in leaves and tubers, respectively, in response to the environments under which the plants are grown. In so far as acrylamide-gel electrophoresis reflects the detailed composition of the bulk protein fractions, these also respond to interacting environmental factors.

4. The morphological responses to environments that regulate growth and form set the milieu within which the nitrogen metabolism operates and is regulated. The outcome is shown by the composition of the nonprotein, ninhydrin-reactive compounds that comprise the soluble-nitrogen fraction of the different organs as grown. The extraction, qualitative and quantitative analysis of the nitrogen compounds, and graphical representation of the results are again those used in earlier studies.

The consequential effects of environments on composition, if attributed directly to effects upon individual gene–enzyme-operated steps, would be very large indeed. Nevertheless, the resultant effects stem from visible changes in the respective shoot growing points and these set in train *both* the morphological *and* the metabolic consequences that ensue.

In potato leaves the nitrogen is principally consolidated in the bulk protein. The PN/SN ratio ranged from high values—83 in the younger leaves on 60-day-old plants under environment III (long days and high temperatures)—down to 10.8 in the younger leaves on 60-day-old plants under environment II (short days and low temperatures), as shown in Table XI. In contrast, the total nitrogen content of tubers is much richer than in leaves in the SN compounds. The PN/SN ratios for tubers ranged from 1.59 for plants at 60 days from planting and under environment II (i.e., short days and low temperatures) to 4.80 at 60 days but under environment III (long days and high temperatures; Table XI). The total nitrogen of leaves, therefore, reflects the balance between leaves and tubers of different ages under different environments conducive to synthesis in, or breakdown and transport from, leaves and to the buildup of nitrogen elsewhere (as in tubers).

TABLE XI

EFFECT OF DIFFERENT ENVIRONMENTS ON THE TOTAL SOLUBLE NITROGEN (SN)
AND RELATIVE PROTEIN NITROGEN (PN) OF TUBERS AND LEAVES ON PLANTS
GROWN UNDER DIFFERENT ENVIRONMENTS[a]

Treatments[b]	Age of plants (days)	Tubers		Older leaves		Younger leaves	
		SN	PN/SN	SN	PN/SN	SN	PN/SN
I	60	652.5	1.79	35.6	15.8	65.1	22.1
II	60	498.7	1.59	113.4	14.4	130.8	10.8
III	60	170.6	4.80	43.0	29.6	27.3	83.4
IV	60	180.0	4.20	91.2	23.3	80.2	40.9

[a] Soluble nitrogen expressed as micrograms of nitrogen per gram fresh weight.
[b] I, Short days (10 hr), high temperature (24°C); II, short days (10 hr), low temperature (12°C); III, long days (14 hr), high temperature (24°C); IV, long days (14 hr), low temperature (12°C).

Thus the regulation of the overall nitrogen economy of potato plants is primarily physiological and developmental (with controls exercised by environments); the biochemical reaction systems (as described earlier), which are necessarily invoked at the different sites, respond but do not constitute the ultimate control.

For the study of the environmental responses in question, use was made of several strains of potato from Peru. Clonal populations of these were raised vegetatively for the research. The data extracted here were all obtained on a hybrid designated KB-165. However, similar responses during growth to the environments that modify morphogenesis and metabolism of potato were obtained with other strains from Peru (233). The purpose here is to show that the very large responses in growth and form of KB-165 to the various environmental conditions imposed during its growth occur concomitantly with and are causally related to a similarly wide range of biochemical effects. Because a fuller account has appeared (233), only short selections from the available data on nitrogen metabolism will be given, sufficient only to illustrate the basis for the conclusions to be drawn.

The environments used involved selections of long and short days coupled with higher and lower temperatures that were either constant throughout or were applied either by day or by night. The environments were maintained throughout an extended period of growth of the plants. The experiments were planned symmetrically, usually in a block of four different combinations of environments. Representative plants were measured, weighed, and photographed in such a way that the

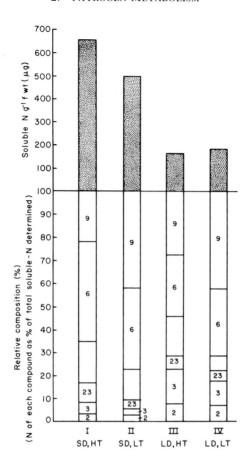

Fig. 8. Soluble nonprotein nitrogen compounds of potato (*Solanum tuberosum*) tubers (KB-165) as grown under short or long days at high (24°C) or low (12°C) temperatures. Key for Figs. 8 and 9: 2, aspartic acid; 3, glutamic acid; 4, serine; 5, glycine; 6, asparagine; 7, threonine; 8, alanine; 9, glutamine; 12, lysine; 13, arginine; 14, methionine; 15, proline; 16, valine; 18, leucine and isoleucine; 19, phenylalanine; 21, tyrosine; 23, γ-aminobutyric acid; 25, unknown. Redrawn from Steward *et al.* (233).

general habit of the plants was revealed in terms of their growth and form, branching habits, leaf form, crops of roots, stolons, tubers, leaves, etc. Finally, the impact of the environments on the form of the growing points of the shoots was determined by sections (233).

A minimum range of data in Table XI and Figs. 8 and 9A,B will show at the outset that the soluble-nitrogen of potato tubers and leaves is responsive both to morphological determinants and to environments that affect development. The illustrations used to present the data on

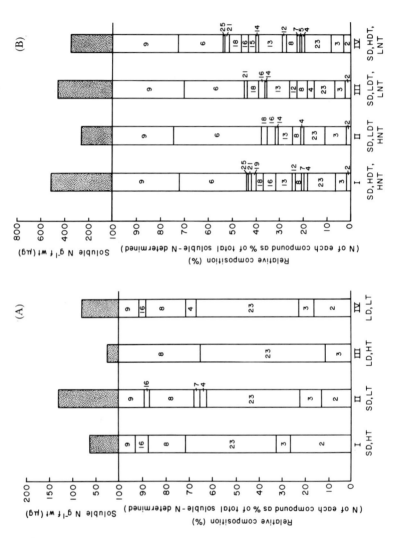

Fig. 9. Soluble nonprotein nitrogen compounds of potato (hybrid KB-165) as grown in different environments. (A) Data for leaves on plants grown for 60 days under short or long days and high (24°C) or low (12°C) temperatures. (B) Data for tubers grown under short days with different thermoperiodic cycles. For

potato are similar to those used in the earlier chapter. These data show that leaves and tubers on the shoots of the same plants differ greatly in their total soluble-nitrogen content, in the relative contributions of specific compounds to the total, and in the relations that obtain between protein- and soluble-nitrogen moieties. The principal points here are as follows.

Table XI shows the broad range in total soluble-nitrogen content in tubers and leaves on cloned plants of the same age and as grown under different environmental conditions. Tubers are distinguished by a greater soluble-nitrogen content than leaves. Short days during growth (environments I and II) are conducive to a high content of nonprotein nitrogen in tubers and to a low PN/SN ratio.

In contrast, long days (environments III and IV) are conducive in tubers to a much lower nonprotein-nitrogen content and a much greater PN/SN ratio. The nonprotein, soluble nitrogen of leaves is invariably less than that in tubers, and on short days (I and II) the lower temperatures (II) favor the soluble nitrogen of leaves; a temperature effect is to be seen also under long days (III and IV), at which it is the higher temperatures that are associated with the lower nonprotein nitrogen in leaves. Overall, the PN/SN ratios are greater in young leaves than in older ones.

The principal features of the composition of the soluble nitrogen fraction of tubers are as follows (see Fig. 8).

Tubers are rich in the amides (Nos. 6 and 9 on the histograms). Those compounds with the dicarboxylic acids [(aspartic, No. 2; and glutamic, No. 3), γ-aminobutyric acid (No. 23), and arginine (No. 13)] represent the bulk of the soluble fraction, although most of the protein amino acids are also present in smaller and varying amounts. That portion of the histograms for the soluble nitrogen of tubers, which is not accounted for by Nos. 2, 3, and 23 or by the amides Nos. 6 and 9, consisted of a general pool of numerous identified amino acids (Nos. 4, 5, 7, 8, 13, 14, 16, 18, 19, and 21) not shown in detail on the figures. In older, more mature tubers (as at 90 days) but under the same environments as those of Fig. 8, this general pool increased relative to the amides and dicarboxylic acids, and then arginine (No. 13) and the leucines (No. 18) became more prominent (see Fig. 14 in 233 for data).

The composition of the soluble nitrogen of potato leaves (see Fig. 9A) is very different from that of tubers, even though they are present concurrently on the same plants. The dicarboxylic acids and γ-aminobutyric acid (Nos. 2, 3, and 23) together with alanine are the predominant nonprotein-nitrogen compounds in leaves, the amides being much less prominent than in tubers and asparagine being virtually absent.

The summary (233) shows that the environments determining growth and form in potato shoots, which as just seen are linked to the composition of leaves and tubers, make their first impact on the apical growing regions. In the transition from a leafy shoot to tuber or stolon (or to a floral apex) brought about by environments, it is as if the environments "mold" the form of the growing points and so set in train trends of development that, as they control form, also control subsequent metabolism. The great contrasts between the formative effects of environments on terminal shoot apices and on those laterals that, low on the same axis, may form stolons and tubers at their tips, have been illustrated (233, Figs. 12a,b). A great contrast was shown between the form of apices at stolon tips and at shoot apices under environment II (short days and low temperatures, where the stolons formed tubers in contrast to the apex remaining as if in transition to flower) and also under environment III (long days and high temperatures, where the stolon tips were fully vegetative and elongated without forming tubers, whereas the main terminal apex of the stem was transformed into a flowering shoot). These transformations—brought about by different redistributions of growth and development throughout the respective apices, in response to the environmental stimuli (Fig. 13 in 233).

Tubers form best on plants grown under short days, so Fig. 9B shows the relative composition of the soluble nitrogen of tubers borne on plants grown during 60 short days with different temperature regimes. Environments I and III used temperatures constant throughout day and night; environments II and IV used different thermoperiodic cycles. These environments, as shown, affected many compounds, but especially the amides and arginine. It is simpler therefore to focus upon a narrower range of compounds for which there is much known about the enzymes concerned in their biosynthesis and metabolism. This is shown on Table XII for glutamine and asparagine.

Glutamine predominated over asparagine in tubers present on the younger plants, but this preference for glutamine over asparagine was reversed with age of the plants. Overall, the asparagine/glutamine ratio for tubers varied with the temperature regimes under short days, over a 10-fold range (0.23 to 2.00), and if this comparison is extended to leaves their general preference for glutamine over asparagine was virtually absolute.

Although asparagine and glutamine are chemical homologs differing only by one CH_2 group, this does not simplify the interpretation of their respective biosynthetic responses in potato to environments. In fact, these different responses reemphasize what has previously been noted, namely, that the two amides often seem to behave physiologically in very

TABLE XII

Effect of Environments on Tubers:
The Asparagine/Glutamine Ratios of Tubers on Plants of Different Ages[a]

| Age of plants (days) | Treatments | | | |
	I HDT/HNT	II LDT/HNT	III LDT/LNT	IV HNT/LNT
45	0.55	0.63	0.23	0.52
60	1.00	1.40	0.84	0.70
90	1.91	2.00	0.84	1.35

[a] All plants grown on short days but with high (H) or low (L) day (D) or night (N) temperatures (T).

different ways. Their respective roles in potato tubers and leaves also suggest this, and their respective places in the metabolic flow of carbon and of nitrogen are, in fact, not homologous but distinctive for they involve different immediate precursors, products, and enzyme systems (cf. Charts 11 and 12).

Nor is the dilemma of asparagine and glutamine accumulations in specific plants dispelled by a later concise discussion by Fowden et al. (88, pp. 155–157) or by Miflin (159). In these interpretations, the amino groups of protein and even of nonprotein amino acids derive from glutamate, whereas the nonamino nitrogen, whether of asparagine, tryptophan, histidine, or arginine, derives from the amide nitrogen of glutamine. But do these details of biosynthesis illuminate the respective roles of asparagine and glutamine in a specific plant or in plants generally? Their respective accumulations in specific organs during development, their varied responses to light and dark and to inorganic nutrients other than nitrogen (especially trace elements and sulphur deficiencies), and the impact upon them of the turnover, translocation, storage, and reutilization of nitrogen are not obvious from such schemes alone.

In the final developmental outcome, environments affect the disposition of carbon and nitrogen between these two compounds in potato plants. Thus the increase in our knowledge of their enzymatically distinctive pathways and reactions (to and from glutamine and asparagine) may have clarified biosynthetic routes that are chemically feasible without prescribing either the carbon or nitrogen traffic along these routes or the points where it may be arrested so that accumulations occur. The ultimate controls seem to be triggered through environments and nutrition operating on inherited pathways of development. This is a chasten-

D. J. Durzan and F. C. Steward

ing thought, for otherwise the descriptive biochemistry *in vitro* seems so conclusive.

2. Compartmentation

Neither in leaves nor in tubers is there any direct correspondence between the soluble-nitrogen constituents present and the amino acid composition of the bulk protein they contain. Therefore the composition of the total soluble nitrogen fraction reflects neither the requirements of soluble-nitrogen compounds to synthesize the protein nor the compounds to be expected when it is broken down in the organs in question. These and similar circumstances encountered in the metabolism of tissue slices and cultured explants (21, 229, 268) had led to ideas of compartmentation, as necessary to interpret nitrogen metabolism, even before the current knowledge of the full complexity of fine structure in cells.

The first ideas of compartmentation involved biochemical pools from which nitrogen and carbon were withdrawn prior to synthesis and to which they were returned after protein breakdown. As these views developed the concept was that certain nitrogen-rich substances (e.g., amides and arginine) were preferred substances for storage, whereas others were more preferentially adapted to link those sites of storage to the active sites of synthesis. Also, the concept of protein turnover in cells provided a link between the paths of carbon in protein metabolism and in respiration, which hitherto had been kept unduly separate.

However, the present cytology of cells describes their organelles and membranes (Volume VII, Chapter 1), and these must surely constitute the surfaces and compartments where the complex reactions of intermediary nitrogen metabolism (see Section III) are performed. To bridge the gap postulated between "pools" of reactants that are biochemical abstractions and cellular inclusions and organelles that are physically demonstrable—still present those who would chart the course of nitrogen and carbon throughout with some formidable problems. Nor have these problems been dispersed by the sorts of elegant physical and electronic devices that as seen in Volume VII, Chapter 2, have enabled the intricate courses of energy transfers to be charted, often without, be it said, the isolation of their enzyme-catalyzed intermediates which may only function in very small amounts and in steps that are of very brief duration. The course of carbon and nitrogen through metabolism has not yet been charted in this way, nor will it be in the foreseeable future.

3. Concluding Remarks

This "case history" describes some far-reaching consequences of environmental effects on potato. These are most obvious in their impact upon morphogenesis. They affect, however, metabolism that is organ specific. Therefore, environments together with developmental determinants regulate how genetic information available to cells of leaves, tubers, etc. is variously expressed. These consequences influence the nitrogen compounds that predominate in the nonprotein moiety of the different organs, but they also influence a large number of metabolites that appear in smaller amounts. Although these final responses to environments are the more evident, their initial effects trace back to effects on apical growing regions where morphogenetic trends are initiated. But why such profound consequences, including such detailed effects on nitrogen metabolism, should flow from interactions between day and night length and alternations of day and night temperatures is still mysterious. Though biochemistry may suffice to show that all the necessary reactions of nitrogen metabolism are gene determined and feasible, this does not answer how, and why, developmental factors and environmental controls intervene to determine the final outcome. If all the final consequences for the complex reactions of nitrogen metabolism of shifts and balances in environments were to be attributed directly to interventions upon each and every ·gene–enzyme-dependent step, then their number and frequency would be very large indeed. Hence the modulation of metabolism may need to occur at higher organizational levels and involve much larger blocks or combinations of information. In fact, the very detail of current knowledge of enzymology tends to obscure the problem and compounds the dilemma of regulation.

Thus far the analysis of nitrogen metabolism in relation to growth and development has been concerned only with angiosperms. The variations on these patterns within flowering plants, and also in other branches of the plant kingdom, are, of course, innumerable. However, some reference to two coniferous plants (*Picea glauca* and *Pinus banksiana*), now to be made, extends the discussion to perennial plants that are gymnosperms and forest trees and that expose their plant body and growing regions to a very different range of environmental stresses.

C. NITROGEN METABOLISM OF CONIFERS: WHITE SPRUCE (*PICEA GLAUCA*) AND JACK PINE (*PINUS BANKSIANA*)

1. Introduction

Conifers differ much from the plants hitherto considered. Potato, the banana (*Musa*), even mint replace their vulnerable aerial foliage and shoot growing points with massive organs of perennation that both store nutrients and give protection to their resting buds. Deciduous trees protect their resting buds by bud scales and by withdrawing water. In contrast, the perennial plant body of conifers develops a persistent aerial complement of green leaves and buds that survive the seasonal extremes of temperature, the stresses of xerophytic conditions (as in winter), and the extreme diurnal fluctuations of temperature and day length that vary greatly with the seasons. The morphogenetic (e.g., 64) and metabolic adaptations by which all this is accomplished permit survival in environments and on soils that would not support most agricultural crops.

How then does nitrogen relate to the normal growth and habit of conifers? The first consideration involves the nitrogen balance of a boreal forest (129, 246). In acidic forest soils, available nitrogen is in short supply. It is the main factor that limits the productivity of the trees. The nitrogen content of the humus, as distinct from the mineral soil below, is largely immobilized in organic form from which it is only slowly released or cycled as available nitrogen. The paucity of available nitrogen is shown by the response of trees to fertilizers and by foliar diagnosis of their nitrogen contents. Nutrient availability is usually fostered naturally by mycorrhizae commonly associated with roots. This symbiotic relationship contributes to the nutritional well-being of the trees, with the tree providing carbohydrate (photosynthate) to the fungus.

2. Seed Composition during Development and Germination

Embryos in dry conifer seeds have been nourished heterotrophically by their maternal gametophytes, which contain high levels of lipid and starch reserves and arginine-rich storage proteins. To examine whether climate and the genetics of the parent tree may predetermine the composition of the seeds, dry seeds from various geographical sources across the natural boreal range of jack pine were separated into embryos (diploid) and female gametophytes (haploid), and their sugars, free and bound amino acids, and soluble proteins were measured (63). The following climatic factors were correlated with seed composition.

Higher temperatures at the seed source generally reduced the levels of most soluble components of seeds, except for the amides, glutamine, and asparagine, but they increased embryo weight and length. Combinations of higher temperatures and greater precipitation at the seed source reduced the arginine content of seeds. In the gametophyte, which nourishes the embryo, it was increased precipitation rather than temperature that caused high amide content, which in turn caused the developing embryo to be heavier and larger at higher temperatures.

When water initiated germination, the amides rapidly increased in the embryo at the expense of the gametophyte, but after germination amide levels again declined as the extractable protein increased. During imbibition and germination the reserve proteins of seeds are mobilized for the synthesis of new proteins and nucleic acids (63, 65, 67–69, 194), especially at the emerging shoot (156) and root apices (155). The turnover and resynthesis of these proteins are accompanied by sequential changes in the composition of the free amino acid pools. These changes are mainly in free arginine, the amides, proline and certain monoamino acids such as serine, alanine and γ-aminobutyrate (69, 194).

3. Pivotal Nitrogenous Compounds in the Soluble-Nitrogen Pool

Two salient aspects of the descriptive nitrogen metabolism of conifers relate to the soluble-nitrogen pools. The first concerns the relationships among the dominant free amino acids (arginine, glutamine, proline, and asparagine) during growth and in response to environmental stress. The second concerns the total level of soluble nitrogen as it relates to the assimilation of nitrogen, the turnover of proteins (yielding α-amino acids and amides) and nucleic acids (yielding β-amino acids) and with the production of various categories of secondary substances characteristic of conifers (e.g., monosubstituted guanidines; Chart 19).

Arginine is a storage product that often accumulates in the soluble pool as growth subsides. The formation of a molecule of arginine requires four nitrogen atoms by transamination from aspartic and glutamic acids. During active growth the latter normally contribute via transamination to the production of protein amino acids. During such periods nitrogen is released from stored arginine via arginase, to produce first urea (as an internal nitrogen source) and amines such as ornithine (Charts 11 and 19), presumably as nitrogen sources for protein and nucleic acid synthesis. In spruce and pine, arginine metabolism is not entirely based on the urea cycle as originally proposed (171). Frequently, arginine nitrogen is diverted to the production of nitrogen-rich monosubstituted guanidines (Appendix III). These compounds have a wide range of properties and strong physiological activities (Appendix IV). The content of free arginine fluctuates with protein synthesis and degradation. In leaves, arginine accumulates in darkness, but in light it is converted to the amides (168). It also accumulates during nutrient deficiency (e.g., low phosphate) and when ammonium is the sole source of nitrogen (71).

Glutamine is a primary product of the assimilation of urea nitrogen (58, 59). Glutamine contributes its nitrogen by transamination to the synthesis of a wide range of protein amino acids (Chart 1). In pine and spruce trees nitrogen is translocated in the form of amides. Glutamine also contributes to the production of purine and pyrimidine bases (Chart 15) for nucleic acid synthesis, and this is supported by studies with [^{14}C]urea. High levels of free glutamine usually anticipate a wave of cell divisions in germinating seeds. Glutamine can be a precursor for arginine and proline and for unusual metabolic products such as α-keto-glutaramate and succinimide (57). Production of these compounds seems also to be related to tree stress under dry weather conditions.

Free proline is typically derived from arginine (especially in the fall) and occasionally from glutamine (especially in spring). Proline also accumulates when growth is arrested by drought. There are significant correlations between hot or dry years and reproduction (cones are rich in proline). In late summer as terminal buds mature, proline is rapidly incorporated into proteins and then becomes hydroxylated *in situ* to hydroxyproline. Hydroxyproline is not usually encountered in the free amino acid pool but has been found in pollen (272). During seed germination proline rapidly enters the newly synthesized isoenzymes of peroxidase in cells that are about to undergo differentation (69).

Asparagine usually becomes prominent in the soluble-nitrogen pool of all organs after growth subsides, when, presumably, more protein is being degraded than synthesized (168). Asparagine, when converted to aspartic acid, may participate in several alternative pathways. The aspartate may transfer its nitrogen to argininosuccinate, to adenylosuccinate, or it may react with carbamoyl phosphate to form arginine, AMP, or carbamoyl asparate, respectively (e.g., Charts 11 and 18). It also contributes to the biosynthesis of certain growth regulators (ethylene and IAA-aspartate).

4. Some Factors Affecting the Soluble-Nitrogen Pool

Under field conditions, the salient features of the seasonal changes in the soluble nitrogen of spruce buds (Fig. 10) are as follows. Over most of the year the developing buds, shoot tips, and leaves of conifers store a substantial complement of soluble nonprotein nitrogen, which is dominated by arginine. At other times of the year asparagine, glutamine, and proline as well as some of the more familiar amino acids such as alanine, serine, and γ-aminobutyrate may replace the otherwise dominant arginine (54). Although conifers are adapted to a nutrition essentially low in nitrogen (246), cells of these evergreen trees will respond to exogenous nitrogen, which may be supplied as fertilizers in nurseries or by aircraft over forests (18). The trees then build up "luxury" levels of nonprotein nitrogen. During the assimilation of ammonia, mediated via the amides, the central role of arginine and its derivatives (urea, amines, and the monosubstituted guanidines) become distinctive features of

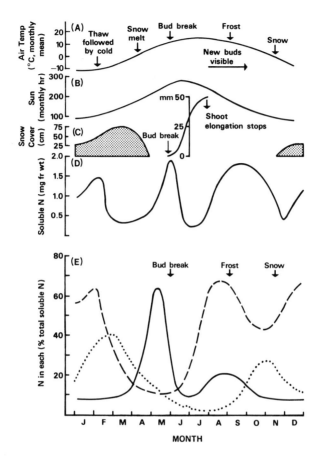

Fig. 10. Effect of seasonal factors on spruce buds and their composition. (A) Mean
monthly air temperature. (B) Mean monthly sunshine. (C) Depth of snow cover. (Inset)
Length of leader shoot from May to July. (D) Total soluble nitrogen of spruce buds. (E)
Annual trends in glutamine (—), arginine (---), and proline (····) content of spruce buds.
Data of D. J. Durzan (unpublished; cf. 55).

their nitrogen metabolism (63, 71). This metabolism supplements but
does not override the amide-oriented metabolism directed toward the
formation of nucleic acids and proteins. Through the activity of arginase
on arginine, the slow internal release of urea in the tree replenishes the
nitrogen for those stages of growth and development otherwise linked to
the sequential appearance of free amides.

After the snow melts in the spring and during the swelling of buds,
urea nitrogen released by arginase finds its way into the soluble-nitrogen
pool, mainly as glutamine. As buds break and new shoots emerge, the

glutamine nitrogen declines as the monoamino acids increase and domi-
nate the soluble nitrogen. The pool then consists mainly of alanine and
glutamic and aspartic acids, all substances that can donate or transfer
their nitrogen to an enlarging pool of precursors for nucleic acids and
proteins or return it to storage products such as arginine. The latter (i.e.,
arginine) again dominates as shoot growth subsides and a new terminal
bud becomes visible. Hence the seasonal responses in soluble nitrogen
are canalized through glutamine or arginine and a complement of protein monamino acids (including proline). Proline levels increase
markedly under conditions of drought (37, 54–57). Patterns of accumulation of these compounds distinguish the metabolic events at
spring, summer, and winter and anticipate developmental trends in
shoot tips under field conditions.

The changes of soluble nitrogen in buds and shoot apexes, though
complex, are related to factors in the environment. The mean monthly
air temperature and the total monthly hours of sunshine display a seasonal course with time, as shown in Fig. 11A. Figure 11B shows the
fluctuations in arginine (the principal storage form of nitrogen, up to
70–75% of the soluble nitrogen) in the buds during a single season. In
the comparison of Figs. 10A–E and 11A,B the complexity of the regulation of nitrogen compounds in the shoot tips is obvious. On the one
hand it is difficult to assign a single physical factor to the operative role
when, in fact, it is clear that their interactions are so important. On the
other hand it is also difficult to visualize single determining biochemical
reactions when, physiologically, the growing shoots need to mobilize
such a wide range of gene-determined enzymatic steps.

Although the effects of temperature, as shown in Fig. 11, may be
substantial, the effect of light treatments at night on flowering (60) or
reduced light intensity on growth caused by shading under a canopy
may be even greater. The light that reaches the forest floor beneath the
tree canopy is a factor in determining the establishment of seedlings and
saplings. Some species are more tolerant of shade (e.g., white spruce)
than others, which are quite intolerant (e.g., jack pine). In one study (53)
seedlings of both species were raised under field conditions for 3 years
under continuous levels of shade based on natural light intensity and
fractions thereof (45, 25, and 13% of natural light). In the fall of the
third year, when leader growth had subsided, the sapling parts (leaves,
stems, and roots) were harvested and evaluated for the impact that continuous shade had on the level, distribution, and composition of the
soluble nitrogen.

Both species (pine and spruce) responded to shading by decreasing
their amide content, especially at the lower levels of light. This trend was

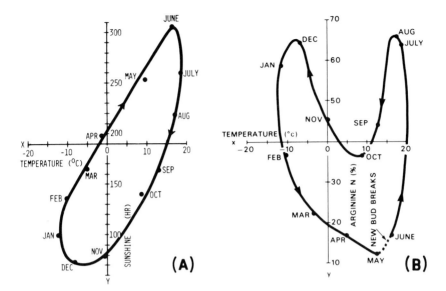

Fig. 11. Seasonal fluctuation in climatic factors and in arginine content of spruce shoot tips. (A) Climatic factors. Temperature and sunshine attain minimum values in winter and maximum values in summer, but the extremes of temperature lag behind those of sunshine by about 1 month, causing the elliptical shape of the curve. (B) Arginine content. There are two maxima in arginine content: during winter, when low temperature causes an accumulation of arginine as a storage product, and during late summer, when arginine accumulation marks the onset of dormancy. The minimum values observed relate to the use of stored arginine by the shoot tips (i) for the formation of glutamine during the onset of growth in May and (ii) in its conversion in October to proline, which accumulates during the onset of dormancy in the bud (cf. Fig. 10). Data of D. J. Durzan (unpublished).

more pronounced in the more shade-intolerant pine, and especially so in its roots, which contained only very low levels of total amide at 13% of natural light. Low amide content may reflect the greatly reduced ability of pine to grow in such deep shade.

5. Intermediary Metabolism as Revealed by Labeled Compounds

Seasonal trends having been detected in the composition of the soluble nitrogen, the question arose whether the development of buds could be related to the metabolic flow of carbon and nitrogen from one compound to another. A common practice is to use ammonium or urea nitrogen to increase the available nitrogen to conifers and thus extend their juvenile growth phase. Urea when so supplied would supplement the endogenous urea derived from arginine.

a. [^{14}C]Urea and ^3H_2O.   Urease is often difficult to detect in seedlings and young shoots. After several hours of exposure to urea, especially in the light, urease can be detected cytochemically in the cells that first encounter the urea (i.e., in the epidermal and subepidermal cells) (58).

Under controlled and aseptic conditions, the urea assimilated by seedlings in the presence of tritiated water yielded free amino acids labeled specifically at the α-amino carbon with covalently attached tritium (59). The first such compound detected was glutamate, consistently with the C_5 port of entry for nitrogen (Chart 1). Unfortunately, this does not distinguish between the operation of glutamate dehydrogenase and the GS–GOGAT system, because if tritium were introduced at the amide group it could be exchangeable.

When [^{14}C]urea was fed to seedlings in active growth, the major early radioactive products of metabolism were free alanine and carbamoylaspartate followed by glutamine, glutamic and aspartic acids, and serine. Radioactive amino acids were also recovered from newly synthesized protein. The early formation of carbamoyl aspartate indicated a carbon-14 flow toward the formation of pyrimidine nucleotides (Chart 15), and when nucleic acids were isolated, radioactivity in the pyrimidine bases was indeed detected. Because a wide range of amino acids was labeled with carbon-14 from urea, it was not surprising that the purine bases in RNA and DNA also contained radioactivity (Charts 15, 17, and 18). In short, carbon and nitrogen from urea contributed to the synthesis of both protein and nucleic acids.

In other plants, such as the legumes, ureides rather than arginine and guanidines serve as major storage forms of nitrogen. Ureides are formed only as products of purine breakdown. Thereafter, for every ureide degraded, 2 equivalents of urea should be released (Chart 18). Urea so released may not, however, accumulate in conifer tissue because urease may be induced, as for example by exposure to high levels of supplied adenine and guanine.

For reasons that are not completely understood, urea fertilization leads to increased wood formation in conifers. Interest in this affect of urea contrasts with the situation in agricultural crops where the goal of nitrogen fertilization is to increase protein nitrogen rather than fiber. In forest trees the production of wood under conditions of even low nitrogen supply must be extremely efficient. This efficiency must be linked with the production of photosynthate by leaves in light and with dark reactions associated with secondary growth in the cambium. Some of these dark reactions undoubtedly require the cofactor biotin, which is required for most carboxylation and transcarboxylation reactions. It has been shown that biotin in spruce can be derived in part from [^{14}C]carbamoyl phosphate and urea (139; D. J. Durzan, *Plant Physiol.*, in press) and this may be the basis for the stimulation of wood production by urea.

b. L-[^{14}C]Arginine and L-[^{14}C]Citrulline.   Metabolic pathways in spruce and pine embrace a far wider range of substances than was earlier recognized by Mothes (168) and more recently by Naylor (171) and Pharis *et al.* (185) (see Appendix III). Many of the identified compounds in the soluble-nitrogen pool are known constituents of plant protein. Others (e.g., γ-aminobutyric acid, β-alanine, and various guanidine compounds) are not found in protein but arise in the soluble fraction from now-familiar sequences of metabolic events. The even wider range of substances formed in white spruce and other conifers relates to the metabolism of arginine.

Experiments with [^{14}C]arginine and [^{14}C]citrulline suggest that the ornithine cycle operates in spruce buds in spring (56) and serves to produce the free arginine found in the buds at this time. However, the large accumulation of free arginine that occurs in fall buds appears not to be formed by the cycle but is the result of protein degradation or turnover (54).

An unexpected observation was that arginine was metabolized to a wide range of monosubstituted guanidines that hitherto had passed unnoticed. One such pathway involved the formation of 2-keto-4-guanidinovaleric acid (70). The occurrence of [^{14}C]guanidine derivatives revealed that the metabolism of arginine in these conifers is incompletely known and probably involves several new pathways. Pathways from arginine to monosubstituted guanidines persist in buds as they break in the spring, during the germination of seeds, and in the production of callus grown from explants taken from various parts of the tree. In these situations at least 18 guanidino compounds (Sakaguchi reactive products) and two *N*-phosphoryl derivatives have been detected (56, 62, 66).

From Appendix IIId it is to be expected that the guanidines may have strong physiological effects and in some cases may give rise to their corresponding amines. This means that each guanidine molecule may be cleaved by an enzyme similar to arginase (see Table X).

The fate of the monosubstituted guanidines derived from arginine remains obscure. No doubt the guanidines may lead to amines such as agmatine, to γ-aminobutyrate, and possibly to polyamines (D. J. Durzan and F. C. Steward, unpublished, and Chart 11). *N*-Carbamoylputrescine, an intermediate in the formation of polyamines, is difficult to detect in conifers. These pathways and others involving monosubstituted guanidines (e.g., agmatine) account more rapidly for the metabolism of arginine in conifers than did the ornithine cycle alone.

Experiments with labeled citrulline, a urea-cycle intermediate, were done in the spring under circumstances in which soil moisture was limiting for bud expression. [This was known because supply of water to the roots promptly released the growth of buds (56, Fig. 2)]. [^{14}C]Citrulline applied to these buds in the early morning was absorbed and metabolized; all of the intermediates of the urea cycle could be detected (Chart 11). Whereas three monosubstituted [^{14}C]guanidines were recovered in spring, many more such compounds were found in the fall.

Radioactivity supplied to spruce buds as either [^{14}C]citrulline or [^{14}C]arginine could be recovered from bud protein principally as arginine, glutamic acid, and proline. The levels of ^{14}C in the bud protein were highest in the dark periods when the buds expand (56, Fig. 2).

c. L-[^{14}C]Proline and L-[^{14}C]-Glutamine.

In north temperate zones, spring, late fall, and winter are seasons of drought and severe cold. When the seasonal pattern is disrupted, as during a thaw in winter or during unusually cold weather in spring, many conifers react by accumulating specific metabolic intermediates.

In late fall, resting spruce buds were excised and fed L-[^{14}C]proline; evidence for the occurrence and accumulation of Δ'-pyrroline-5-carboxylic acid and glutamic-γ-semialdehyde was then found (57). The atypical accumulation and persistence of these intermediates from proline is to be regarded as a feature of the onset of winter rest.

When the protein of buds exposed to [^{14}C]proline and [^{14}C]glutamine was examined, radioactive proline glutamate, aspartate, arginine, and hydroxyproline were all recovered although in different proportions from each labeled source. In earlier studies the same products were recovered from bud protein after [^{14}C]arginine was fed. Interconversions of the carbon of proline and glutamine suggested that they had a common metabolic intermediate, namely, γ-glutamyl phosphate, which is believed to be enzyme-bound (260). This compound is connected to L-proline via glutamic-γ-semialdehyde and is an intermediate in the biosynthesis of glutamine and pyrrolidone carboxylic acid (pyroglutamate or 5-ketoproline).

A feature of glutamine metabolism in resting buds is the extent of removal of amide nitrogen in the formation of glutamic acid and α-ketoglutarate. The formation of another keto acid, α-ketoglutaramic acid, either by transamination or oxidation, could not be ruled out, but radioactivity in the ketoamino product never accounted for more than 6% of the total radioactivity derived from glutamine. Decarboxylation of this ketoamino acid could account for the occurrence of compounds tentatively identified by coincident chromatography in two different systems as succinamic acid (by decarboxylation) and succinimide (by ring closure).

In summary, the studies of the changes in the soluble-nitrogen components of spruce, as shown by seasonal and climatic factors (including shading) were all followed over the relatively long term, even extending over annual cycles. In contrast, changes in the soluble nitrogen as shown by the use of ^{14}C-labeled substrates in the form of urea, arginine, citrulline, proline, and glutamine were all observed over short-term periods and on buds, and consequently they were markedly responsive to the activity of the buds and the season of the year. These different approaches and distinctive results emphasize again how much the nitrogen metabolism and biochemistry of spruce is responsive to its status in the alternating cycles of growth, dormancy, and perennation which, though endogenously manifested, are environmentally tuned and regulated.

D. SOME DETERMINANTS OF NITROGEN METABOLISM

The ability of plants through their genetic constitution to carry out specific enzyme-catalyzed reactions will be determined, on conventional ideas, by various sequences of genes → proteins → enzymes, etc. But the mere fact that these abilities exist does not prescribe how and where and to what extent they may be invoked. A veritable "library" of inherited information is selectively deployed in organogenesis by the circumstances that give each organ its morphological identity. Moreover, these events are often modulated by environmental factors. Thus biochemical regulation, as of nitrogen metabolism, cannot yet be specified solely in biochemical–genetic terms for it is also a problem of morphogenesis, which, from zygote to adult, is a progression of highly improbable events. Despite this improbability, plants emerge as stable physical entities within their environments. Plants, therefore, really need to be described in physical as well as biochemical terms and in compliance with, not defiance of, physical principles that will determine their development and stability. The adjustments and correlations of form and composition of plants are part of these considerations.

Even the perception and directional response of plants to that most tangible of physical stimuli, gravity, have nevertheless proved to be both complex and incompletely understood (11). This is even more true of such complex responses as those of nitrogen metabolism that require multiple stimuli. Developmental responses to light (whether morphological or metabolic) may utilize the phytochrome system as the initial perception (to be discussed in a later volume), but, where the subtle interactions between daily duration of light and darkness and diurnal fluctuations of temperature are involved, the problems of perception of the stimuli and the mechanisms of response are compounded. This is particularly relevent to nitrogen metabolism because it is as these physical factors in the environment "mold" shoot growing points that they anticipate the composition (as of nitrogen compounds) of the developed organs (see 233). Relationships between growth, development, and composition have already been discussed (Volume IVA, Chapter 4, pp. 566 ff., and Volume VIC, Chapter 7), and work on cells in culture has led to progress in relating metabolism and cell cycles (291; cf. 237, 238).

The dilemma is that self-duplicating cells and their organelles, which divide equationally, nevertheless give rise to the diversity both of form and composition of their derivative cells and organs in vivo. DNA in the form of isolated fragments (i.e., plasmids) may now endow bacteria with biosynthetic properties, drawn from higher organisms, that they did not possess hitherto, through the techniques of recombinant DNA and genetic engineering. Also, the self-reproducing organelles of higher plants (e.g., mitochondria and plastids) have their own distinctive complements of DNA. Nevertheless, the diversity acquired during development, whether in terms of form or composition of cells, is not explained away by changes induced by the composition of the DNA of the cells of developed organs. On the contrary, the decisive "choices" or decisions during development are taken when cells assume distinctive positions and roles in the plant body. But, despite all that has been said and written about "patterns" that emerge during development and the "strategy of the genes" (93, 100) and the need for their being "turned on or off," the ultimate explanations of how all this comes about are not yet known. What is apparent, however, is that morphogenesis and responses in composition go "hand in hand," and in higher plants exogenous environmental factors interact with endogenous developmental trends to determine both form and composition (especially with respect to nitrogen compounds). Indeed, it is because higher plants may maintain potentially unlimited growth and repetitive organ formation in their growing points and totipotency in their developed somatic cells that they present so acutely the problems of diversity during development. The scope of the integration here needed should be conceived holistically; notwithstanding that, currently, the "ideal" investigations are those that may be pursued through progressive reductionism!

Ideally, one should be able to trace the problems of regulation throughout development from zygotes and, in the context of this chapter, do so in terms of nitrogen metabolism. The act of fertilization imparts to the egg in an ovule a "built-in capacity to grow." The information now in the zygotic nucleus, together with the heterotrophic nutrition of the embryo sac, permits the fertilized egg to synthesize all the nitrogen compounds required by its growth and further development. Seed formation involves the prior development of an embryo, complete with growing regions capable of forming tissues and organs but also able to store products that will suffice until the plantlet becomes a completely autonomous entity. Storage in cotyledons, in endosperm, or in enlarged parts of the axis provides for future needs for organic carbon, nitrogen, phosphorus, and sulfur from which to fabricate complex structures and for the release of needed energy for their fabrication. Therefore, specifically for nitrogen, seed formation involves the synthesis of particular seed proteins; seed dormancy, their retention in storage organs; and germination, protein breakdown, translocation, and reuse.

Seeds generally (unless viviparous) enter on a prolonged period of dormancy and provide in their storage organs (whether cotyledons or endosperm) a distinctive morphological setting for their stored products in regions in which the cells have ceased to multiply but may still enlarge. Thus the sequential events here are (i) synthesis of specific proteins, (ii) the cessation of synthesis with provision for storage, and finally, (iii) their ultimate breakdown, translocation, and reuse (as in the apical growing regions). These topics properly overlap many considerations of growth and development that will be discussed in other volumes. Suffice it to say, however, that this sequence of events invokes all the complexity of the genetic information prescribing how the biochemistry will work with all the variables capable, at the critical periods, of controlling in the cells the pace of its progress (or even its reversal). Among these variables, water supply or stress and growth-promoting agents or their antagonists have been invoked.

Sussex and colleagues (241, 242) have attributed the onset of storage seed proteins in the cotyledons of legumes (pea) to a critical stage in their development (20 days after fertilization) after cell division has subsided though cell enlargement ensues. Signals then received transform a prior high water content and low protein content in the cells into lowered water content and higher content of specific proteins. Thus the initial bulk synthesis of seed proteins involves cells in a particular morphological setting constrained by balanced endogenous growth factors to enlarge and not divide. Benzyladenine, as the cytokinin, and abscisic acid, its inhibitor, are plausible exogenous counterparts of the regulatory endogenous growth factors. Thereafter in development, nitrogen metabolism, differentiation, and morphogenesis proceed concomitantly, and, en route, new centers of synthesis and storage of nitrogen compounds will arise.

Sutcliffe and Pate (242a) have collated articles on selected topics related to the physiology and growth of the garden pea (*Pisum sativum*). Work on its nitrogenous nutrition and metabolism still looms large (242a, Chapters 3, 6, 9, 10, 12, 13, and 15), particularly with respect to the diversity of nonprotein-nitrogen compounds that flows from use of nitrate and in response to the consequential stimulus to nitrate reductase activity. These nonprotein-nitrogen compounds are speedily involved in the events of translocation (242a, Fig. 13.4 for the prominence of amides, mainly asparagine, and of ureides, especially allantoic acid) as in the xylem sap. The thoughts developed here for the pea plant are amplified and extended to other legumes in Chapter 4 of this volume.

Legumes, with their several prominent centers of nitrogen metabolism (nodules, roots, leaves, apices, fruits, etc.), have classically been favored as experimental subjects for the study of nitrogen metabolism. This is particularly so because of their nitrogen-fixing capacity (see Chapter 1) and their ability to reveal in bleeding saps the nature and quantities of compounds being translocated (see Chapter 4, Section III) between centers of metabolism and synthesis. All this is currently facilitated by use of isotopically labeled nitrogen and carbon (see Chapter 4, Section V).

Although Sutcliffe and Pate (242a) compiled a very comprehensive picture of the diverse events that occur in pea plants throughout their development (amplified by Pate, Chapter 4 of this volume), the problems of their correlation and the regulatory control of inherited information remain. A summary statement, applicable alike to metabolism and morphogenesis (242a, p. 287), is that "all of the correlations mentioned above are under the control of complex interactions between hormones and nutrients, the levels of which are affected by environmental conditions." As in the case of other plants considered here (Section IV,B–D), one concludes again that it is from centers of growth (as in the apices of shoot and root, in buds, and in developing organs) that these regulatory controls originate.

E. Nitrogen Metabolism in the Integrated Responses of Cultured Cells

1. Quiescent versus Activated Cells

Chapter 4 in Volume II, (see pp. 335–363) invoked the contrast between actively growing and mature quiescent cells in relation to different interrelated physiological functions. This was done specifically with respect to nitrogen metabolism and the synthesis of protein, to water and the uptake and accumulation of solutes, and to respiration as the ultimate source of energy for cells. The preceding sections have stressed how nitrogen metabolism changes through morphogenesis and the impact on these events of environmental controls. Later work on cultured carrot (*Daucus carota*) cells now illuminates these problems at the cellular level, uncomplicated by limitations imposed by morphogenesis, in a system in which all of the information available to the zygote is free to work in activated carrot cells, as shown by the incidence of their totipotency as revealed by somatic embryogenesis.

2. Nitrogen Compounds in Activated Cultured Explants

The nitrogen compounds of the storage pool of the quiescent cells of carrot roots do not correspond, qualitatively or quantitatively, to the hydrolytic products of the protein in the growing cells of cultured explants. Therefore, from storage cellular pools to sites of synthesis and from sites of breakdown to the reuse of the breakdown products, the nitrogen compounds of the cultured explants need to be reworked under the influence of the exogenous growth-regulatory factors that stimulate protein synthesis, respiration, and turnover. To interpret these events in terms of specific metabolites and postulated pools, special means had to be devised. These involved the exogenous supply of specifically ^{14}C-labeled metabolites, the chromatographic analysis of the nitrogen compounds in the nonprotein nitrogen, and the total protein nitrogen of the cells, together with the specific activities of the respired carbon dioxide. From these data metabolic schemes were devised (229).

Thus the carrot cells, as activated by the growth factors, were conceived in terms of relatively inert storage pools or compartments of nitrogen compounds and of protected pools of metabolites through which they were passed en route to protein, with compounds (like glutamine and alanine) being substances through which the traffic in carbon and nitrogen was canalized (21) at key points. This intensely dynamic picture of the carrot cells recognized that entering exogenous sugar, whether from an ambient medium or via translocation, passed through the tricarboxylic acid cycle to provide ports of entry for inorganic nitrogen or for nitrogen transferred from nitrogen-rich sources already in the cells (Chart 9). Moreover, the tricarboxylic acid cycle could also receive, and even respire away, carbon from protein breakdown products, leaving nitrogen-rich residues to reenter the storage pools (Fig. 3). Since these ideas were formulated, the evident cytological complexity of cells in general (Volume VII, Chapter 1) and of cultured carrot in particular (115, 116) now place these concepts of functional storage pools, compartments, and sites of active metabolism and synthesis in an even more heterogeneous organizational setting than hitherto.

3. Nitrogen Metabolism during Growth Induction

The impact of powerful, composite, growth-promoting systems, naturally occurring as, for example, in coconut milk or a fluid from immature *Aesculus* fruits, on the nitrogen metabolism and growth induction of quiescent carrot cells has been much investigated. The stimulus has been dissected into partial systems, designated I and II (mediated by inositol and indoleacetic acid, respectively). The distinctive synergistic effects of the component parts of each system and the way they produce mutually complementary effects (which are also supplemented by casein hydrolysate added to the culture medium) have been tested. The tests have been made with reference to aspects of growth of the carrot explants (in fresh weight, in cell number, and in average cell size), but also with respect to their impact upon nitrogen metabolism (i.e., total soluble nitrogen, protein nitrogen, and nucleic acid content per 1000 cells or per explant). In addition, the soluble (nonprotein, ninhydrin reactive) nitrogen compounds were separated and quantitatively determined by the methods used in the study of potato plants already described; thus the effects of the growth-factor treatments on the relative composition of the nonprotein nitrogen of the cells were illustrated in a similar manner (228).

4. Trace Elements and Growth-Factor Effects in Nitrogen Metabolism

The effectiveness of the organic growth factors (i.e., the component parts of systems I and II) and the effects of the partial systems (231, pp. 63–71) also presupposes a proper balance of the several inorganic trace elements. Without this balance of trace elements even otherwise competent complements of growth factors in nutrient media will not suffice for maximum growth. Furthermore, various combinations of trace elements and exogenous growth-promoting substances and their cofactors have evoked a wide range of metabolic responses in uniform clones of cultured carrot explants as seen especially in their content of nitrogen compounds. This illustrates the complexity of the regulatory control of nitrogen metabolism in the growing cells and shows how their many responses may be correlated. Examples of the results will be cited (the data are to be found in 234).

To study the effects of trace elements, the basal nutrient media, the coconut milk supplements and any other growth-promoting additives, were prepared with limitingly low content of the trace elements and each of the growth-factor combinations was then tested, against the appropriate controls, for the effects of trace elements added singly and in combinations. The responses of the nitrogenous components of the cultured tissue to the various growth-factor–trace-element regimes were revealed in terms of the total nonprotein-nitrogen fraction and especially of its content of the amides (glutamine and asparagine), of alanine, of the dicarboxylic acids (aspartic and glutamic), and of arginine. The effects on the protein nitrogen were shown by its total content and its percentage of the total nitrogen of the tissue and its relationship to the total nucleic acid present.

The great range of responses in the nitrogen metabolism within genetically identical clones of carrot explants to treatments of the same duration, under standard environmental conditions, as brought about by growth-factor–trace-element regimes, need not be discussed individually; they are best seen collectively in the figures in Steward and Rao (235, 236; see also Table XIII). All of these figures depict the effects of various parallel treatments that show the composition of the soluble nitrogen of the explants after they had been subjected to various parallel treatments. The data also show the effects of the treatments on total soluble and protein nitrogen per gram fresh weight, as a percent of total nitrogen, and in relation to total nucleic acid and their effects on the growth responses of the explants as shown by the number of cells and the total weight of the explants. The two papers cited (235, 236) together describe 57 different growth-factor–trace-element regimes that produced distinctive patterns of nitrogenous compounds in the explants as cultured. To deal with these data in summary form liberal extracts from these papers are therefore made as follows:

The growth-promoting agents employed by Steward and Rao (235) consisted of (i) the balanced complex of factors found in coconut milk, (ii) an active isolate from *Aesculus* (AF_{aesc}), which is one of a class of growth factors (AF_1) that interact with inositol (AF_1 + inositol) and which in this sense comprise growth-promoting System I, and (iii) the substance zeatin (Zeat.), which is typical of a class of active factors (AF_2) that interact with indoleacetic acid (AF_2 + IAA) and function, therefore, as a growth-promoting complex termed System II in the culture of carrot tissue.

The carrot explants stimulated by coconut milk grew better than those stimulated by the other combinations of growth factors, and they converted their soluble nitrogen more effectively to protein. The growth, whether it was induced by coconut milk or by System I or II, and other specific effects attributable to the growth factors employed, were markedly affected also by the elements iron and molybdenum.[11]

---

[11]Although the whole range of micronutrients has been examined, iron and molybdenum were selected for detailed study because of their distinctive effects and interactions.

TABLE XIII

Responses of Carrot Explants during 18 Days of Culture[a]

				Treatment			
Medium[a]	Avg. fr. wt. per plant (mg)	Cells per explant ($\times 10^3$)	Cell size (μg/cell)	Soluble nitrogen per explant (μg)	Protein nitrogen per explant (μg)	Total nitrogen per explant (μg)	Nucleic acids per explant (μg)
1. B**	3.8	22.8	0.1841	2.8	9.1	11.9	16.3
43. B** + CM + Fe + Mo + CH	167.4	1316.5	0.1174	42.7	67.5	110.2	60.3

[a] In a basal nutrient medium free of trace elements (No. 1) and in a basal medium replenished with Fe and Mo and also supplemented with a full complement of added growth-promoting factors (No. 43). After (236), Table 1, Series (d) controls. B**, Basal medium free of trace elements; CM**, coconut milk freed from trace elements; CH, enzymatic casein hydrolyzate; Fe, iron; Mo, molybdenum.

The carrot explants that had responded to coconut milk emphasized alanine in their soluble, nonprotein, nitrogenous pool, whereas those subjected to the active components of System I or II as clearly emphasized glutamine as the prominent nonprotein, nitrogen-rich compound.

The partial effects due to the component parts of System I (AF$_{aesc}$ or inositol) and to the component parts of System II (Zeat. or IAA), as these interacted also with iron and molybdenum in an otherwise trace-element-free basal medium (B**), revealed a pattern of interlocking effects, due to trace elements and to growth factors, on the metabolism (especially the nitrogen metabolism) of the aseptically cultured carrot explants.

Steward and Rao (236) performed another symmetrically designed set of 36 treatments on parallel samples of carrot explants. The resultant effects on the soluble-nitrogen fraction of the cultures as grown were illustrated by histograms (236, Fig. 1a–c, p. 244).

Three sets of growth factor combinations were used, and superimposed on each set of these combinations were treatments that tested for these growth-inducing environments, the effects of molybdenum, iron, and these two trace elements in combination. The three distinctive sets of growth-factor combinations were

1. A trace-element-free basal medium plus (i) the active component of System I (AF$_{aesc.}$), (ii) the same basal medium plus inositol, and (iii) the same basal medium plus AF$_{aesc.}$ and inositol in combination.
2. A trace-element-free basal medium, plus (i) zeatin as an active component of System II, (ii) the same basal medium plus IAA, and (iii) the same basal medium plus IAA and zeatin in combination. That is, Combination 2 recaptiulates Combination 1 except that it invokes the corresponding growth-factor stimuli of System II instead of those that operate in System I.
3. A trace-element-free basal medium plus trace-element-free coconut milk (i) without further additives, (ii) with added IAA, and (iii) with added inositol.

These treatments produced (236, Fig. 1a–c) a wide range of responses in growth of the explants and especially so in the size and composition of the soluble-nitrogen component in the cultures as grown. Although initially all of the components of the soluble-nitrogen could be said to respond to these treatments, the dominant effects involved either the amides (especially glutamine) or alanine, although in situations where neither of these were as predominant, many of the other constituents were relatively more prominent. The general concept to which all these experiments lead is as follows.

The relatively large, quiescent cells of the original carrot explants possessed large pools of soluble nitrogenous reserves and contained also the genetic information that conveys both morphological and biochemical totipotency (227); nevertheless, in order to grow, these cells require much more than a complete range of the obvious organic and inorganic nutrients. If these cells receive the appropriate exogenous stimuli, then they may use the supplied nutrients; but it is the combined effects of the growth-factor systems and of the trace elements that furnish these stimuli. Therefore, the organic growth regulators and the inorganic trace elements must converge on the several sites at which metabolism (especially nitrogen metabolism), cell division, and growth are directed within the cells. Many chemical agents participate distinctively in these events; these include Fe and Mo, the constituents of coconut milk, zeatin-like compounds, IAA, compounds like the *Aesculus* factor, inositol, and even the constituents of casein hydolysate—and the effects they all produce are very diverse. Soluble nitrogen may be diverted into alternative channels as alanine or glutamine, and in some cases arginine, form larger or smaller pools of soluble compounds; ribosomal nucleic acids are synthesized and protein levels per cell or per explant are regulated. Finally, cells may be caused to divide faster than they can store soluble-nitrogen compounds under some treatments, or to expand and build up reserves of soluble substances under others. One should not, therefore, attempt to localize all of these stimuli and responses at one or even a few intracellular sites. The machinery of growth and cell division and the compartmented metabolism that supports it is, within the limits determined by genetic constitution, either set in motion or repressed by the balanced array of exogenous growth factors and the special nutrients (trace elements) that the ambient medium must provide. Thus the sequence of events when growth subsides (as when parenchyma cells mature) and the converse events (when quiescent cells are stimulated *in situ* to grow again) may be seen as mediated by the interplay between exogenous stimuli and special nutrients (especially Fe and Mo) and with the highly compartmented cellular organization.

5. Nitrogenous Solutes: An Aspect of the Osmotica of Cultured Cells

The discussion of nitrogen metabolism of cultured cells is incomplete without considering it in relation to the entire complex of solutes (ions and molecules), which, being accumulated as osmotica in cells, are implicated in their growth and development. (This task fulfilled the objective of the development of the aseptic carrot system in its intended use to study metabolism ion uptake and solute accumulation in growing cells.) Five papers dealt with the problems of solute accumulations that develop in cultured carrot cells as they divide, enlarge and mature (169 and references cited therein). Essentially the position is as follows.

When quiescent carrot cells in aseptic culture embark on renewed activity and growth, they follow a typical growth curve when they are in an ambient medium that contains salts at low levels with nitrate as the form of nitrogen and a carbohydrate. Due to the growth-factor stimuli, which have been discussed, cells move toward the rapid division that may ensue, and, as they do so, they acquire solutes (organic and inorganic), which have been identified (196). At the outset, K^+ is the preferred cation, and it tends then not to be accompanied stoichiometrically by chloride, reaching only relatively low degrees of accumulation over the ambient medium. At this stage, when cells remain small it is the nonelectrolytes (carbohydrate, organic acids, and soluble-nitrogen compounds) that preferentially contribute to the osmotica of the cells. Later, however, as cells enlarge and after sugars have accumulated, nonelectrolytes may be reversibly replaced on transfer to a salt (KCl) solution when K^+ and Cl^- enter the cells in near stoichiometric amounts, and K^+ may then reach greater degrees of ionic accumulation. The preferential formation of amide nitrogen or of alanine in response to different component parts of the overall growth-factor stimulus is typical of an ambient medium in which KNO_3 is the source of K^+ and of nitrogen. If, however, the total growth-promoting complex (CM) operates on tissue subjected to graduated concentrations of KNO_3 contrasted with graduated concentrations of NH_4Cl or NH_4NO_3, very dramatic contrasts ensue. The balance of soluble-nitrogen compounds in carrot cultures grown on media that contain from 6 to 18 mM KNO_3, in spite of their increased total soluble nitrogen, remains very stable in the proportions of the various amino acids, especially of alanine, glutamine, asparagine, etc. But at 6 mM NH_4Cl in lieu of 6 mM KNO_3 there is a great accumulation of soluble nitrogen that is not turned into protein, and with this a dramatic shift occurs from the alanine-dominated composition to one dominated by glutamine–arginine, and this shift persists in solutions that contain up to 18 mM NH_4NO_3. (42b, Fig. 8).

The last of the papers on solutes in cultured cells summarizes (169, Table 1) the consequences of 20 different treatments of carrot explants for their content of inorganic ions and osmotically active solutes; it then deals with the implications of the entire study as an aspect of nutrition and development. The last of another group of five papers (234) is especially concerned with the overall consequences of the cultural procedures for the composition of the solutes accumulated in carrot cells.

The wide range of effects on the composition of the organic, nitrogenous solutes of cultured carrot cells that are associated with the growth factors and the trace elements that modulate their activity finds a parallel in the consequences of similar treatments for the content of inorganic ions (K^+, Na^+, and Cl^-) in the cells and for the total organic solutes that comprise their osmotica (see 234, Table 1, pp. 222–223, for these data as they relate to eight growth-factor regimes under four different trace-element regimes).

Thus the responses of proliferating carrot tissue in aseptic cultures reveal the extent to which their cells can utilize their genetically determined propensities for cell growth and nitrogen metabolism as affected by an array of both organic and inorganic interacting regulatory factors. There is more, however. In spite of the range of responses of given clones of carrot explants under *constant* environmental regimes of light and temperature, they will respond still further to diurnal cycles of light and darkness and to different temperatures applied by day and night.

6. Interacting Factors in the Composition and Nutrition of Cultured Cells

But what does this all mean? After genetics has endowed cells (whether they are free to proliferate or are more controlled in their growth in tissues and organs as they develop from growing points) with specific, biochemically feasible attributes, their actual expression is determined by a great array of interacting factors. These factors have in part endogenous developmental origins, in part they have an exogenous nutritional (trace element) basis, and they are also subject to overall environmental regulation. These considerations raise questions of how complete an understanding is to be expected solely from the specificity of gene–enzyme biochemistry in the approach to nitrogen metabolism; they also pose the challenge inherent in a full understanding of the means by which genetic information is controlled and used during growth and development and how overall regulation is achieved within cellular organizations.

Autotrophic green plants are familiarly regarded as unique in their nutritional uses of carbon and nitrogen. Plant cells have, however, other less obvious but distinctive characteristics.

For example, multiplying cells that are primarily engaged, anabolically, in the building up of complexity budget a portion of their energy in the form of reduced entropy. Cells as physical systems use metabolic energy, capturing a part of it in the creation and maintenance of their organization built out of complex nitrogen compounds. Cells that accumulate high concentrations of solutes, organic and inorganic, and retain these, dynamically, against adverse concentration gradients also budget a portion of their energy to increase solute concentrations (i.e., free energy), and this requires that osmotic work be done.

Thus, by the use of the carrot cultures, the time course of solute content has been correlated with the time course of growth, with the onset of cell division, on the one hand, and cell enlargement on the other. The soluble-nitrogen compounds are to be seen as part of this time course of solute content, which flows from the growth-induction stimuli. The data show that the partial and complete growth-promoting systems may regulate the composition in terms of both organic solutes, including nitrogen compounds, and the inorganic ions.

7. Sources and Sinks for Carbohydrates and Nitrogen Compounds

Green leaves in the light are the source of carbohydrates, over and above their immediate use there, and, in the configuration of the sugars, they tie up solar energy that enables plants to "feed on negative entropy." But, just as clearly, reduced nitrogen from nitrate via the roots is needed to build up proteins in leaves and to permit energy and carbon to be incorporated into the complexity of structure that enables plants and their cells to function physically (see Chapter 4). And once the levels of osmotica to maintain turgor have been established, reciprocal "flows" of carbohydrate from shoot to root and of reduced nitrogen compounds and inorganic ions from roots to shoots preserve nutritional balance as shoots are "sources" for carbohydrates and "sinks" for nitrogen compounds; in roots the reverse is true. But when external nitrogen and nutrients are deprived and limiting, the carbohydrates may build up as osmotica in roots, a condition that, when extreme, produced the "low salt–high sugar" barley (*Hordeum vulgare*) roots that when excised were able to replace their organic solutes rapidly with accumulated alkali halides, provided their aerobic metabolism was maintained by adequate aeration.

V. Summary: Perspectives on Nitrogen Metabolism of Selected Plants

Obviously, the pursuit of problems of nitrogen metabolism through the growth and development of selected members of any of the great divisions of the plant kingdom would reveal points of both similarity and distinction. Contrasts between the selected examples drawn from angiosperms and gymnosperms are therefore only indicative of these variations on the central theme. The examples chosen have ranged from proliferating, unorganized, but nevertheless potentially totipotent cells to the organs of a dicotyledon, as they develop from shoot apices and as they respond to environments, and to forest trees, which cover a large part of the earth's surface and survive against seemingly hostile environments. Therefore, the present study has range even though it cannot claim to be comprehensive.

It would be satisfying if one could prescribe all of the chemical rules that limit the capacity of plants to fabricate form and substance from nitrogen. Obviously this is not so. The days when an empirical analysis of a pure plant protein, in terms of its amino acids, and the search in plants for the protein amino acids in the free state seemed to be laudable research objectives have passed even within a contemporary life span. What then can one say in conclusion?

What autotrophic plants do with nitrogen is still a feature that distinguishes higher plants from higher animals; without the former, the latter could neither have evolved nor long exist. Nevertheless, at the chemical level plants utilize essentially similar genetic mechanisms to preserve and use their inherited information and to build the large molecules (proteins and enzymes) that make them work as biosynthetic machines. But plants and animals diverge when biochemistry becomes subservient to morphogenesis, and this is even apparent in their respective uses of the simpler molecules.

Plants conserve and reuse their breakdown products, keeping them as part of their internal complement of solutes. However, plants also utilize to a far greater degree than animals their external environments to modify the biochemistry of their simpler soluble compounds (albeit within limits set by heredity). Plants also modify (through the morphogenetic responses of their growing points) both the course of their development and the great diversity in the form and composition of the organs they produce. While these differences emphasize the contrasts between plants and animals, they are also the basis of the great diversity in plants, exemplified by the different plants discussed.

All of the plants referred to (whether they develop from zygotes in an embryo sac, from seeds arrested in their development, from vegetative shoot apices, or from cells explanted from mature organs) employ their unique complement of genetic information. This information prescribes limits within which biochemistry and physiology can operate, but the expression of these inherited potentialities is, in plants, also modulated in many ways. External sources of essential elements, especially of trace elements (singly and in combinations) greatly affect the intrinsic biochemical propensities of the cells to fabricate and store nitrogen compounds. Although not yet fully understood, exogenous growth factors (acting no doubt in lieu of those that operate endogenously during normal development) working in synergistic combinations and along with the trace elements make their additional impact upon the responses of explanted cells as they grow. But even this does not exhaust the degrees of freedom that the cells possess, for, even lacking organized growing points, they may also respond metabolically to the external variables of darkness versus light, to day and night length, and to interactions with temperature, etc. But once the cells form part of an organized growing region, the way they express their inherited information becomes subject in organ-specific ways to limitations that morphology prescribes, and in turn the morphology may be drastically modulated by external environments, especially in the behavior of perennial plants of temperate climates. But what can be concluded from the angiosperm–gymnosperm contrasts as they are here revealed?

The main convergent pathways of nitrogen and carbon metabolism for angiosperms have been summarized in the metabolic charts. The key metabolites (whether amino acids; amides, especially glutamine; or their corresponding keto acids) and the ever more complex array of enzyme systems through which their metabolism is negotiated are by now familiar. Nitrate or ammonium nitrogen when fixed, and urea when supplied, may be exogenous nitrogen sources. But the compounds, especially when nitrogen-rich, that accumulate when soluble nitrogen is not being metabolically consumed do not reflect the composition of the proteins to which they may give rise or from which they may have been released on breakdown. A great range of so-called secondary metabolites may accumulate in ways that are seemingly unrelated to the normal course of essential intermediary metabolism. The problems of regulation, whether developmental, nutritional, or environmental, largely concern how all this chemical machinery, genetically determined, is put to work.

In principle a range of metabolic potentialities exists in the coniferous trees that is similar to that in the flowering plants, although their regulation and detailed expression may seem to be very different. In particular, the regulation by environments seems to be even more demanding in the conifers in their responses to climatic, seasonal, and even diurnal fluctuations. Although many of the broad pathways may be similar, there are points of difference, especially in the soluble-nitrogen compounds that accumulate. This is notably so in the greater stress on arginine and its derivatives in the conifers (integrating somewhat with members of the urea cycle) rather than on amides (especially glutamine) in the angiosperms (integrating with intermediates of the tricarboxylic acid cycle, etc.). But understanding of the nitrogen metabolism and physiology of the conifers still requires deeper excursions into lesser known areas involving ureides (though urea itself may not be prominent), involving proline, and, especially, involving arginine and a wide range of guanidines. The problems still remain to know how far the conventional arginine–citrulline cycle is really involved as a main route of metabolism and to know how conifers manage to steer their buds so successfully through the various seasonal hazards and fluctuations to which they are exposed. Nevertheless, conifers, which often face such seemingly hostile environments on soils that are often nitrogen-poor, consitute a very successful part of the earth's green cover and are notable as producers of cellulose and wood.

Because nitrogen is of such overriding importance in every aspect of the metabolism of plants, this chapter has inevitably presented large amounts of information impinging on plant biochemistry, physiology, and development. In the outcome, virtually no study of the biology of plants seems complete without taking account of the involvement of nitrogen, not only through its specific metabolic implications, but because nitrogen serves to integrate the many diverse aspects of plant biology into a coherent whole.

References

1. Abelson, J. (1979). RNA processing and the intervening sequence problem. *Annu. Rev. Biochm.* **48,** 1035–1069.
2. Aberhart, D. J. (1979). Stereochemical studies on the metabolism of amino acids. *Recent Adv. Phytochem.* **13,** 29–54.
3. Adams, D. O., and Yang, S. F. (1979). Ethylene biosynthesis: identification of 1-aminocyclopropane-1-carboxylic acid as an intermediate in the conversion of methionine to ethylene. *Proc. Natl. Acad. Sci. USA* **76,** 170–174.
4. Allen, G. M., and Jones, M. E. (1964). Decomposition of carbamyl phosphate in aqueous solutions. *Biochemistry* **3,** 1238–1247.
5. Altschul, A. M. (1969). Food: proteins for humans. *Chem. Eng. News* **47,** 68–81.
6. Altschul, A. M. (1976). The protein–calorie trade-off. *In* "Genetic Improvement of Seed Proteins," pp. 5–16. Proc. Workshop 18–20 March 1974. Natl. Res. Counc. & Natl. Acad. Sci., Washington, D.C.
7. Amzel, M. L., and Poljak, R. J. (1979). Three-dimensional structure of immunoglobins. *Annu. Rev. Biochem.* **48,** 961–997.
8. Anderson, N. G. (1976). Interactive macromolecular sites. I. Basic theory. *J. Theor. Biol.* **60,** 401–412.
9. Anderson, N. G. (1976). Interactive macromolecular sites. II. Role in prebiotic macromolecular selection and early cellular differentiation. *J. Theor. Biol.* **60,** 413–419.
10. Atkinson, D. E. (1977). "Cellular Energy Metabolism and Its Regulation." Academic Press, New York.
11. Audus, L. J. (1959). "Plant Growth Substances," 2nd ed. Wiley (Interscience), New York.
12. Battersby, A. R., Fookes, C. J. R., Matcham, G. W. J., and McDonald, E. (1980). Biosynthesis of the pigments of life: formation of the macrocycle. *Nature (London)* **285,** 17–21.

13. Beale, S. I. (1978). δ-Aminolevulinic acid in plants: its biosynthesis, regulation, and role in plastid development. *Annu. Rev. Plant Physiol.* **29**, 95–120.

14. Beevers, H. (1979). Conceptual developments in metabolic control 1924–1974. *Plant Physiol.* **54**, 437–442.

15. Beevers, L. (1976). "Nitrogen Metabolism in Plants." Arnold, London.

16. Bell, E. A. (1973). Amino acids of natural origin. *MTP Int. Rev. Sci. Org. Chem. Ser. One* **6**, 1–26.

17. Bell, E. A. (1980). Non-protein amino acids in plants. *Encycl. Plant Physiol. New Ser.* **8**, 403–432.

18. Bengston, G. W. (1973). Fertilizer use in forestry: material and methods of application. *Proc. FAO-IUFRO Int. Symp. Forest Fert.* pp. 97–168.

19. Benkovic, S. J. (1980). On the mechanism of action of folate and biopterin-requiring enzymes. *Annu. Rev. Biochm.* **49**, 227–251.

20. Bernardi, G. (1978). Intervening sequences in the mitochondrial genome. *Nature (London)* **276**, 558–559.

21. Bidwell, R. G. S., Barr, R. A., and Steward, F. C. (1964). Protein synthesis and turnover in cultured plant tissue: sources of carbon for synthesis and the fate of the protein breakdown products. *Nature (London)* **203**, 367–373.

22. Bidwell, R. G. S., and Durzan, D. J. (1975). Some recent aspects of nitrogen metabolism. *In* "Dynamic Aspects of Plant Physiology: A Symposium Honoring F. C. Steward" (P. J. Davies, ed.), pp. 152–225. Cornell Univ. Press, Ithaca, New York.

23. Bisswanger, H., and Schmincke, O. H. E. (1979). "Multifunctional Proteins." Wiley, New York.

24. Boulter, D. (1979). Structure and biosynthesis of legume storage proteins. Seed protein improvement: cereals, grain, legumes. *Proc. IAEA Int. Symp.* **1**, 125–136.

25. Bowman, C. M., and Dyer, T. A. (1979). 4.5S Ribonucleic acid, a novel ribosome component in the chloroplasts of flowering plants. *Biochem. J.* **183**, 605–613.

26. Bradbeer, W. J., Atkinson, Y. E., Börner, T., and Hageman, R. (1979). Cytoplasmic synthesis of plasmic polypeptides may be controlled by plastid synthesized RNA. *Nature (London)* **279**, 816–817.

27. Braun, A. C. (1980). Genetic and biochemical studies on the suppression of and recovery from the tumorous state in higher plants. *In Vitro* **16**, 38–48.

28. Brouk, B. (1975). "Plants Consumed by Man." Academic Press, New York.

29. Brown, R., and Dyer, A. F. (1972). Cell division in higher plants. *In* "Plant Physiology" (F. C. Steward, ed.), Vol. 6c, pp. 49–90. Academic Press, New York.

30. Bryan, P. A., Crawley, R. D., Brunner, C. E., and Bryan, J. K. (1970). Isolation and characterization of a lysine-sensitive aspartokinase from a multicellular plant. *Biochem. Biophys. Res. Commun.* **41**, 1211–1217.

31. Buchanan, J. M. (1973). The amidotransferases. *Adv. Enzymol.* **39**, 91–183.

32. Buchanan, B. B., Wolosiuk, R. A., and Schürmann, P. (1979). Thioredoxin and enzyme regulation. *Trends Biochem. Sci.* **4**, 93–96.

33. Burr, F. A. (1979). Zein synthesis and processing on zein protein body membranes. *Proc. IAEA Int. Symp.* **1**, 159–164.

34. Butler, L. G., and Reithel, F. J. (1977). Urease-catalyzed urea synthesis. *Arch. Biochem. Biophys.* **178**, 43–50.

35. Chen, Y. M., and Walker, J. B. (1977). Transaminations involving keto- and amino-inositols and glutamine in *Actinomyces* which produce gentamicin and necinicin. *Biochem. Biophys. Res. Commun.* **77**, 688–692.

36. Chibnall, A. C. (1939). "Protein Metabolism in the Plant." Yale Univ. Press, New Haven, Connecticut.

37. Chalupa, V., and Durzan, D. J. (1973). Growth and development of resting buds of conifers *in vitro. Can. J. For. Res.* **3**, 196–208.
38. Chapman, D. J., and Ragan, M. A. (1980). Evolution of biochemical pathways: evidence from comparative biochemistry. *Annu. Rev. Physiol.* **31**, 639–678.
39. Conn, E. C. (1973). Biosynthesis of cyanogenic glycosides. *Biochem. Soc. Symp.* **38**, 277–302.
40. Conn, E. C. (1973). Cyanogenetic glycosides. *In* "Toxicants Occurring Naturally in Foods" (pp. 299–308), Natl. Res. Counc. & Natl. Acad. Sci., Washington, D.C.
41. Conn, E. E. (1980). Cyanogenic glycosides. *Encycl. Plant Physiol. New Ser.* **8**, 460–492.
42. Cooke, R. J., Oliver, J., and Davies, D. D. (1979). Stress and protein turnover in *Lemna minor. Plant Physiol.* **64**, 1103–1109.
42a. Cozzarelli, N. R. (1980). DNA gyrase and supercoiling of DNA. *Science (Washington, D.C.)* **207**, 953–960.
42b. Craven, G. H., Mott, R. L., and Steward, F. C. (1972). Solute accumulation in plant cells. IV. Effects of ammonium ions on growth and solute content. *Ann. Bot. (London)* **36**, 897–914.
43. Crick, F. (1979). Split genes and RNA splicing. *Science (Washington, D.C.)* **204**, 264–271.
44. Crick, F. (1970). Central dogma of molecular biology. *Nature (London)* **227**, 561–567.
44a. Cullimore, J. V., and Sims, A. P. (1980). An association between photorespiration and protein catabolism: studies with *Chlamydomonas. Planta* **150**, 392–396.
45. Davies, D. D. (1968). The metabolism of amino acids in plants. *In* "Recent Aspects of Nitrogen Metabolism in Plants." (E. F. Hewitt and C. V. Cutting, eds.), pp. 125–135. Academic Press, New York.
46. Davies, D. D. (1979). The central role of phosphoenolpyruvate in plant metabolism. *Annu. Rev. Plant Physiol.* **30**, 131–158.
47. Davies, J. S. (1977). Occurrence and biosynthesis of D-amino acids. *Chem. Biochem. Amino Acids Pept. Proteins* **4**, 1–27.
48. Dixon, N. E., Gazzolo, C., Blakeley, R. C., and Zerner, B. (1976). Metal ions in enzymes using ammonia or amides. *Science (Washington, D.C.)* **191**, 1144–1150.
49. Drey, C. N. C. (1977). The chemistry and biochemistry of beta-amino acids. *Chem. Biochem. Amino Acids Pept. Proteins* **4**, 241–299.
50. Drummond, M. (1979). Crown gall disease. *Nature (London)* **281**, 343–347.
51. Duda, C. T. (1976). Plant RNA polymerases. *Annu. Rev. Plant Physiol.* **27**, 119–132.
52. Dure, L. S., III (1976). Messenger RNA synthesis and utilization in seed development and germination. *In* "Genetic Improvement of Seed Proteins," pp. 329–340. Natl. Res. Counc. & Natl. Acad. Sci., Washington, D.C.
53. Durzan, D. J. (1971). Free amino acids as affected by light intensity and the relation of responses to the shade tolerance of white spruce and shade intolerance of jack pine. *Can. J. For. Res.* **1**, 131–140.
54. Durzan, D. J. (1968). Nitrogen metabolism of *Picea glauca*. I. Seasonal changes of free amino acids in buds, shoot apices and leaves, and the metabolism of uniformly labelled ^{14}C-L-arginine by buds during the onset of dormancy. *Can. J. Bot.* **46**, 909–919.
55. Durzan, D. J. (1968). Nitrogen metabolism of *Picea glauca*. III. Diurnal changes of amino acids, amides, protein and chlorophyll in leaves of expanding buds. *Can. J. Bot.* **46**, 929–937.
56. Durzan, D. J. (1969). Nitrogen metabolsim of *Picea glauca*. IV. Metabolism of uniformly labelled ^{14}C-L-arginine, (carbamyl-^{14}C)-L-citrulline and 1,2,3,4-^{14}C-γ-guanidinobutyric acid during diurnal changes in the soluble and protein nitrogen associated with the onset of expansion of spruce buds. *Can. J. Biochem.* **47**, 778–783.

57. Durzan, D. J. (1973). Nitrogen metabolism of *Picea glauca*. V. Metabolism of uniformly labelled ^{14}C-L-proline and ^{14}C-L-glutamine by dormant buds in late fall. *Can. J. Bot.* **51,** 359–369.

58. Durzan, D. J. (1973). The metabolism of ^{14}C-urea by white spruce seedlings as influenced by light and darkness. *Can. J. Bot.* **51,** 1197–1211.

59. Durzan, D. J. (1973). The incorporation of tritiated water into amino acids in the presence of urea by white spruce seedlings in light and darkness. *Can. J. Bot.* **41,** 351–358.

60. Durzan, D. J., Campbell, R. A., and Wilson, A. (1979). Inhibition of female cone production in white spruce by red light treatment during night under field conditions. *Environ. Exp. Bot.* **19,** 133–144.

61. Durzan, D. J., and Chalupa, V. (1976). Growth and metabolism of cells and tissue of jack pine (*Pinus banksiana*). I. The establishment and some characteristics of a proliferated callus from jack pine seedlings. *Can. J. Bot.* **54,** 437–445.

62. Durzan, D. J., and Chalupa, V. (1976). Growth and metabolism of cells and tissue of jack pine (*Pinus banksiana*). V. Changes in free arginine and Sakaguchi-reactive compounds during callus growth and in germinating seedlings. *Can. J. Bot.* **54,** 483–495.

63. Durzan, D. J., and Chalupa, V. (1968). Free amino acids and soluble proteins in the embryo and female gametophyte of jack pine as related to climate at the seed source. *Can. J. Bot.* **46,** 417–428.

64. Durzan, D. J., Chafe, S. C., and Lopushanski, S. M. (1973). Effects of environmental changes on sugars, tannins, and organized growth in cell suspension cultures of white spruce. *Planta* **113,** 241–249.

65. Durzan, D. J., Mia, A. J., and Wang, B. S. P. (1971). Effects of tritiated water on the metabolism and germination of jack pine seeds. *Can. J. Bot.* **49,** 2139–2149.

66. Durzan, D. J., and Pitel, J. (1977). The occurrence of *N*-phosphorylarginine in the spruce budworm (*Choristoneura fumiferana* Clem.). *J. Insect Biochem.* **7,** 11–13.

67. Durzan, D. J., Pitel, J., Mia, A. J., and Ramaiah, P. K. (1973). Metabolism of uracil by germinating jack pine seedlings. *Can. J. For. Res.* **3,** 209–221.

68. Durzan, D. J., Pitel, J., and Ramaiah, P. K. (1972). Acid-soluble nucleotides and ribonucleic acids from germinating jack pine seeds. *Can. J. For. Res.* **2,** 206–216.

69. Durzan, D. J., and Ramaiah, P. K. (1971). The metabolism of L-proline by jack pine seedlings. *Can. J. Bot.* **49,** 2163–2173.

70. Durzan, D. J., and Richardson, R. G. (1966). The occurrence and role of α-keto-δ-guanidinovaleric acid in *Picea glauca* (Moench) Voss. *Can. J. Biochem.* **44,** 141–143.

71. Durzan, D. J., and Steward, F. C. (1967). The nitrogen metabolism of *Picea glauca* (Moench) Voss. and *Pinus banksiana* Lamb. as influenced by mineral nutrition. *Can. J. Bot.* **45,** 695–710.

72. Dyer, T. A., and Bowman, C. M. (1979). Nucleotide sequences of chloroplast 5S ribosomal ribonucleic acid in flowering plants. *Biochem. J.* **183,** 595–604.

73. Ellis, R. J. (1979). The most abundant protein in the world. *Trends Biochem. Sci.* **4,** 241–244.

74. Farquhar, G. D., Wetselaar, R., and Firth, P. M. (1979). Ammonia volatilization from scenescing leaves of maize. *Science (Washington, D.C.)* **203,** 1257–1258.

75. Fedec, P., and Cossins, E. A. (1976). Effect of gibberellic acid and L-methionine on C-1 metabolism in aerated carrot discs. *Phytochemistry* **15,** 1819–1823.

76. Fink, D. J., Falb, R. D., and Bean, M. K. (1978). Production of α-keto acids by immobilized enzymes. *Am. Inst. Chem. Eng. Symp. Ser.,* No. 172, **74,** 18–24.

77. Fisher, H. W., and Williams, R. C. (1979). Electron microscopic visualization of nucleic acids and of their complexes with proteins. *Annu. Rev. Biochem.* **48,** 649–679.

78. Folk, J. E. (1980). Transglutaminases. *Annu. Rev. Biochem.* **49,** 517–531.

79. Fowden, L. (1976). Amino acids occurrence, biosynthesis and analogue behavior in plants. *Perspect. Exp. Biol.* **2**, 263–272.

80. Fowden, L. (1976). The world of a chemist among plants. *Interdiscip. Sci. Rev.* **1**, 63–71.

81. Fowden, L. (1974). Nonprotein amino acids from plants: distribution, biosynthesis, and analog function. *Recent Adv. Phytochem.* **8**, 95–122.

82. Fowden, L. (1972). Amino acid complement of plants. *Phytochemistry* **11**, 2271–2276.

83. Fowden, L. (1972). Fluoroamino acids and protein synthesis. *Ciba Found. Symp.*, 141–159.

84. Fowden, L. (1970). The non-protein amino acids of plants. *Prog. Phytochem.* **2**, 203–266.

85. Fowden, L. (1965). Amino acid biosynthesis. *In* "Biosynthetic Pathways in Higher Plants" (J. B. Pridham and T. Swain, eds.), pp. 73–99. Academic Press, New York.

86. Fowden, L., and Frankton, J. B. (1968). Specificity of amino acyl-sRNA synthetases with special reference to arginine activation. *Phytochemistry* **7**, 1077–1086.

87. Fowden, L., Lea, P. J., and Norris, R. D. (1972). Amino acid analogues and protein synthesis in plants. *Symp. Biol. Hung.* **13**, 137–145.

88. Fowden, L., Lea, P. J., and Bell, E. A. (1979). The nonprotein amino acids of plants. *Adv. Enzymol.* **50**, 117–175.

89. Fowden, L., Smith, I. K., and Dunnill, P. M. (1968). Some observations on the specificity of amino acid biosynthesis and incorporation into proteins. *In* "Recent Aspects of Nitrogen Metabolism in Plants" (E. J. Hewitt and C. V. Cutting, eds.) pp. 165–177. Academic Press, New York.

90. Gehrke, C. W., Zumwalt, R. W., Stalling, D. L., Roach, D., and Aue, W. A. (1971). A search for amino acids in Apollo 11 and 12 lunar fines. *J. Chromatogs.* **59**, 305–319.

91. Giovanelli, J., Mudd, H. S., and Datko, H. S. (1974). Homoserine esterification in green plants. *Plant Physiol.* **54**, 725–736.

92. Gordon, A. H. (1979). Electrophoresis and chromatography of amino acids and proteins. *Ann. N.Y. Acad. Sci.* **325**, 95–105.

93. Grant, V. (1975). "Genetics of Flowering Plants." Columbia Univ. Press, New York.

94. Greenstein, J. P., and Winitz, M. (1961). "Chemistry of the Amino Acids," 3 vol. Wiley, New York.

95. Granroth, B. (1970). Biosynthesis and decomposition of cysteine derivatives in onion and other *Allium* species. *Ann. Acad. Sci. Fenn. Ser. A2* **154**, 71.

96. Guitton, Y. (1963). The nitrogen metabolism of gymnosperms. Enzymatic biosynthesis of asparagine from α-ketosuccinamic acid in *Pinus pinea. C. R. Hebd. Seances Acad. Sci. Ser. D* **257**, 506–507.

97. Guitton, Y. (1957). Sur le métabolisme azoté des gymnospermes. Présence de l'arginase dans les graines. *C.R. Hebd. Seances Acad. Sci. Ser. D* **245**, 1157–1160.

98. Hahlbrock, K., and Brisebach, H. (1979). Enzymic controls in the biosynthesis of lignin and flavonoids. *Annu. Rev. Plant Physiol.* **30**, 105–130.

99. Halsam, E. (1974). "The Shikimate Pathway." Butterworth, London.

100. Hanawalt, P. C. (1980). Molecules to living cells. *Read. Sci. Am.*

101. Hangarter, R. P., Peterson, M. D., and Good, N. E. (1980). Biological activities of indoleacetylamino acids and their use as auxins in tissue culture. *Plant Physiol.* **65**, 761–767.

102. Hartman, C. S. (1970). Purines and pyrimidines. *In* "Metabolic Pathways" (D. M. Greenberg, ed.), Vol. 4, pp. 1–68. Academic Press, New York.

103. Hatch, M. D. (1971). Mechanism and function of the C_4 pathway of photosynthesis. *In* "Photosynthesis and Photorespiration" (M. D. Hatch, C. B. Osmond, and R. O. Slayter, eds.), pp. 139–152. Wiley (Interscience), New York.

104. Heber, U. (1974). Metabolite exchange between chloroplasts and cytoplasm. *Annu. Rev. Plant Physiol.* **25**, 393–421.
105. Hegarty, M. P. (1978). Toxic amino acids of plant origin. *In* "Effects of Poisonous Plants on Livestock" (R. F. Keeler, K. R. Van Kampen, and L. R. James, eds.), pp. 575–585. Academic Press, New York.
105a. Hellebust, J. A., and Bidwell, R. G. S. (1964). Protein turnover in attached wheat and tobacco leaves. *Can. J. Bot.* **42**, 1–12.
105b. Hellebust, J. A., and Bidwell, R. G. S. (1963). Protein turnover in wheat and snapdragon leaves: preliminary investigations. *Can. J. Bot.* **41**, 969–983.
106. Heller, W. (1973). Paläobiochemische Aspekte der Evolution. *Naturwissenschaften* **60**, 460–468.
107. Hewitt, E. J. (1974). Aspects of trace element requirements in plants and microorganisms: the metalloenzymes of nitrate and nitrite reduction. *MTP Int. Rev. Sci. Biochem. Ser. One* **11**, 199–245.
108. Hewitt, E. J., Hucklesby, D. P., and Notton, B. A. (1976). Nitrate metabolism. *In* "Plant Biochemistry" (J. Bonner and J. Varner, eds.), pp. 633–681. Academic Press, New York.
109. Hodson, R. C., Williams, S. K., and Davidson, W. R. (1975). Metabolic control of urea catabolism in *Chlamydomonas reinhardi* and *Chlorella pyrenoidosa*. *J. Bacteriol.* **121**, 1022–1035.
110. Hoffmann, D., Hecht, S. S., Ornaf, R. M., and Wynders, E. L. (1974). N'-Nitrosonornicotine in tobacco. *Science (Washington, D.C.)* **186**, 265–267.
111. Holley, R. J., Apgar, J., Everett, G., Madison, J., Marquisee, M., Merrill, S., Penswick, J., and Zamir, A. (1965). Structure of a ribonucleic acid. *Science (Washington, D.C.)* **147**, 1462–1465.
111a. Hoffman, N. E., Fu, J.-R., and Young, S. F. (1983). Identification and metabolism of 1-(malonylamino)cyclopropane-1-carboxylic acid in germinating peanut seeds. *Plant Physiol.* **71**, 197–199.
112. Holzer, H. (1980). Control of proteolysis. *Annu. Rev. Biochem.* **49**, 63–91.
113. Horowitz, N. H. (1979). Genetics and the synthesis of proteins. *Ann. N.Y. Acad. Sci.* **325**, 253–266.
114. Isenberg, I. (1979). Histones. *Annu. Rev. Plant Physiol.* **48**, 159–191.
115. Israel, H. W., and Steward, F. C. (1966). The fine structure of quiescent and growing cells: its relation to growth induction. *Ann. Bot. (London)* N.S. **30**, 63–69.
116. Israel, H. W., and Steward, F. C. (1967). The fine structure and development of plastids in cultured cells of *Daucus carota*. *Ann. Bot. (London)* N.S. **31**, 1–18.
117. Itakura, K., Hirose, T., Crea, R., Riggs, A. D., Heynacker, H. L., Bolivar, F., and Boyer, H. W. (1978). Expression in *Escherichia coli* of a chemically synthesized gene. *Science (Washington, D.C.)* **198**, 1056–1063.
118. Jencks, W. P. (1971). Infrared measurements in aqueous media. *Methods Enzymol.* **6**, 918–919.
119. Jones, M. E. (1971). Regulation of pyrimidine and arginine biosynthesis in mammals. *Adv. Enzyme Regul.* **9**, 19–49.
120. Jones, M. E. (1963). Carbamyl phosphate. *Science (Washington, D.C.)* **140**, 1373–1379.
121. Jukes, T. H. (1974). The "intruder" hypothesis and selection against arginine. *Biochem. Biophys. Res. Commun.* **58**, 80–84.
122. Jukes, T. H. (1978). The amino acid code. *Adv. Enzymol.* **47**, 375–432.
123. Khorana, H. G. (1968). Synthesis in the study of nucleic acids. *Biochem. J.* **109**, 709–725.
124. Khorana, H. G. (1979). Total synthesis of a gene. *Science (Washington, D.C.)* **203**, 614–625.

125. Kim, S.-H. (1976). Three-dimensional structure of transfer RNA. *Prog. Nucleic Acid Res. Mol. Biol.* **17**, 181–216.

126. Kleczkowski, K., and Wielgat, B. (1968). Carbamoylation of putrescine in plant material. *Bull. Acad. Sci. Pol. Cl. 2* **16**, 521–526.

127. Klee, C. B., Crouch, T. H., and Richman, P. G. (1980). Calmodulin. *Annu. Rev. Biochem.* **49**, 489–515.

128. Kleindienst, M. R., Clark, J. D., and Lee, C. (1977). Amino acids in fossil woods. *Nature (London)* **267**, 468.

129. Knowles, R. (1969). Microorganisms and nitrogen in the raw humus of a black spruce forest. *Trend* **15**, 13–17.

130. Kornberg, R. D. (1974). Chromatin structure: a repeating unit of histones and DNA. *Science (Washington, D.C.)* **184**, 868–871.

131. Krebs, H. A. (1973). Pyridine nucleotides and rate control. *Symp. Soc. Exp. Biol.* **27**, 299–318.

132. Krebs, E. G., and Beavo, J. A. (1979). Phosphorylation–dephosphorylation of enzymes. *Annu. Rev. Biochem.* **48**, 923–959.

133. Kupchan, S. M. (1975). Advances in the chemistry of tumor-inhibitory natural products. *Recent Adv. Phytochem.* **9**, 167–188.

134. Kurland, C. G. (1977). Structure and function of the bacterial ribosome. *Annu. Rev. Biochem.* **46**, 173–200.

135. Lange, L. G., III, Riordan, J. F., and Vallee, B. F. (1974). Functional arginyl residues as NADH binding sites of alcohol dehydrogenases. *Biochemistry* **13**, 4361–4370.

136. Lark, K. G., and Cress, D. E. (1978). Cell division and DNA synthesis in plant cells. *In* "Frontiers of Plant Tissue Culture 1978" (T. A. Thorpe, ed.), pp. 179–189. Int. Assoc. Plant Tissue Culture Univ. of Calgary, Alberta, Canada.

137. Larkins, B. A., Pedersen, K., Handa, A. K., Hurkman, W. J., and Smith, L. D. (1979). Synthesis and processing of maize storage proteins in *Xenopus laevis* oocytes. *Proc. Natl. Acad. Sci. USA* **76**, 6448–6452.

138. Lea, P. J., and Fowden, L. (1975). The purification and properties of glutamine-dependent asparagine synthetase isolated from *Lupinus albus. Proc. R. Soc. London Ser. B* **192**, 13–26.

139. Lezius, A., Ringelmann, E., and Lynen, F. (1963). Zur biochemischen Funktion des Biotins. IV. Die biosynthese des Biotins. *Biochem. Z.* **336**, 510–525.

140. Lijinsky, W. (1976). Health problems associated with nitrites and nitrosomines. *Ambio* **5**, 67–72.

141. Lijinsky, W. (1976). Interaction with nucleic acids of carcinogenic and mutagenic *N*-nitroso compounds. *Prog. Nucleic Acid. Res. Mol. Biol.* **17**, 247–269.

142. Liu, K. D. F., and Huang, A. H. C. (1977). Subcellular localization and developmental changes of asparatate-α-ketoglutarate transaminase isoenzymes in the cotyledons of cucumber seedlings. *Plant Physiol.* **59**, 777–782.

143. Lodish, H. (1976). Translation control of protein synthesis. *Annu. Rev. Biochem.* **45**, 39–72.

144. Löffelhardt, W. (1977). The biosynthesis of phenylacetic acids in the blue-green alga *Anacystis nidulans:* evidence for the involvement of a thylakoid-bound L-amino acid oxidase. *Z. Naturforsch.* **32**, 345–350.

145. Long, E. O., and Dawid, I. B. (1980). Repeated genes in eukaryotes. *Annu. Rev. Biochem.* **49**, 727–764.

146. Luckner, M. (1980). Expression and control of secondary metabolism. *In* "Secondary Plant Products" (E. A. Bell and B. V. Charlwood, eds.), pp. 23–63. Springer-Verlag, New York.

147. Lueck, J. D., Herrman, J. L., and Nordlie, R. C. (1972). General kinetic mechanism of microsomal carbamyl phosphate: glucose phosphotransferase, glucose-6-phosphatase and other activities. *Biochemistry* **11**, 2792–2799.

147a. Mangat, B. S., Levin, W. B., and Bidwell, R. G. S. (1974). The extent of respiration in illuminated leaves and its control by ATP levels. *Can. J. Bot.* **52**, 673–681.

148. Margna, U. (1977). Control at the level of substrate supply—an alternative in the regulation of phenylpropanoid accumulation in plant cells. *Phytochemistry* **16**, 419–426.

149. Marx, J. L. (1980). Newly made proteins zip through the cell. *Science (Washington, D.C.)* **207**, 164–167.

150. Mauger, A. B., and Witkop, B. (1966). Analogs and homologs of proline and hydroxyproline. *Chem. Rev.* **66**, 47–86.

151. McKee, H. S. (1962). "Nitrogen Metabolism in Plants." Oxford Univ. Press (Clarendon), London and New York.

152. Mebs, D. (1973). Chemistry of animal venoms, poisons and toxins. *Experientia* **29/11**, 1328–1334.

153. Meister, A. (1975). Biochemistry of glutathione. In "Metabolic Pathways" (D. M. Greenberg, ed.), Vol. 7, pp. 101–188. Academic Press, New York.

154. Meister, A. (1965). "Biochemistry of the Amino Acids," 2nd ed. Academic Press, New York.

155. Mia, A. J., and Durzan, D. J. (1977). Cytochemical and subcellular organization of root apical meristems of dry and germinating jack pine embryos. *Can. J. For. Res.* **7**, 263–276.

156. Mia, A. J., and Durzan, D. J. (1974). Cytochemical and subcellular organization of the shoot apical meristem of dry and germinating jack pine embryos. *Can. J. For. Res.* **4**, 39–54.

157. Miflin, B. J. (1973). Amino acid biosynthesis and its control in plants. In "Biosynthesis and Its Control in Plants" (B. V. Milborrow, ed.), pp. 49–68. Academic Press, New York.

158. Miflin, B. J. (1976). Metabolic control of biosynthesis of nutritionally essential amino acids. In "Genetic Improvement of Seed Proteins," pp. 135–155. Natl. Res. Counc. & Natl. Acad. Sci. USA.

159. Miflin, B. J., (ed.) (1980). "Amino Acids and Derivatives." Academic Press, New York.

160. Miflin, B. J., and Lea, P. J. (1977). Amino acid metabolism. *Annu. Rev. Plant Physiol.* **28**, 299–329.

161. Miller, J. A. (1973). Naturally occurring substances that can induce tumors. In "Toxicants Occurring Naturally in Foods," pp. 508–549. Natl. Res. Counc. & Natl. Acad. Sci. USA, Washington, D.C.

162. Miller, O. L., Jr. (1972). Visualization of RNA synthesis on chromosomes. *Int. Rev. Cytol.* **33**, 1–25.

163. Miller, O. L., Jr., and Beatty, B. R. (1969). Visualization of nucleolar genes. *Science (Washington, D.C.)* **164**, 955–957.

164. Mitchell, D. J., and Bidwell, R. G. S. (1970). Synthesis and metabolism of homoserine in developing pea seedlings. *Can. J. Bot.* **48**, 2037–2042.

165. Moreno, U. (1970). "Physiological Investigations on the Potato Plant with Special Reference to the Effects of Different Environments." Ph.D. Thesis, Cornell Univ., Ithaca, New York.

166. Mothes, K. (1961). The metabolism of urea and ureides. *Can. J. Bot.* **39**, 1785–1807.

167. Mothes, K. (1972). Some remarks about cytokinins. *Fiziol. Rast. (Moscow)* **19**, 1011–1022.

168. Mothes, K. (1929). Physiologische Untersuchungen über das Asparagin und das Arginin in Coniferen. *Planta* **7**, 584–649.

169. Mott, R. L., and Steward, F. C. (1972). Solute accumulation in plant cells. V. An aspect of nutrition and development. *Ann. Bot. (London) N.S.* **36**, 915–937.

170. Murakoshi, I., and Hatanaka, S. I. (1977). Sulfur containing amino acids in nature. *Yuki Gosei Kagaku (J. Synth. Org. Chem. Jpn.)* **35**, 343–353.

171. Naylor, A. W. (1959). Interrelations of ornithine, citrulline and arginine in plants. *Symp. Soc. Exp. Biol.* **13**, 193–209.

172. Nelson, O. E. (1979). Inheritance of amino acid content in cereals. *Proc. IAEA Int. Symp.* **1** 79–88.

173. Nevers, P., and Saedler, H. (1977). Transposable genetic elements as agents of gene instability and chromosomal rearrangements. *Nature (London)* **268**, 109–115.

174. Northcote, D. H. (1974). Plant biochemistry. *MTP Int. Rev. Sci. Biochem. Ser. One* **2**, 1–287.

175. Oaks, A., and Bidwell, R. G. S. (1970). Compartmentation of intermediary metabolites. *Annu. Rev. Plant Physiol.* **21**, 43–66.

176. Ochoa, S., and de Haro, S. (1979). Regulation of protein synthesis in eukaryotes. *Annu. Rev. Biochem.* **48**, 549–580.

177. Ogawa, T., and Okazaki, T. (1980). Discontinuous DNA replication. *Annu. Rev. Biochem.* **49**, 421–457.

178. Olins, A. L., Carlson, R. D., and Olins, D. E. (1975). Visualization of chromatin substructure: *v* bodies. *J. Cell Biol.* **64**, 528–537.

179. O'Neal, T. D., and Naylor, A. W. (1976). Some regulatory properties of pea leaf carbamyl phosphate synthetase. *Plant Physiol.* **63**, 23–28.

180. Paik, W. K., and Kim, S. (1975). Protein methylation; chemical, enzymological and biological significance. *Adv. Enzymol.* **42**, 227–286.

181. Païs, M., Jorreau, F.-X., Sierra, M. G., Moscaretti, O. H., Ruveda, E. A., Chang, C.-J., Hagaman, E. W., and Wenker, E. (1979). Carbon-13 NMR analysis of cyclic peptide alkaloids. *Phytochemistry* **18**, 1869–1872.

182. Palade, G. (1975). Intracellular aspects of the process of protein synthesis. *Science (Washington, D.C.)* **189**, 347–358.

183. Penning de Vries, F. W. T. (1975). The cost of maintenance processes in plant cells. *Ann. Bot. (London)* **39**, 77–92.

184. Pestka, S. (1976). Insights into protein biosynthesis and ribosome function through inhibitors. *Prog. Nucleic Acid. Res. Mol. Biol.* **17**, 217–245.

185. Pharis, R. P., Barnes, R. L., and Naylor, A. W. (1964). Effects of the nitrogen level calcium level and nitrogen source upon the growth and composition of *Pinus taeda* L. *Physiol. Plant.* **17**, 560–572.

186. Pirie, N. W. (1973). Plants as sources of unconventional protein foods. *In* "The Biological Efficiency of Protein Production" (J. G. W. Jones, ed.), pp. 101–118. Cambridge Univ. Press, London.

187. Pitel, J. A., and Durzan, D. J. (1980). Chromosomal proteins of conifers. III. Metabolism of histones and non-histone chromosomal proteins in jack pine (*Pinus banksiana* Lamb.) during germination. *Physiol. Plant.* **8**, 137–194.

188. Pitel, J. A., and Durzan, D. J. (1978). Chromosomal proteins of conifers. I. Extraction and characterization of histones and nonhistone chromosomal proteins from dry seeds of conifers. *Can. J. Bot.* **56**, 1915–1927.

189. Pitel, J. A., and Durzan, D. J. (1978). Chromosomal proteins of conifers. II. Tissue

specificity of the chromosomal proteins of jack pine (*Pinus banksiana* Lamb.). *Can. J. Bot.* **56**, 1928–1931.

190. Pitel, J. A., and Durzan, D. J. (1978). Changes in the composition of the soluble and chromosomal proteins of jack pine (*Pinus banksiana* Lamb.) seedlings induced by fenitrothion. *Environ. Exp. Bot.* **18**, 153–162.

191. Pitel, J. A., and Durzan, D. J. (1975). Pyrimidine metabolism in seeds and seedlings of jack pine (*Pinus banksiana*). *Can. J. Bot.* **53**, 673–686.

192. Price, H. J., Sparrow, A. H., and Nauman, A. F. (1973). Evolutionary and developmental consideration of the variability of nuclear parameters in higher plants. I. Genome volume, interphase chromosome volume, and estimated DNA content of 236 Gymnosperms. *Brookhaven Symp. Biol.* **25**, 390–421.

193. Pusheva, M. A., and Khoreva, S. L. (1977). Amino acid composition of polynucleolide complexes isolated from algae. *Microbiology (Engl. Transl. of Mikrobiologiya)* **46**, 49–52.

193a. Quigley, G. J., and Rich, A. (1976). Structural domains of transfer RNA molecules. *Science (Washington, D.C.)* **194**, 796–806.

194. Ramaiah, P. K., Durzan, D. J., and Mia, A. J. (1971). Amino acid, soluble protein, and isoenzyme patterns of peroxidase during the germination of jack pine. *Can. J. Bot.* **50**, 2151–2161.

195. Ratner, S. (1973). Enzymes of arginine and urea synthesis. *Adv. Enzymol.* **39**, 1–90.

196. Rehr, S. S., Bell, E. A., Janzen, D. H., and Feeny, P. P. (1973). Insecticidal amino acids in legume seeds. *Biochem. Syst.* **1**, 63–67.

197. Reinbothe, H., and Mothes, K. (1962). Urea, ureides and guanidines in plants. *Annu. Rev. Plant Physiol.* **13**, 129–150.

198. Ressler, C. (1975). Plant neurotoxins (lathyrogens and cyanogens). *Recent Adv. Phytochem.* **9**, 151–166.

199. Rhodes, D., Sims, A. P., and Folkes, B. F. (1980). Pathway of ammonia assimilation in illuminated *Lemna minor*. *Phytochemistry* **19**, 357–365.

199a. Rich, A. (1976). Structural domains of transfer RNA molecules. *Science (Washington, D.C.)* **194**, 796–806.

200. Richter, D., and Isono, K. (1977). The mechanism of protein synthesis—initiation, elongation and termination in translation of genetic messages. *Curr. Top. Microbiol. Immunol.* **11**, 83–125.

201. Robinson, A. B., and Rudd, C. J. (1974). Deamidation of glutaminyl and asparaginyl residues in peptides and proteins. *Curr. Top. Cell. Regul.* **8**, 247–295.

202. Roca, W. M. (1972). "The Development of Certain Crop Plants as Affected by Environments." Ph.D. Thesis, Cornell Univ., Ithaca, New York.

203. Rognes, S. E. (1972). A glutamine-dependent asparagine synthetase from yellow lupine seedlings. *PEBS Lett.* **10**, 62–66.

204. Rognes, S. E. (1975). Glutamine-dependent asparagine synthetase from *Lupinus luteus*. *Phytochemistry* **14**, 1975–1981.

205. Ross, C. (1965). Comparison of incorporation metabolism of RNA pyrimidine nucleotide precursors in leaf tissues. *Plant Physiol.* **40**, 65–73.

206. Ross, C., and Cole, C. V. (1968). Metabolism of cytidine and uridine in bean leaves. *Plant Physiol.* **43**, 1227–1231.

207. Ross, C., and Murray, M. G. (1917). Development of pyrimidine-metabolizing enzymes in cotyledons of germinating peas. *Plant Physiol.* **48**, 626–630.

208. Roublelakis, K. A., and Kliewer, W. M., (1978). Enzymes of the Krebs–Henseleit cycle in *Vitus vinfera* L. I. Ornithine carbamoyl-transferase: isolation and some properties. II. Argininosuccinate synthetase and lyase. III. *In vivo* and *in vitro* studies of arginase. *Plant Physiol.* **62**, 337–339, 340–343, 344–347.

209. Ruggieri, G. D. (1976). Drugs from the sea. *Science (Washington, D.C.)* **194,** 491–500.
210. Schilperoort, R. H., and Bomhoff, G. H. (1974). Crown gall. A model from tumor research and genetic engineering. *In* "Genetic Manipulations with Plant Material" (L. Ledoux, ed.), pp. 141–162. Plenum, New York.
211. Schimmel, P. R., and Söll, D. (1979). Aminoacyl-tRNA synthetases: general features and recognition of transfer RNAs. *Annu. Rev. Biochem.* **48,** 601–648.
212. Schwarz, Z., and Kössel, H. (1980). The primary structure of 16S rRNA from *Zea mays* chloroplast is homologous to *E. coli* 16S rRNA. *Nature (London)* **283,** 739–742.
213. Shargool, P. D. (1975). Degradation of argininosuccinate lyase by a protease synthesized in soybean cell suspension cultures. *Plant Physiol.* **55,** 632–635.
214. Shargool, P. D. (1973). The response of soybean argininosuccinate synthetase to different energy charge values. *FEBS Lett.* **33,** 348–350.
215. Sharon, N. (1979). Lectins. In "Glycoconjugate Research" (J. D. Gregory and R. W. Jeanloz, eds.), Vol. 1, pp. 459–491. Academic Press, New York.
216. Shrift, A. (1973). Metabolism of selenium by plants and microorganisms. *In* "Organic Selenium Compounds, Their Chemistry and Biology" (L. Klayman and W. H. H. Gunther, eds.), pp. 763–814. Wiley (Interscience), New York.
217. Sinsheimer, R. L. (1977). Recombinant DNA. *Annu. Rev. Biochem.* **46,** 415–438.
218. Skokut, T. A., Wolk, P. C., Thomas, J., Meeks, J. C., and Shaffer, P. W. (1978). Initial organic products of assimilation of [^{13}N]ammonium and [^{13}N]nitrate by tobacco cells cultured on different sources of nitrogen. *Plant Physiol.* **62,** 299–304.
219. Smith, E. L. (1979). Amino acid sequences of proteins—the beginnings. *Ann. N.Y. Acad. Sci.* **325,** 107–119.
220. Smith, T. A. (1980). Plant amines. *Encycl. Plant Physiol. New Ser.* **8,** 433–460.
221. Smith, T. A. (1977). Phenethylamine and related compounds in plants. *Phytochemistry* **16,** 9–18.
222. Smith, T. A. (1971). The occurrence, metabolism and functions of amines in plants. *Biol. Rev. Cambridge Philos. Soc.* **46,** 201–241.
223. Smith, T. A. (1975). Recent advances in the biochemistry of plant amines. *Phytochemistry* **14,** 865–890.
224. Solomonson, L. P., and Spehar, A. M. (1977). Model for the regulation of nitrate assimilation. *Nature (London)* **265,** 373–375.
224a. Somerville, C. R., and Ogren, W. L. (1980). Inhibition of photosynthesis in *Arabidopsis* mutants lacking leaf glutamate synthase activity. *Nature (London)* **286,** 257–259.
225. Sparrow, A. H., and Nauman, A. F. (1973). Evolutionary changes in genome and chromosome sizes and in DNA content in the grasses. *Brookhaven Symp. Biol.* **25,** 367–389.
226. Stafford, H. A. (1974). The metabolism of aromatic compounds. *Annu. Rev. Plant Physiol.* **25,** 459–486.
227. Steward, F. C. (1968). "Growth and Organization in Plants." Addison-Wesley, Reading, Massachusetts.
228. Steward, F. C. (1970). From cultured cells to whole plants; the induction and control of their growth and morphogenesis. The Croonian Lecture 1969. *Proc. R. Soc. London Ser. B* **175,** 1–30.
229. Steward, F. C., Bidwell, R. G. S., and Yemm, E. W. (1958). Nitrogen metabolism, respiration and growth of cultured plant tissue. IV. The impact of growth on protein metabolism and respiration of carrot tissue explants. General discussion of results. *J. Exp. Bot.* **9,** 285–305.
230. Steward, F. C., and Durzan, D. J. (1965). Metabolism of organic nitrogenous compounds. *In* "Plant Physiology" (F. C. Steward, ed.), Vol. 4A, pp. 379–686. Academic Press, New York.

231. Steward, F. C., and Krikorian, A. D. (1971). "Plants, Chemicals and Chemicals and Growth." Academic Press, New York.

232. Steward, F. C., and Krikorian, A. D. (1979). Problems and potentialities of cultured cells in retrospect and prospect. *In* "Plant Cell and Tissue Culture: Principles and Applications" (W. R. Sharp, P. O. Larsen, E. F. Paddock, and V. Raghaven, eds.), pp. 221–262. Ohio State Univ. Press, Columbus.

233. Steward, F. C., Moreno, U., and Roca, W. M. (1981). Growth, form and composition of potato plants as affected by environment. *Ann. Bot. (London)* **48** (Suppl 2.), 1–44.

234. Steward, F. C., Mott, R. L., and Rao, K. V. N. (1973). Investigations on the growth and metabolism of cultured explants of *Daucus carota*. V. Effects of trace elements and growth factors on the solutes accumulated. *Planta* **111**, 219–243.

235. Steward, F. C., and Rao, K. V. N. (1970). Investigations on the growth and metabolism of cultured explants of *Daucus carota*. III. The range of responses induced in carrot explants by exogenous growth factors and by trace elements. *Planta* **91**, 129–145.

236. Steward, F. C., and Rao, K. V. N. (1971). Investigations on the growth and metabolism of cultured explants of *Daucus carota*. IV. Effects of iron, molybdenum and the components of growth promoting systems and their interactions. *Planta* **99**, 240–264.

237. Street, H. E. (1979). Embryogenesis and chemically induced organogenesis. *In* "Plant Cell and Tissue Culture" (W. R. Sharp, P. O. Larsen, E. F. Paddock, and V. Raghaven, eds.), pp. 123–153. Ohio State Univ. Press, Columbus.

238. Street, H. E., Gould, A. R., and King, J. (1976). Nitrogen assimilation and protein synthesis in plant cell cultures. *In* "Perspectives in Experimental Biology" (N. Sunderland, ed.), pp. 337–356. Pergamon, Oxford.

239. Strong, F. M. (1966). Naturally occurring toxic factors in plants and animals used as food. *Can. Med. Assoc. J.* **94**, 568–573.

240. Sumner, J. B., and Sommers, G. F. (1953). "Chemistry and Methods of Enzymes," 3rd ed. Academic Press, New York.

241. Sussex, I. M., and Dale, R. M. K. (1980). Hormonal control of storage protein synthesis in *Phaseolus vulgaris*. *In* "The Plant Seed" (I. Rubenstein, B. G. Gengenbach, R. L. Phillips, and C. E. Green, Jr., eds.), pp. 129–141. Academic Press, New York.

242. Sussex, I. M., Dale, R. M. K., and Couch, M. L. (1980). Developmental regulation of storage protein in seeds. *In* "Genome Organization and Expression in Plants" (C. J. Leaver, ed.), pp. 283–289. Plenum, New York.

242a. Sutcliffe, J. F., and Pate, J. S. (eds.) (1977). "The Physiology of the Garden Pea." Academic Press, New York and London.

243. Suzuki, T., and Takahashi, E. (1975). Metabolism of xanthine and hypoxanthine in the tea plant (*Thea sinensis* L.). *Biochem. J.* **146**, 79–85.

244. Suzuki, T., and Takahashi, E. (1975). Biosynthesis of caffeine by tea-leaf extracts. *Biochem. J.* **146**, 87–96.

245. Synge, R. L. M. (1968). Occurrence in plants of amino acid residues chemically bound otherwise than in proteins. *Annu. Rev. Plant Physiol.* **19**, 113–136.

246. Tamm, C. O. (1982). Nitrogen cycling in undisturbed and manipulated boreal forests. *Proc. R. Soc. London Ser. B* **296**, 419–425.

247. Taylor, J. M. (1979). The isolation of eukaryotic messenger RNA. *Annu. Rev. Biochem.* **48**, 681–717.

248. Terano, S., and Suzuki, T. (1978). Formation of β-alanine from spermine and spermidine in maize shoots. *Phytochemistry* **17**, 148–149.

249. Terano, S., and Suzuki, Y. (1978). Biosynthesis of γ-aminobutyrate acid from spermine in maize seedlings. *Phytochemistry* **17**, 550–551.

250. Thatcher, R. C., and Weaver, T. L. (1976). Carbon–nitrogen cycling through micro-bial formamide metabolism. *Science (Washington, D.C.)* **192**, 1234–1235.

251. Thelander, L., and Reichard, P. (1979). Reduction of ribonucleotides. *Annu. Rev. Biochem.* **48**, 133–158.

252. Thoma, F., Koller, T., and Klug, A. (1979). Involvement of histone H1 in the organi-zation of the nucleosome and of the salt dependent superstructures of chromatin. *J. Cell Biol.* **83**, 403–427.

253. Thompson, J. F., Morris, C. J., and Smith, I. K. (1969). New naturally occurring amino acids. *Annu. Rev. Biochem.* **38**, 137–158.

254. Thompson, J. F., and Muenster, A. M. (1971). Separation of the *Chlorella* ATP:urea amido-lyase into two components. *Biochem. Biophys. Res. Commun.* **43**, 1049–1055.

255. Tolbert, N. E. (1973). Compartmentation and control in microbodies. *Symp. Soc. Exp. Biol.* **27**, 215–239.

256. Trewavas, A. (1976). Post-translational modification of proteins by phosphorylation. *Annu. Rev. Plant Physiol.* **27**, 349–374.

257. Trewavas, A. (1977). The control of plant growth by protein kinases. *NATO Adv. Study Inst. Ser. A* **12**, 309–319.

258. Tsai, C. S., and Axelrod, B. (1965). Catabolism of pyrimidines in rape seedlings. *Plant Physiol.* **40**, 39–44.

259. Tsang, M. L.-S., and Schiff, J. A. (1976). Sulfate-reducing pathway in *Escherichia coli* involving bound intermediates. *J. Bacteriol.* **125**, 923–933.

260. Tsuda, Y., Stephani, R. A., and Meister, A. (1971). Direct evidence for the formation of an acyl phosphate by glutamine synthetase. *Biochemistry* **10**, 3186–3189.

261. Tyler, B. (1978). Regulation of the assimilation of nitrogen compounds. *Annu. Rev. Biochem.* **47**, 1127–1162.

262. Tzagoloff, A., Macino, G., and Sebald, W. (1979). Mitochondrial genes and transla-tion products. *Annu. Rev. Biochem.* **48**, 419–441.

263. Umbarger, H. E. (1969). Regulation of amino acid metabolism. *Annu. Rev. Biochem.* **38**, 323–370.

264. Umbarger, H. E. (1978). Amino acid biosynthesis and its regulation. *Annu. Rev. Biochem.* **47**, 533–606.

265. Uy, R., and Wold, F. (1977). Posttranslational covalent modification of proteins. *Science (Washington, D.C.)* **198**, 890–896.

266. Van Thoai, N. (1965). Nitrogenous bases. *In* "Comprehensive Biochemistry" (M. Florkin and E. H. Stotz, eds.), Vol. 6, pp. 208–253. Elsevier, New York.

267. Varner, J. E. (1974). Deuterium oxide as a tool for the study of amino acid metabo-lism. *What's New Plant Physiol.* **6**, 1–5.

268. Varner, J. E. (1977). Hormonal control of protein synthesis. *In* "Nucleic Acids and Protein Synthesis in Plants" (L. Bogorad, ed.), Vol. 12, pp. 293–307. Plenum, New York.

269. Vickery, H. B. (1972). The history of the discovery of the amino acids. II. A review of amino acids described since 1931 as components of native protein. *Adv. Protein Chem.* **26**, 82–171.

270. Viets, F. G. (1975). The environmental impact of fertilizers. *CRC Crit. Rev. Environ. Control* **5**, 423–453.

271. Virtanen, A. I. (1961). Some aspects of amino acid synthesis in plants and related subjects. *Annu. Rev. Plant Physiol.* **12**, 1–12.

272. Virtanen, A. I., and Kari, S. (1955). Free amino acids in pollen. *Acta Chem. Scand.* **9**, 1548–1551.

273. Vogel, J. H., and Vogel, R. H. (1974). Enzymes of arginine biosynthesis and their repressive control. *Adv. Enzymol.* **40**, 65–90.

274. Volpe, P. (1976). The gene expression during the life cycle. *Horiz. Biochem. Biophys.* **2,** 285–340.
275. Waley, S. G. (1966). Naturally occurring peptides. *Adv. Protein Chem.* **21,** 1–112.
276. Walker, J. M., Gooderham, K., and Johns, E. W. (1979). The isolation and partial sequence of peptides produced by cyanogen bromide cleavage of carp-thymus non-histone chromosomal high-mobility-group protein 2. *Biochem. J.* **181,** 659–665.
277. Wallis, M. (1974). On the frequency of arginine in proteins and its implications for molecular evolution. *Biochem. Biophys. Res. Commun.* **56,** 711–716.
278. Wallsgrove, R. M., Lea, P. J., and Miflin, B. J. (1979). Distribution of the enzymes of nitrogen assimilation within the pea leaf cell. *Plant physiol.* **63,** 232–236.
279. Weber, G., Prajda, N., and Williams, J. C. (1974). Regulation of key enzymes: strategy in reprogramming of gene expression. *Symp. Biol. Hung.* **18,** 123–142.
280. Weinstein, B. (ed.) (1974). "Chemistry and Biochemistry of Amino Acids, Peptides, and Proteins: A Survey of Recent Developments," 3 vols. Dekker, New York.
281. Wessels, B. W., McKean, D. J., Lien, N. C., Shinnick, C., DeLuca, P. M., and Smithies, O. (1978). Amino acid sequence determination of proteins labeled in tritium gas by microwave discharge. *Radiat. Res.* **74,** 35–50.
282. Whitney, P. A., Cooper, T. G., and Magasanik, B. (1973). The induction of urea carboxylase and allophanate hydrolase in *Saccharomyces cerevisiae. J. Biol. Chem.* **248,** 6203–6209.
283. Widholm, J. M. (1972). Anthranilate synthetase from 5-methyl-tryptophan-suscepti-ble and resistant cultured *Daucus carota* cells. *Biochim. Biophys. Acta* **279,** 48–57.
284. Widholm, J. M. (1972). Cultured *Nicotiana tabacum* cells with an altered anthranilate synthetase which is less sensitive to feedback inhibition. *Biochim. Biophys. Acta* **261,** 52–58.
285. Widholm, J. M. (1977). Selection and charterization of amino acid analogue resistant plant-cell cultures. *Crop Sci.* **17,** 597–600.
286. Woese, C. R. (1973). Evolution of the genetic code. *Naturwissenschaften* **60,** 447–459.
287. Wong, K. F., and Dennis, D. T. (1973). Aspartokinase in *Lemna minor* L. *Plant Physiol.* **51,** 327–331.
288. Wood, H. G., and Zwolinski, G. K. (1976). Transcarboxylase: role of biotin, metals, and subunits in the reaction and its quaternary structure. *CRC Crit. Rev. Biochem.* **4,** 47–121.
289. Wool, I. G. (1979). The structure and function of eukaryotic ribosomes. *Annu. Rev. Biochem.* **48,** 719–754.
290. Yang, E.-M., Montoya, A. L., Nester, E. W., and Gordon, M. P. (1980). Plant tumor reversal associated with the loss of foreign DNA. *In Vitro* **16,** 87–92.
291. Yeoman, M. M. (ed.) (1976). "Cell Division in Higher Plants." Academic Press, New York.
292. Yeoman, M. M., and Aitchison, P. A. (1976). Molecular events of the cell cycle: a preparation for division. *In* "Cell Division in Higher Plants" (M. M. Yeoman, ed.), pp. 111–133. Academic Press, New York.
293. Yon, R. J. (1972). Wheat-germ aspartate transcarbamoylase. Kinetic behaviour sug-gesting an allosteric mechanism of regulation. *Biochem. J.* **128,** 311–320.
294. Zelitch, I. (1975). Improving the efficiency of photosynthesis. *Science (Washington, D.C.)* **188,** 676–633.
295. Kolata, G. (1981). Z-DNA: from the crystal to the fly. *Science (Washington, D.C.)* **214,** 1108–1110.
296. Marcus, A. (1981). "Proteins and Nucleic Acids." Academic Press, New York.
297. Steward, F. C., and Thompson, J. F. (1950). Photosynthesis and respiration: a rein-terpretation of recent work with radioactive carbon. *Nature (London)* **166,** 593–596.

APPENDIX I

New Amino Acids and Amides Related to the Protein Amino Acids of Plants and Their Free Occurrence[a]

Name and structure	Occurrence, recognition, etc.	References
Alanine derivatives		
N-Acetylalanine $CH_3CH(NHCOCH_3)COOH$	Amino terminal residue of wheat-germ cytochrome c	1967. F. C. Stevens et al. J. Biol. Chem. **242**, 2764.
D-Alanine	Pisum sativum	1973. M. Fukuda et al. Phytochemistry **12**, 2593.
D-Alanyl-D-alanine	Phalaris tuberosa	1975. J. L. Frahm and R. J. Illman. Phytochemistry **14**, 859.
	Tobacco (Nicotiana) leaves	1973. M. Noma et al. Agric. Biol. Chem. **37**, 2439.
Serine derivatives		
Galactoserine	In medium from tomato (Lycopersicon esculentum) cell suspensions (glycoprotein)	1973. D. T. A. Lamport et al. Biochem. J. **133**, 125.
N-Acetylserine	In pea (Pisum) histones	1969. R. J. Delange et al. J. Biol. Chem. **244**, 5669.
Glutamate derivatives		
Pyroglutamyl dipeptides	Mushroom [Agaricus bisporus (A. campestris)]	1970. M. R. Altamura et al. J. Food Sci. **35**, 134.
Lysine derivatives		
N^ε-Methyllysine	Salmonella flagellin	1959. R. P. Ambler and M. W. Rees. Nature (London) **184**, 56.
	Wheat (Triticum) germ	1964. K. Murray. Biochemistry **3**, 10.
	In Sedum acre	1973. E. Leistner and I. D. Spencer, J. Am. Chem. Soc. **95**, 4715.
N^ε-Dimethyllysine	In histone	1967. W. K. Paik and S. Kim. Biochem. Biophys. Res. Commun. **27**, 479.

[a] Acceptance of the occurrence of amino acids actually in proteins should follow rigorous criteria (see 269). Compounds reported in this appendix do not necessarily meet all of these criteria and should therefore be considered as substances "related to" the established protein amino acids, but not as integral parts of the protein molecule, except where this is specifically substantiated.

APPENDIX I   *(Continued)*

Name and structure	Occurrence, recognition, etc.	References
N^{ε}-Trimethyllysine	In histone	1968. K. Hempel *et al.* *Naturwissenschaften* **55,** 37.
	Wheat germ histone	1969. R. Delange *et al. J. Biol. Chem.* **244,** 1385.
	Free in plants	1964. T. Takomoto *et al. Yakugaku Zasshi (J. Pharm. Soc. Jpn.)* **84,** 1176.
N^{ε}-Acetyllysine	Seeds of *Reseda odorata*	1968. P. O. Larsen. *Acta Chem. Scand.* **22,** 1369.
	Two sites in pea (*Pisum*) histone	1969. R. J. DeLange *et al. J. Biol. Chem.* **244,** 5669.
Arginine derivatives L-Citrullinyl-L-arginine	Red algae	1974. K. Miyazawa and K. Ho. *Nippon Suisan Gakkaishi (Bull. Jpn. Soc. Sci. Fish.)* **40,** 815.
	Broad bean (*Vicia faba*) seeds	1976. T. Kasai *et al. Agric. Biol. Chem.* **40,** 2449.
Arginylglutamine	*Cladophora*	1959. S. Makisumi. *J. Biochem.* **46,** 63.
Phenylalanine derivatives *N*-Feruloylglycyl-L-phenylalanine	Sequence in barley (*Hordeum vulgare*) globulins; *N*-feruloylglycine suggested as a "starter" in protein biosynthesis	1973. C. F. Van Sumere. *Phytochemistry* **12,** 407.
	In *Medicago sativa* bulk protein as N-terminal sequence	1980. C. F. Van Sumere *et al. Phytochemistry* **19,** 704.
Proline derivatives Dimethylproline	N-Terminal residue of cytochromes	1977. G. W. Pettigrew and G. M. Smith. *Nature (London)* **265,** 661.
Hydroxyproline derivatives 3-Hydroxyproline	Collagen	1966. J. S. Wolf III *et al. Fed. Proc. Fed. Am. Soc. Exp. Biol.* **25,** 862.

(continued)

APPENDIX I   (*Continued*)

Name and structure	Occurrence, recognition, etc.	References
cis-3,4,*trans*-3,4-Dihydroxy-L-proline	Hydrolysis of diatom cell walls *Navicula pelliculosa*	1969. T. Nakajima and B. E. Volcani. *Science* (Washington, D.C.) **164,** 1400.
Free *cis*- and bound *trans*-4-hydroxy-L-proline	*Santalum album*	1970. R. Kuttan and A. N. Radhakrishnan. *Biochem. J.* **119,** 651.
	Bound trans form in *Santalum*	1976. U. V. Mani and A. N. Radhakrishnan. *Indian J. Biochem. Biophys.* **13,** 13.
Methionine derivative *N*-Formylmethionine	Initiates polypeptide synthesis in bacteria	1964. K. Marcker and F. Sanger. *J. Mol. Biol.* **8,** 835.
	In chloroplasts	1967. J. H. Schwartz *et al. J. Mol. Biol.* **30,** 309. 1968. A. E. Smith and K. Marcker. *J. Mol. Biol.* **38,** 241.
	Wheat (*Triticum*) transformylase	1971. J. P. Leis and E. B. Keller. *Biochemistry* **10,** 889.
Tyrosine derivative Isodityrosine	Cross-linking amino acid in cell-wall glycoprotein	1982. S. C. Fry. *Biochem. J.* **204,** 449.
Other compounds claimed α-Aminobutyric acid	Seed proteins of *Salvia officinalis*	1964. C. H. Breiskorn and J. Glasz. *Naturwissenschaften* **51,** 216.
Ornithine	In peptide from Irish Moss (*Chondrus crispus*)	1958. E. G. Young and D. G. Smith. *J. Biol. Chem.* **233,** 406.
	Bound in flexibacteria and algae	1965. O. Holm-Hansen and R. A. Lewin. *Physiol. Plant.* **18,** 418.
L-α-Amino-δ-hydroxy-amino-valeric acid	Hydrolysates of iron-free ferrichrome	1961. T. Emory and J. B. Neilands. *J. Am. Chem. Soc.* **83,** 1626.

APPENDIX I   (*Continued*)

Name and structure	Occurrence, recognition, etc.	References
S-Farnesylcysteine	At C-terminus of lipopeptide from yeast	1979. A. Sakurai *et al. Proc. Int. Congr. Plant Growth Regul. 10th,* Abstr. No. 505, p. 25.

The literature also reports a number of amino acids other than the well-established protein amino acids that occur in antibiotics and in antimetabolites. These substances arouse claims for new protein amino acids of which at least 50 have been noted; these include β-alanylhistidine, (+) α-aminobutyric acid, D (−) β-aminoisobutyric acid, L-amino-2-methylaminopropionic acid (bound in *Cycas*), L-citrulline, *N*-formylglycine, 3-hydroxy-5-methylproline, sarcosine, and *O*-phosphoserine.

Repeated claims of D-amino acids that occur in nature, usually combined in peptides, especially in bacteria, have been made. Although a list of at least 25 such D-amino acids was compiled, they are not given here in detail because their significance for, or relevance to, normal protein metabolism in plants has not been established. The more recent references to new D-amino acids may be obtained from Davies (47).

Arginine derivatives occur in animal protein (180).

APPENDIX II

AMINO AND IMINO ACIDS AND THEIR DERIVATIVES THAT OCCUR FREE IN PLANTS[a,b,c]

Name and structure	Occurrence, recognition, etc.	References
a. Derivatives of monoaminomonocarboxylic acids		
[¹⁴C]Carbamoyl-β-alanine	Precursor of pyrimidines in plants	1963. J. Buckowicz et al. Acta Biochim. Pol. **10**, 157.
N-Malonyl-D-alanine	Pisum sativum	1973. T. Ogawa et al. Biochim. Biophys. Acta **297**, 60.
D-Vinylglycine; (R)-2-Aminobut-3-enoic acid	In Rhodophyllus nidorosus in an optically impure form	1974. G. Dardenne et al. Phytochemistry **13**, 1897.
N-Feruloylglycine and N-ferulolyglycyl peptides	In globulins of barley (Hordeum vulgare) seeds	1973. C. F. Van Sumere et al. Phytochemistry **12**, 407.
3-(2-Feruloyl) alanine	Together with L-aspartic acid, a hydrolysis product (hot acid) of ascorbalamic acid in Brassica oleracea	1973. R. Couchman et al. Phytochemistry **12**, 707.
β-Acetamido-L-alanine	Acacia armata	1968. A. S. Seneviratne and L. Fowden. Phytochemistry **7**, 1039.
Cyclopentenylglycine	Seeds of Hydnocarpus anthelminthicus	1977. U. Cramer and F. Spener. Eur. J. Biochem. **74**, 495.
Homoisoleucine	Aesculus californica	1968. L. Fowden and A. Smith. Phytochemistry **7**, 809.
b. Derivatives of unsaturated monoaminomonomonocarboxylic acids and their derivatives		
2-Amino-4-methylhex-5-ynoic acid (**Ia**)	Seeds of Euphorbia longan	1969. M. Sung et al. Phytochemistry **8**, 1227.
2-Amino-4-hydroxymethylhex-5-ynoic acid (**Ib**)		

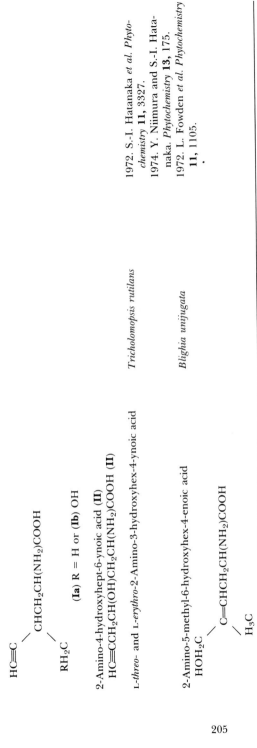

HC≡C
　　＼
　　　CHCH₂CH(NH₂)COOH
　　／
RH₂C

(Ia) R = H or **(Ib)** OH

2-Amino-4-hydroxyhept-6-ynoic acid **(II)**
HC≡CCH₂CH(OH)CH₂CH(NH₂)COOH **(II)**

L-*threo*- and L-*erythro*-2-Amino-3-hydroxyhex-4-ynoic acid

2-Amino-5-methyl-6-hydroxyhex-4-enoic acid

HOH₂C
　　　＼
　　　　C=CHCH₂CH(NH₂)COOH
　　　／
H₃C

Tricholomopsis rutilans

Blighia unijugata

1972. S.-I. Hatanaka et al. *Phytochemistry* **11**, 3327.
1974. Y. Niimura and S.-I. Hatanaka. *Phytochemistry* **13**, 175.
1972. L. Fowden et al. *Phytochemistry* **11**, 1105.

---

a This appendix excludes substances already listed in Appendix I as possible constituents of protein. It includes substituted derivatives of established free amino acids in plants. For historical references and details of occurrence, see earlier appendixes in Volume IVA.
b A large number of named amino and imino acids have been detected in urine. Even though not yet reported in plants, these may be found to have botanical origin.
c Well over 100 compounds in these categories have been isolated and named as occurring in microorganisms, fungi, and animals. For these, reference should be made to *Specialist Periodical Reports* (The Chemical Society).

(*continued*)

APPENDIX II (Continued)

Name and structure	Occurrence, recognition, etc.	References
2-Amino-4-methylhex-4-enoic acid $$\underset{\text{H}_3\text{C}}{}\overset{\text{H}}{\underset{\text{C}}{}}=\overset{}{\underset{\text{C}}{}}\overset{\text{CH}_2\text{CH(NH}_2)\text{COOH}}{\underset{\text{H}_3\text{C}}{}}$$ and its γ-glutamylpeptide, homoisoleucine, 2-amino-4-methyl-6-hydroxyhex-4-enoic acid	*Aesculus californica* seeds Spectroscopic examination	1968. L. Fowden and A. Smith. *Phytochemistry* **7**, 809 1968. D. S. Millington and R. C. Sheppard. *Phytochemistry* **7**, 1027.
L-(−)-5-Methyl-2-amino-4-hexenoic acid	*Leucocortinarius bulbiger*, a mushroom	1968. G. Dardenne and J. Casimir. *Phytochemistry* **7**, 1401.
δ-Hydroxyleucenine $$\underset{\text{H}_3\text{C}}{\overset{\text{HOH}_2\text{C}}{}}\text{C}=\text{CHCHCOOH}$$ $$\qquad\qquad\quad\underset{\text{NH}_2}{|}$$	Hydrolysates of phalloidine Suggested precursor for 4-hydroxygalegine	1955. T. Weiland and W. Schön. *Angew. Chem.* **593**, 157. 1963. N. V. Thoai and G. Desvages. *Bull. Soc. Chim. Biol.* **45**, 413.
c. Derivatives of monoaminodicarboxylic acids and their amides		
3-Aminoglutaric acid $$\text{H}_2\text{NCHCH}_2\text{COOH}$$ $$\quad\;\;|$$ $$\quad\text{CH}_2\text{COOH}$$	*Chondria armata*	1965. T. Takemoto and T. Sai. *Yakugaku Zasshi* (*J. Pharm. Soc. Jpn.*) **85**, 33.

Compound	Source / Notes	Reference
γ-Methylglutamic acid HOOCCH(CH₃)CH₂CH(NH₂)COOH	2S,3S Analog of earlier reported compound; in *Gleditsia*	1974. G. A. Dardenne *et al. Phytochemistry* **13**, 1295.
γ-Methyl-γ-hydroxyglutamic acid HOOCC(OH)CH₂CH(NH₂)COOH | CH₃	(2S,4S), and (2S,4R)-γ-Hydroxy-γ-methylglutamic acid; in *Pandanus veitchii* Erythro form in *Lathyrus maritimus* (2S,3S,4R)-3-hydroxy analog in *Gleditsia*	1967. J. Jadot *et al. Biochim. Biophys. Acta* **136**, 79. 1968. J. Przybylska and F. Strong. *Phytochemistry* **7**, 471. 1974. G. A. Dardenne *et al. Phytochemistry* **13**, 2195.
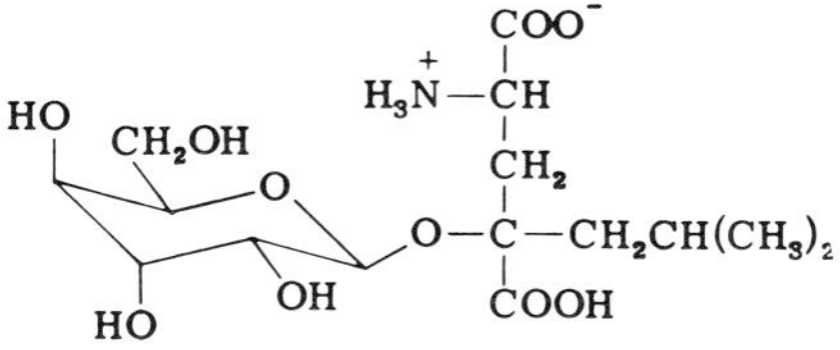 (2S,4R)-4-(β-D-Galactopyranosyloxy)-4-isobutylglutamic acid 	Flowers of *Reseda odorata*	1973. P. O. Larsen *et al. Phytochemistry* **12**, 1713.
γ-Ethylidineglutamic acid 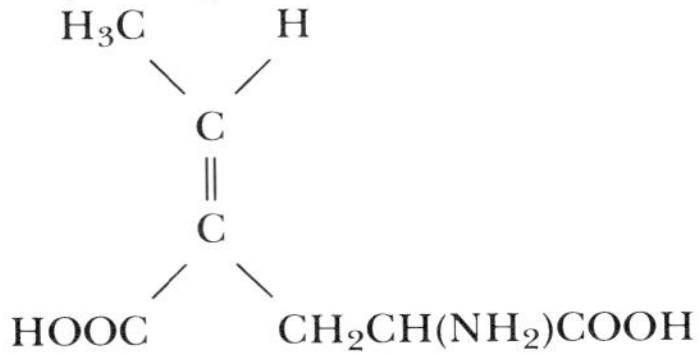 	*Tulipa*	1966. L. Fowden. *Biochem. J.* **98**, 57.
N⁵-Isopropyl-L-glutamine HOOC CH₂CH(NH₂)COOH	Seeds of *Lunaria annua*	1965. P. O. Larsen. *Acta Chem. Scand.* **19**, 1071.
γ-Propylidene-L-glutamic acid and γ-ethylidene and γ-methylene derivatives	*Mycena pura*	1975. S. Hatanaka and H. Katayama. *Phytochemistry* **14**, 1434.
(2S)-2-Aminoadipic acid	*Reseda odorata*	1976. H. Sørensen. *Phytochemistry* **15**, 1527.

(continued)

207

APPENDIX II (Continued)

Name and structure	Occurrence, recognition, etc.	References
2-Hydroxysuccinamic acid $NH_2CCH_2CHCOOH$ (with O double bond and OH)	Growing pea (*Pisum*) leaves	1978. N. D. H. Lloyd and K. W. Joy. *Biochem. Biophys. Res. Commun.* **81**, 186.
N^α-(1-Carboxyethyl)aspartate, N^α-(1-carboxyethyl)glutamate, N^α-(1,3-dicarboxyethyl)aspartate, N^α-(1,3-dicarboxypropyl)-glutamate	Claimed, but tentative identification and occurrence in plant tumors induced by *Agrobacterium tumefaciens*	1978. C.-C. Chang and B.-Y. Lin. *Plant Physiol. (Suppl.)*, p. 73.
Amides of leucine, tyrosine and phenylalanine	Ladino clover (*Trifolium repens* f. *lodigense*) seeds	1976. T. Kasai *et al. Agric. Biol. Chem.* **40**, 2489.
4-Ethylamide of glutamic acid (L-ethanine)	Detected in tea [*Camellia* (*Thea*) *sinensis*] leaves and synthesis of L-theanine	1964. L. Furuyama *et al. Bull. Chem. Soc. Jpn.* **37**, 1078.
d. Diamino-, mono-, di- and tricarboxylic acids and their derivatives		
$C_{12}H_{17}NO_7$	*Lactarius helvus* (amino acid B)	1965. E. Honkanen and A. I. Virtanen. *Acta Chem. Scand.* **19**, 1010.
	Proposed 2-methylene cycloheptene-1,3-diglycine	1964. E. Honkanen and A. I. Virtanen. *Acta Chem. Scand.* **18**, 1319.
L-Amino-2-methylaminopropionic acid	Seeds of *Cycas circinalis*	1967. A. Vega and E. A. Bell. *Phytochemistry* **6**, 759.
	L-Isomer	1968. A. Vega *et al. Phytochemistry* **7**, 1885.
	Free and bound in 9 species	1973. S. F. Dossaji and E. A. Bell. *Phytochemistry* **12**, 143.

208

(continued)

Compound	Occurrence / Notes	References
β-N-Oxalyl-L-α,β-diaminopropionic acid $HOOCCONHCH_2CHOOH$ $\quad\quad\quad\quad\quad\vert$ $\quad\quad\quad\quad NH_2$	Seeds of *Lathyrus sativus*	1964. S. L. N. Rao *et al. Biochemistry* **3**, 432.
α- and γ-Oxalyl derivatives of α,γ-diaminobutyric acid	Seeds of *Lathyrus latifolius* Chemical synthesis of D-2-oxalyl-amino-3-aminopropionic acid Distribution of α-amino-β-oxalyl-aminopropionate Earlier reports of occurrence in pine (*Pinus strobus*) pollen not verified	1966. E. A. Bell and J. P. O'Donovan. *Phytochemistry* **5**, 1211. 1970. G. Wu *et al. Phytochemistry* **15**, 1257. 1977. M. Y. Quereshi *et al. Phytochemistry* **16**, 477. 1976. P. O. Larsen and F. Norris. *Phytochemistry* **15**, 1761.
α,ε-Diaminopimelic acid $HOOCCH(CH_2)_3CHCOOH$ $\quad\quad\vert\quad\quad\quad\quad\vert$ $\quad\; NH_2\quad\quad\; NH_2$	*Phaseolus vulgaris*	1970. R. M. Zacharius. *Phytochemistry* **9**, 2047.
Nδ-Acetylornithine		1965. A. Ménagé and G. Morel. *C. R. Hebd. Seances Acad. Sci. Ser. D* **261**, 2001.
N²-(D-1-Carboxyethyl)-L-ornithine (octopinic acid)	Crown galls	1970. A. Goldmann-Ménagé. *Ann. Sci. Nat. Bot. Biol. Veg.* **11**, 223.
N²-(1,3-Dicarboxypropyl)-L-ornithine (ornaline)	In crown-gall tumors	1977. J. L. Firmin and R. G. Fenwick. *Phytochemistry* **16**, 761.
Nγ-Acetyl-α,γ-diaminobutyric acid Optical form not stated	*In vivo* synthesis by crown-gall-specific *Agrobacterium tumefaciens* Latex of *Euphorbia pulcherrima* Synthesis	1978. J. D. Kemp. *Plant Physiol.* **62**, 26. 1962. I. Liss. *Phytochemistry* **1**, 87. 1963. H. R. Schutte and W. Shütz. *Ann. Chem. (Warsaw)* **665**, 203.

Name and structure	Occurrence, recognition, etc.	References
γ-N-Acetyl-α,γ-diaminobutyric acid	Beta vulgaris	1972. L. Fowden. Phytochemistry 11, 227.
γ-N-Lactyl-α,γ-diaminobutyric acid ε-N-Acetyllysine ε-N-Acetyl-allo-γ-hydroxylysine γ-L-Glutamyl-γ-aminobutyric acid		
(2S,2′S)-N⁶-(2′-Glutaryl)lysine (saccaropine)	Reseda odorata	1976. H. Sørensen. Phytochemistry 15, 1527.
	Distribution in higher plants	1977. R. Nawaz and H. Sørensen. Phytochemistry 16, 599.
N²-(1-Carboxyethyl)-L-lysine	In crown galls	1960. K. Biemann et al. Bull. Soc. Chim. Biol. 42, 979.
	Sunflower (Helianthus) gall	1977. E. Hack and J. D. Kemp. Biochem. Biophys. Res. Commun. 78, 785.
e. Hydroxyamino acids and their derivatives		
(2S,3S,4R)- and (2S,3R,4R)-β-Hydroxy-γ-methyl-glutamic acid	Gymnocladus dioicus	1972. G. A. Dardenne et al. Phytochemistry 11, 787, 791.
(2S,3S)-3-Hydroxy-4-methyleneglutamic acid	Gleditsia caspica	1974. G. Dardenne and J. Casimir. Phytochemistry 13, 2195.
(2S,4R)-4-Methylglutamic acid (2S,3S,4R)-3-Hydroxy-4-methylglutamic acid β-D-Galactoside of 4-hydroxy-4-isobutylglutamic acid	Reseda odorata	1973. P. O. Larsen et al. Phytochemistry 12, 1713.
Correction to (2S,4S)-4-(β-D-Galactopyranosyl)-4-isobutyl-glutamic acid		1977. K. Kaas and H. Sørensen. Acta Chem. Scand. Ser. A 31, 364.
Pinnatanine	European bladder nut (Staphylea pinnata); hydrolysis yields L-allo-γ-hydroxyglutamic acid	1971. M. D. Grove et al. Tetrahedron Lett., p. 4477.

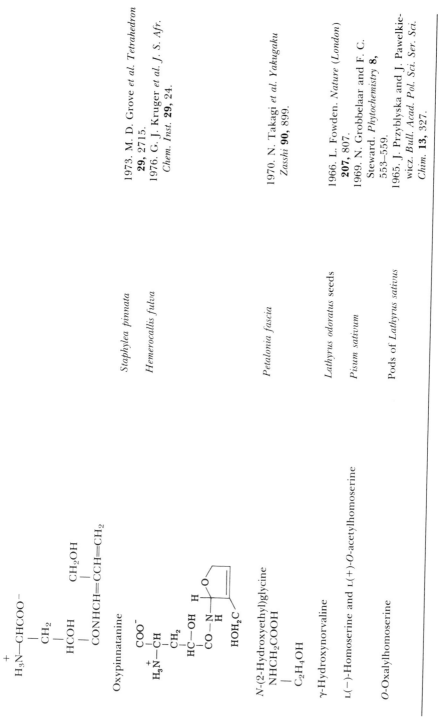

$\overset{+}{H_3N}-CHCOO^-$		
Oxypinnatanine	*Staphylea pinnata*	1973. M. D. Grove et al. *Tetrahedron* **29**, 2715.
	Hemerocallis fulva	1976. G. J. Kruger et al. *J. S. Afr. Chem. Inst.* **29**, 24.
N-(2-Hydroxyethyl)glycine $NHCH_2COOH$ C_2H_4OH	*Petalonia fascia*	1970. N. Takagi et al. *Yakugaku Zasshi* **90**, 899.
γ-Hydroxynorvaline	*Lathyrus odoratus* seeds	1966. L. Fowden. *Nature (London)* **207**, 807.
L(−)-Homoserine and L(+)-O-acetylhomoserine	*Pisum sativum*	1969. N. Grobbelaar and F. C. Steward. *Phytochemistry* **8**, 553–559.
O-Oxalylhomoserine	Pods of *Lathyrus sativus*	1965. J. Przyblyska and J. Pawelkiewicz. *Bull. Acad. Pol. Sci. Ser. Sci. Chim.* **13**, 327.

(continued)

211

APPENDIX II (Continued)

Name and structure	Occurrence, recognition, etc.	References
8-Hydroxynorleucine,(5-hydroxy-2-aminohexanoic acid)	Seeds of *Crotalaria juncea*	1974. R. Pant and H. M. Fales. *Phytochemistry* **13**, 1626.
(2S,3R,4R)-4-Hydroxyisoleucine (2S,3R,4R)-4-hydroxyisoleucine	Seeds of *Trigonella*	1973. L. Fowden *et al. Phytochemistry* **12**, 1707.
Caulerpicin (n = 23, 24, and 25; Me = methyl) Me(CH$_2$)$_{13}$CHCHCH$_2$OH │ NHCO(CH$_2$)$_n$Me **1**	(1) From *Caulerpa*, green alga From *Caulerpa*	1966. M. S. Doty and G. Aguilar-Santos. *Nature (London)* **211**, 990. 1979. M. Mahendran *et al. Phytochemistry* **18**, 1885.
OH │ Me(CH$_2$)$_{12}$HC=CHCH CHCH$_2$OH ^{t} │ NHCO(CH$_2$)$_n$Me **2** (n = 12, 14, 20, 22; t = trans; i.e., mixture of N-acylspingosines)	Spingosine and dihydrosphingosine in *Amansia glomerata* and *Laurencia nidifica* Revised structure of caulerpin, a pigment from green algae	1978. J. H. Cardellina and R. E. Moore. *Phytochemistry* **17**, 554. 1978. B. C. Maiti *et al. J. Chem. Res.* (S). p. 126.
f. Aromatic amino acids and their derivatives Phenylglycine	Phloem sap of *Fagus*	1967. H. H. Dietrichs and H. Funke. *Holzforschung* **21**, 102.
α-(3-Hydroxyphenyl)glycine; α-(3,5-dihydroxyphenyl)glycine; 1-methyl-6-hydroxy-1,2,3,4-tetrahydroisoquinoline-3-carboxylic acid	*Euphorbia helioscopa*, *E. myrsinites*	1968. P. Müller and H. R. Schütte. *Z. Naturforsch.* **236**, 491. 659.
Mono-O-β-D-glucoside of dopamine	Seeds of *Entada gigas* (*E. scandens*)	1972. P. O. Larsen *et al. Int. Symp. Biochem. Physiol. Alkaloide Halle* 1969, p. 113.

3-(4-Hydroxy-3-hydroxymethylphenyl)alanine and 3-(3-hydroxymethylphenyl)alanine

Seeds of *Caesalpina spinosa* (*C. tinctoria*)

1973. R. Watson and L. Fowden. *Phytochemistry* **12**, 617.

3-(3-Hydroxymethylphenyl)-L-alanine in *Iris* spp.

1978. P. O. Larsen *et al. Phytochemistry* **17**, 549.

O-(β-D-Glucopyranosyl)-L-tyrosine and 2-[3-(β-D-glucopyranosyloxy)-4-hydroxyphenyl]ethylamine

Seeds of *Entada pursaetha*

1973. P. O. Larsen *et al. Phytochemistry* **12**, 2243.

N-Hydroxytyrosine

Intermediate in dhurrin biosynthesis in *Sorghum*

1977. B. Møller and E. C. Conn. *Plant Physiol.* **59**, Abstr. No. 451.

2,4-Dihydroxy-6-methylphenylalanine

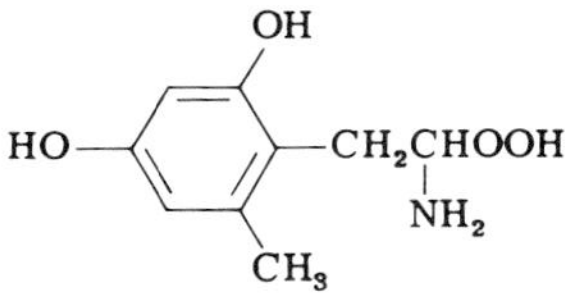

Biosynthesis from serine and acetate

1965. L. A. Hadwiger *et al. Phytochemistry* **4**, 825.

L-(+)-*p*-Aminophenylalanine

Vigna vexillata

1972. G. A. Dardenne *et al. Phytochemistry* **11**, 2567.

L-3-(3-Aminomethylphenyl)alanine and 3-hydroxymethylphenyl and 3-carboxyphenyl analog

Seeds of *Combretum zeyheri*

1975. K. Mwauluka *et al. Biochem. Physiol. Pflanz.* **168**, 15.

4-Hydroxy-3-hydroxymethylphenylalanine and 3-hydroxymethylphenylalanine

1973. R. Watson and L. Fowden. *Phytochemistry* **12**, 617.

3-(3-Carboxyphenyl)alanine, (3-carboxyphenyl)glycine, 3-(3-carboxy-4-hydroxyphenyl)alanine, (3-carboxy-4-hydroxyphenyl)glycine, 2(S)-4-hydroxy-2-aminopimelic acid, lactones of 4-hydroxy-2-aminopimelic acid, γ-glutamylglutamic acid

Caesalpina spinosa (*C. tinctoria*) and related species *Reseda luteola*

1979. L. K. Meier *et al. Phytochemistry* **18**, 1505.

N-Malonylphenylalanine

Barley (*Hordeum vulgare*) seedlings (evidence for D-isomer)

1968. N. Rosa and A. C. Neish. *Can. J. Biochem.* **46**, 797.

N-trans-Caffeoyl-3-(3,4-dihydroxyphenyl)-L-alanine (*trans*-clovamide) and *N-cis*-caffeoyl-3-(dihydrooxyphenyl)-L-alanine

In *Trifolium pratense*

1977. T. Yoshihara *et al. Agric. Biol. Chem.* **41**, 1679.

(continued)

213

Name and structure	Occurrence, recognition, etc.	References
Pretyrosine (structure: cyclohexadiene ring with HO, COOH, CH₂CHCOO₂⁻ / NH₂) $CH_2CHCOO_2^-$, NH_2	Blue-green algae	1977. N. Patel *et al. J. Biol. Chem.* **252**, 5839.
4-Hydroxy-3-methoxy-L-phenylalanine	*Cortinarius brunneus*	1977. G. Dardenne *et al. Phytochemistry* **16**, 1822.
L-6-Bromohypaphorine (6-Bromo-L-tryptophan N^α-trimethyl-betaine)	*Pachymatisma johstoni*, a sponge	1977. W. D. Raverty *et al. J. Chem. Soc. Perkin Trans.* **1**, p. 1204.
g. Heterocyclic compounds with amino or imino groups[d]		
1-Aminocyclopropane-1-carboxylic acid (cyclopropane with COOH, NH₂)	Pears (*Pyrus*) and apples (*Malus*) Precursor for ethylene Ethylene production stimulated by the amino acid *N*-Malonyl conjugate	1957. L. F. Burroughs. *Nature (London)* **179**, 360. 1979. D. O. Adams *et al. Proc. Int. Conf. Plant Growth Regul. 10th,* Abstr. No. 303, p. 18. 1979. D. O. Adams and S. Yang. *Proc. Natl. Acad. Sci. USA* **76**, 170. 1982. N. E. Hoffman *et al. Biochem. Biophys. Res. Commun.* **104**, 756.
β-(Methylenecyclopropyl)β-methylalanine H_2C (methylenecyclopropyl ring) with CH_3, $CHCH(NH_2)COOH$	Seeds of *Aesculus californica*	1968. L. Fowden and A. Smith. *Phytochemistry* **7**, 809.
	Glycine derivative in *Billia hippocastanum*	1970. J. N. Eloff and L. Fowden. *Phytochemistry* **9**, 2423.

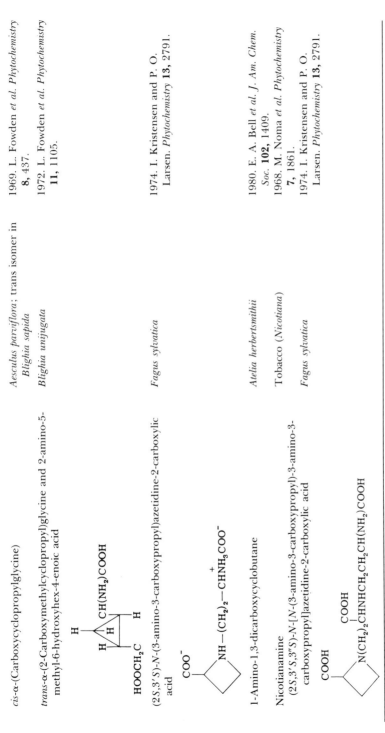

cis-α-(Carboxycyclopropyl)glycine)

trans-α-(2-Carboxymethylcyclopropyl)glycine and 2-amino-5-methyl-6-hydroxyhex-4-enoic acid

(2*S*,3′*S*)-*N*-(3-amino-3-carboxypropyl)azetidine-2-carboxylic acid

1-Amino-1,3-dicarboxycyclobutane

Nicotianamine

(2*S*,3′*S*,3″*S*)-*N*-[*N*-(3-amino-3-carboxypropyl)-3-amino-3-carboxypropyl]azetidine-2-carboxylic acid

Aesculus parviflora; trans isomer in *Blighia sapida*

Blighia unijugata

Fagus sylvatica

Atelia herbertsmithii

Tobacco (*Nicotiana*)

Fagus sylvatica

1969. L. Fowden *et al. Phytochemistry* **8**, 437.

1972. L. Fowden *et al. Phytochemistry* **11**, 1105.

1974. I. Kristensen and P. O. Larsen. *Phytochemistry* **13**, 2791.

1980. E. A. Bell *et al. J. Am. Chem. Soc.* **102**, 1409.

1968. M. Noma *et al. Phytochemistry* **7**, 1861.

1974. I. Kristensen and P. O. Larsen. *Phytochemistry* **13**, 2791.

215

<hr/>

[a] The great diversity of configurations on compounds listed in this category precludes any idea of their overall significance. However, the frequency with which they are derivatives of other well-known simpler substances (e.g., protein amino acids such as alanine, amides, or their derivatives, etc.) warrants their inclusion here for their possible physiological role or advantage in natural selection.

(continued)

Name and structure	Occurrence, recognition, etc.	References
2-Pyrrolidine acetic acid (structure: pyrrolidine ring with N–H and CH$_2$COOH substituent)	Cured tobacco (*Nicotiana*) leaves	1964. H. Tomita *et al. Agric. Biol. Chem.* **28**, 451.
Pyrrolidine-2,5-dicarboxylic acid	Red algae	1975. G. Impellizerri *et al. Phytochemistry* **14**, 1549.
2,5-Dihydroxymethyl-3,4-dihydroxypyrrolidine (structure: pyrrolidine ring with OH, CH$_2$OH, HO, HOH$_2$C, N–H substituents)	Leaves of *Derris elliptica*	1976. A. Welter *et al. Phytochemistry* **15**, 747.
β-(2-Furoyl)-L-alanine 3-(2-Furoyl)alanine (structure: furan ring with C=O, CHCHCOO$^-$, NH$_3^+$)	In hydrolyzed extract of *Koelreuteria paniculata* *Vigna radiata* (*Phaseolus radiatus*) seeds (green gram) Hydrolysis product from ascorbamic acid Buckwheat (*Fagopyrum esculentum*) seeds; L-configuration	1975. F. Irrevere *et al. Lloydia* **38**, 178. 1973. T. Kasai *et al. Agric. Biol. Chem.* **37**, 2923. 1973. R. Couchman *et al. Phytochemistry* **12**, 707. 1973. A. Ichihara *et al. Tetrahedron Lett.*, p. 37.
(3R)-[(1'S)-Aminocarboxymethyl]-2-pyrrolidone-5(S)-carboxylic acid (penmacric acid) (structure: CH$_2$NHCOOH, pyrrolidone ring with O, N–H, HOOC substituents)	Seeds of *Pentaclethra macrophylla*	1975. A. Welter *et al. Bull. Soc. Chim. Belg.* **84**, 243; *Phytochemistry* **14**, 1347. 1975. E. I. Mbadiwe. *Phytochemistry* **14**, 1351.

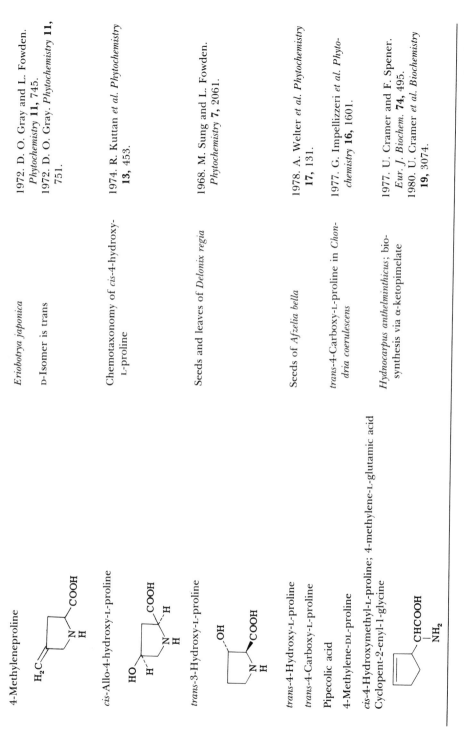

4-Methyleneproline

Eriobotrya japonica
D-Isomer is trans

1972. D. O. Gray and L. Fowden. *Phytochemistry* **11**, 745.
1972. D. O. Gray. *Phytochemistry* **11**, 751.

cis-Allo-4-hydroxy-L-proline

Chemotaxonomy of *cis*-4-hydroxy-L-proline

1974. R. Kuttan *et al. Phytochemistry* **13**, 453.

trans-3-Hydroxy-L-proline

Seeds and leaves of *Delonix regia*

1968. M. Sung and L. Fowden. *Phytochemistry* **7**, 2061.

trans-4-Hydroxy-L-proline
trans-4-Carboxy-L-proline

Seeds of *Afzelia bella*

1978. A. Welter *et al. Phytochemistry* **17**, 131.

Pipecolic acid
4-Methylene-DL-proline

trans-4-Carboxy-L-proline in *Chondria coerulescens*

1977. G. Impellizzeri *et al. Phytochemistry* **16**, 1601.

cis-4-Hydroxymethyl-L-proline; 4-methylene-L-glutamic acid
Cyclopent-2-enyl-1-glycine

Hydnocarpus anthelminthicus; biosynthesis via α-ketopimelate

1977. U. Cramer and F. Spener. *Eur. J. Biochem.* **74**, 495.
1980. U. Cramer *et al. Biochemistry* **19**, 3074.

(continued)

APPENDIX II (Continued)

Name and structure	Occurrence, recognition, etc.	References
cis-4-Hydroxymethyl-L-proline HOH₂C—[pyrrolidine ring]—COOH, N—H	Apple (*Malus*) fruit	1966. A. B. Mauger and B. Witkop. *Chem. Rev.* **66**, 47.
DL-*trans*-4-Hydroxymethylproline HO—[pyrrolidine ring]—COOH, N—H	*Eriobotyra japonica*	1972. D. O. Gray. *Phytochemistry* **11**, 751.
L-Pipecolic acid [piperidine ring]—COOH, N—H	D-Lysine precursor for pipecolate	1970. D. R. Aldag and J. L. Young. *Planta* **95**, 187.
L-(−)-4-Hydroxypipecolic acid OH—[piperidine ring]—COOH, N—H	L(−) in *Peganum harmala* seeds	1971. V. U. Amhad and M. A. Kahn. *Phytochemistry* **10**, 3339.

218

Compound	Source/Notes	Reference
exo-(cis)-3,4-Methanoproline 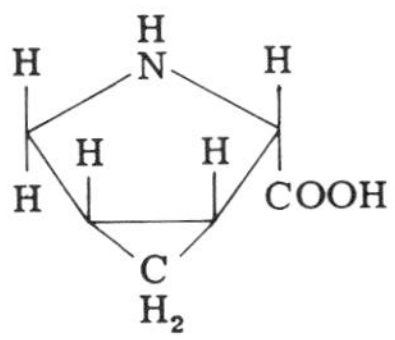	Aesculus parviflora	1969. L. Fowden et al. Phytochemistry **8**, 437.
L-Proline-L-leucine anhydride	Bitter compound in aged sake	1974. K. Takahashi et al. Agric. Biol. Chem. **38**, 927.
L-cis-5-Hydroxypipecolate	Seeds of Gymnocladus dioicus	1977. J. Despontin et al. Phytochemistry **16**, 387.
(2S,5S,6S)-5-Hydroxy-6-methylpipecolic acid	Fagus sylvatica	1974. I. Kristensen et al. Phytochemistry **13**, 2803.
(2S,5S,6S)-5-Hydroxy-6-methylpipecolic acid (2S,5S,6S)-2-Carboxy-5-hydroxy-6-methylpiperidine and its 5R-epimer 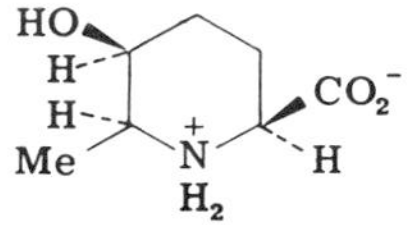	Seeds of Fagus sylvatica	1976. I. Kristensen et al. Tetrahedron **32**, 2799.
4,5-Dihydroxy-L-pipecolic acid 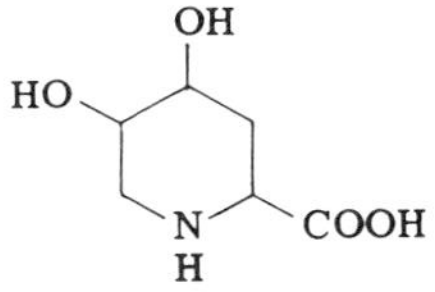	2,4-trans-4,5-cis Form in Calliandra haematocephala 2,4-cis-4,5-trans Form in Derris elliptica 2,4-trans-4,5-trans Form in seeds of Julbernardia, Isoberlinia, and Brachystegia [(2S)-carboxy-(4R,5R)-dihydroxypiperidine]	1972. M. Marlier et al. Phytochemistry **11**, 2597. 1975. G. A. Dardenne. Phytochemistry **14**, 860. 1976. A. Welter et al. Phytochemistry **15**, 747. 1976. P. R. Shewry and L. Fowden. Phytochemistry **15**, 1981.

(continued)

219

Name and structure	Occurrence, recognition, etc.	References
trans-4-Acetylamino-L-pipecolic acid NHCOCH$_3$ (piperidine ring with —COOH)	*Calliandra haematocephala*	1975. G. A. Dardenne. *Phytochemistry* **14**, 860.
Shinorine OCH$_3$ NHCH$_2$COOH HOOC HOCH$_2$CHN CH$_2$OH HO	Red alga *Chondrus yendoi*	1980. I. Tsujino *et al. Bot. Mar.* **23**, 65.
Mimosine or leucenol OH (pyridinone ring N) HOOCHCH$_2$C—N NH$_2$	O-β-D-Glucoside	1971. I. Murakoshi *et al. Chem. Pharm. Bull.* **19**, 2655
Isowillardine; β-(2,4-dihydropyrimidin-3-yl)alanine	Pea (*Pisum*) seedlings (revised structure of 1966 report: *Biochem. Biophys. Acta* **199**, 1)	1968. F. Lambein and R. Van Parijs. *Biochem. Biophys. Res. Commun.* **32**, 474. 1969. E. G. Brown and B. S. Man-

Lupinic acid or β-[6-(4-hydroxy-3-methylbut-*trans*-2-enyl-amino)purine-9-yl]alanine

$$NHCH_2C=CMeCH_2OH$$

Zeatin metabolite in *Lupinus angustifolius*
Enzymatic synthesis

gat. *Biochim. Biophys. Acta* **177**, 427.
1975. J. K. MacLeod *et al. J. Chem. Soc. Chem. Commun.*, p. 809.
1979. C. W. Parker and B. Entsch. *Proc. Int. Conf. Plant Growth Regul. 10th*, Abstr. No. 258, p. 45.

Isoxazolinone derivatives

Substituents and names

	R[1]	R[2]	
1.	H	—CHCHCOOH, NH$_2$	β-(Isoxazolin-5-one-2-yl)alanine
2.	—CH$_2$CHCOOH, NH$_2$	—β-D-Glucopyranosyl	β-(2-β-D-glucopyranosylisoxazolin-5-one-4-yl)alanine

β-D-glucopyranosylisoxazolin-5-one-4-yl)alanine

1 and **2** in *Pisum sativum* seedlings

2, **4**, and **5** in *Lathyrus odoratus* seedlings

1969. F. Lambein *et al. Biochem. Biophys. Res. Commun.* **37**, 375.
1970. F. Lambein and R. Van Parijs. *Biochem. Biophys. Res. Commun.* **40**, 557.
1974. F. Lambein and R. Van Parijs. *Biochem. Biophys. Res. Commun.* **61**, 155.

(*continued*)

APPENDIX II *(Continued)*

	R¹	R²	Occurrence, recognition, etc.	References
3.	H	—CH₂CH₂NH-γ-Glutamyl 2-γ-Glutaminoethylisoxazolin-5-one		

4.	H	—CH₂CH₂CHCOOH
		            │
		            NH₂
		α-Amino-γ-(isoxazolin-5-one-2-yl)butyric acid
				1973. F. Lambein. *Arch. Int. Physiol. Biochim.* **81**, 380.

5.	H	—CH₂CH₂NH₂ 2-Aminoethyl-isoxazolin-5-one	**6**, **7**, and **8** in *Pisum sativum*	
6.	H	—CH₂CH₂CN Nitrile		
7.	H	—β-D-Glucopyranosyl Glucoside		
8.	H	—CH₂COOH Carboxylic acid		

Name and structure	Occurrence, recognition, etc.	References
β-(3,5-Dioxo-1,2,4-oxadiazolidin-2-yl)-l-alanine (quisqualic acid) $$HN{-}C(=O){-}N({-}CH_2CHNH_3^+{-}COO^-){-}C(=O){-}O$$ (ring)	2-(2-Cyanoethyl)-3-isoxazolin-5-one in *Lathyrus odoratus* *Quisqualis*	1974. L. Van Rompuy *et al. Experientia* **30**, 1379. 1975. T. Takemoto *et al. Yakugaku Zasshi* (*J. Pharm. Soc. Jpn.*) **95**, 326, 448.
3-Amino-1,2,4-triazol-1-ylalanine	Product of herbicide 3-amino-1,2,4-triazole Enzymatic production in plants	1967. D. M. Frisch *et al. Phytochemistry* **6**, 921. 1974. I. Murakoshi. *Chem. Pharm. Bull.* **22**, 480.

Ascorbalamic acid 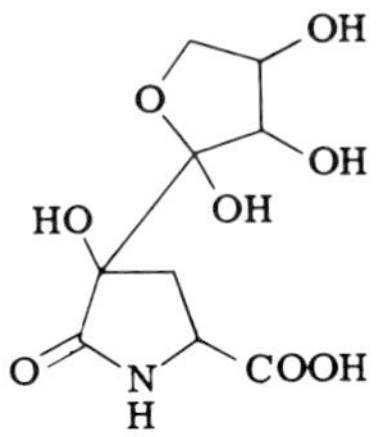	In several plants and yields 3-(2-furoyl)alanine with acid	1973. R. Couchman et al. Phytochemistry **12**, 707.
Histopine; N^2-(1-carboxyethyl)histidine	Sunflower (*Helianthus*) crown gall	1977. J. Kemp. *Plant Physiol.* **59**, Abstr. No. 589; *Biochem. Biophys. Res. Commun.* **74**, 862.
Dimethylhistidine 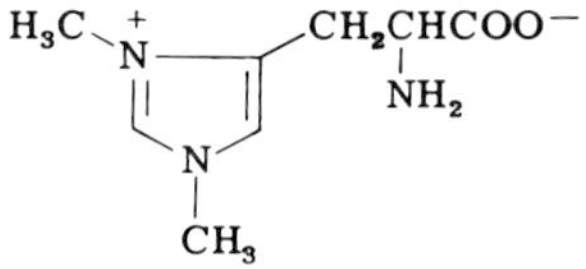	Australian seaweeds	1970. J. C. Madgwick et al. *Arch. Biochem. Biophys.* **141**, 766.
Isosuccinimide-β-glucoside	Diphenylamine sugar reaction in pea (*Pisum*) Pisatoside in pea	1964. V. Vitek. *Biochim. Biophys. Acta* **93**, 429. 1967. J. Kocourek et al. *Arch. Biochem. Biophys.* **121**, 531.
	Glucosyl donor in synthesis of ethyl-β-glucoside	1968. T. Y. Liu and P. Castelfranco. *Arch. Biochem. Biophys.* **123**, 645.
	Enzyme	1970. T. Y. Liu and P. Castelfranco. *Plant Physiol.* **45**, 424.
Succinimide and succinamic acid	Formation from ʟ-glutamine in pine (*Pinus*) buds	1973. D. J. Durzan. *Can. J. Bot.* **51**, 359.

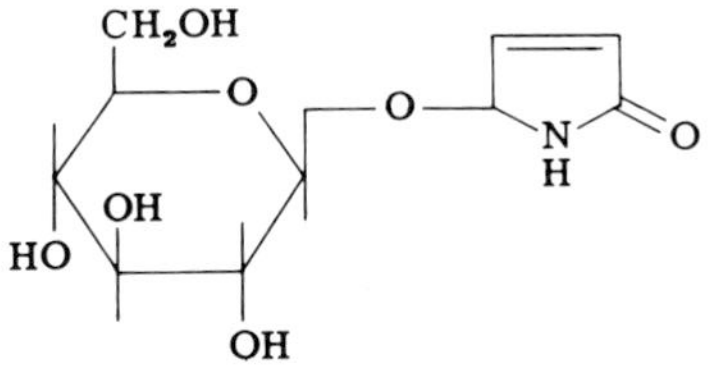

(continued)

223

APPENDIX II (*Continued*)

Name and structure	Occurrence, recognition, etc.	References
Domoic acid	*Chondria armata*	1966. T. Takemoto et al. *Yakugaku Zasshi* (*J. Pharm. Soc. Jpn.*) **86**, 874.
2-Aminoimidazole	Seeds of *Mundulea sericea*	1977. L. E. Fellows et al. *Phytochemistry* **16**, 1399.
3-[2-Amino-2-imidazolin-4(5)-yl]alanine (enduracididine)	In seeds of *Lonchocarpus sericeus*	1977. L. E. Fellows et al. *Phytochemistry* **16**, 1957.
2-[2-Amino-2-imidazolin-4(5)-yl]acetic acid	Distribution in plants	1978. L. E. Fellows et al. *Biochem. Syst. Ecol.* **6**, 213.
cis-4-Cyclohexene-1,2-dicarboximide	Ether extracts of sugar beets (*Beta vulgaris*); inhibitor of germination	1968. E. D. Mitchell and N. E. Tolbert. *Biochemistry* **7**, 1019.

Compound	Occurrence	Reference
5-Hydroxytryptophan	Free in plants	1966. E. A. Bell and L. E. Fellows. *Nature (London)* **210**, 529.
6-Hydroxykynurenic acid	Tobacco (*Nicotiana*) leaves	1968. P. K. MacNicol. *Biochem. J.* **107**, 473.
Indole-3-acetylamino acid conjugates	ε-L-Lysine derivative in culture filtrates of *Pseudomonas*	1968. O. Hutzinger and T. Kosuge. *Biochemistry* **7**, 601. References to earlier work in (230).
	Aspartate, glutamate, glycine, alanine, and valine conjugates in crown-gall callus of *Parthenocissus tricuspidata*	1977. C.-S. Feung *et al. Plant Physiol.* **58**, 666.
	1-*O*-(Indole-3-acetyl)-β-D-glucose in *Pinus pinea*	1980. J. Riov and H. E. Gottlieb. *Physiol. Plant.* **50**, 347.
	β-D-Glucose and aspartyl conjugates in conifers (*Pinus* sp.)	1979. J. Riov *et al. Physiol. Plant.* **46**, 133. References to earlier work in (230).
Naphthaleneacetic acid amino acid conjugates	D-Tryptophan in *Brassica hirta* (*Sinapis alba*)	1965. H. Schraudolf and F. Bergmann. *Planta* **67**, 75.
α-*N*-Malonyl-D-tryptophan	Pea	1968. S. Marumo *et al. Nature (London)* **219**, 959.
4-Chloroindoleacetic acid	Pea	1970. S. Marumo and H. Hattori. *Planta* **90**, 208.
N-Carbomethoxyacetyl and *N*-carboethoxyacetyl-D-4-chlorotryptophan	Soybean (*Glycine max*) plants	1971. D. I. Chkanikov *et al. Fiziol Rast. (Moscow)* **19**, 436.
N-(2,4-Dichlorophenoxyacetyl)L-aspartic acid	Conversion to *N*-(2-methyl-4-chlorophenoxyacetyl acetyl)-L-aspartic acid	1970. D. J. Collins and J. K. Gaunt. *Biochem. J.* **118**, 54.

(continued)

APPENDIX II (*Continued*)

Name and structure	Occurrence, recognition, etc.	References
N-(2,4-Dichlorophenoxyacetyl)-L-glutamic acid	Soybean callus	1971. C. S. Feung *et al. J. Agric. Food. Chem.* **19**, 475.
		1971. D. I. Chkanikov *et al. Fiziol. Rast. (Moscow)* **19**, 436.
Glutamic and aspartic conjugates of 2,4,5-trichlorophenoxy-acetic acid	Soybean callus	1978. M. Arjmand *et al. J. Agric. Food chem.* **26**, 1125.
Alanine conjugate of benzylaminopurine, β-(6-benzylamino-purin-9-yl)alanine	*Phaseolus vulgaris* shoots	1979. D. S. Letham *et al. Plant* **146**, 71.
N^5-(2'-Hydroxybenzyl)-allo-4-hydroxy-L-glutamine and N^5-(4'-hydroxybenzyl)-L-glutamine	Buckwheat (*Fagopyrum esculentum*) seeds	1973. M. Koyama *et al. Agric. Biol. Chem.* **37**, 2749.
L-3-Carboxy-6,7-dihydroxy-1,2,3,4-tetrahydroisoquinoline and L-3,4-dihydroxyphenylalanine	Seeds of *Mucuna mutisiana*	1971. E. A. Bell *et al. Phytochemistry* **10**, 2191.
1-Methyl derivative of the quinoline compound		1972. M. E. Daxenbichler *et al. Tetrahedron Lett.*, p. 1801.
1,2,3,4-Tetrahydro-8-hydroxy-6,7-dimethoxisoquinoline-1-carboxylic acid and its 1-methyl derivative	Peyote (*Lophophora williamsii*) cactus	1970. G. J. Kapadia *et al. J. Am. Chem. Soc.* **92**, 6943.

(+)-5-Hydroxydioxindole-3-acetic acid (**1**) and 5-hydroxy-oxindole-3-acetic acid

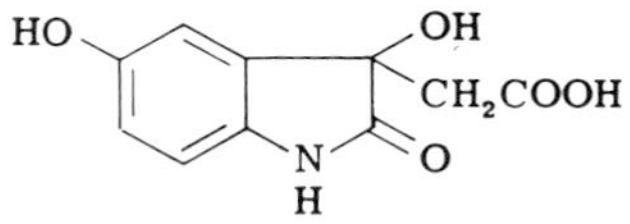

Rice (*Oryza sativa*) bran

1977. Y. Suzuki et al. *Phytochemistry* **16**, 635.

1

N-(1-Carboxy-2-methylbutyl)lachnanthopyridone

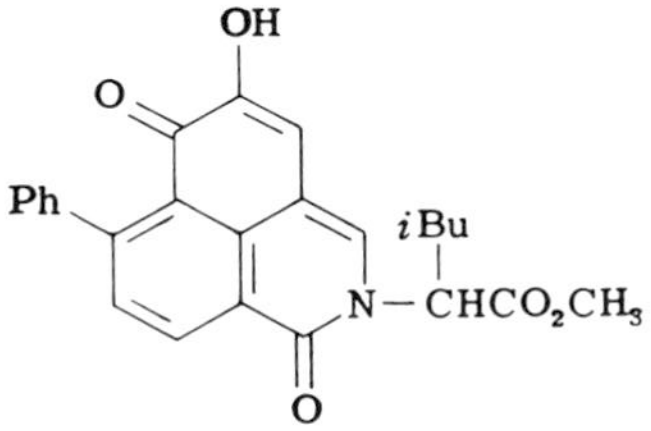

Flowers of *Lachnanthes caroliana* (*L. tinctoria*)

1976. A. C. Bazan and J. M. Edwards. *Phytochemistry* **15**, 1413.

(N of isoleucine at 5 position of ring; *i*Bu = isobutyl; Ph = phenyl)

3-Carboxy-1,2,3,4-tetrahydro-β-carboline

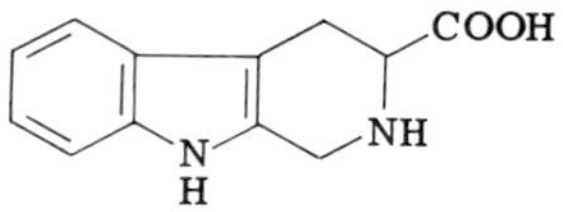

Seeds of *Aleurites fordii*

1975. T. Okuda et al. *Phytochemistry* **14**, 2304.

h. *N*-Methylated amino and imino acids (exclusive of guanidine derivatives)

γ-Methylaminobutyraldehyde (*N*-methylpyrroline)

Nicotiana rustica var. *brasilia*; precursor of nicotine

1968. S. Mizusaki et al. *Plant Physiol.* **43**, 93.

(continued)

Name and structure	Occurrence, recognition, etc.	References
N-Methyl-L-alanine	Leaves of *Dichapetalum cymosum* (up to 5.6% of the dry wt.)	1967. J. N. Eloff and N. Grobbelaar. *J. S. Afr. Chem. Inst.* **20**, 190.
N-Methyl-L-serine	Leaves of *Dichapetalum cymosum*	1969. J. N. Eloff and N. Grobbelaar. *Phytochemistry* **8**, 2201.
Carnitine (betaine of β-hydroxy-γ-aminobutyrate) $\overset{\displaystyle +}{(CH_3)_3}NCH_2\underset{\underset{\displaystyle OH}{\mid}}{CH}CH_2COO^-$	*Pisum sativum*	1969. R. A. Panter and B. Mudd. *FEBS Lett.* **5**, 169. 1975. P. H. McNeil and D. R. Thomas. *Phytochemistry* **14**, 2335.
2-Trimethylaminopropionic acid and 2-trimethylamino-6-oxo-heptanoic acid	Branches of *Limonium vulgare*	1975. F. Larher and J. Hamelin. *Phytochemistry* **14**, 205, 1789.
L-Isoleucine betaine	*Cannabis* seeds	1973. C. A. L. Bercht *et al. Phytochemistry* **12**, 2457.
N,N'-Dimethyl-L-phenylalanine and L-*threo*-β-phenylserine	In peptide of *Canthium euryoides*	1970. G. Boulvin *et al. Bull. Soc. Chim. Belg.* **78**, 583.
L(+)-N-Methyltyrosine	Seeds of *Combretum zeyheri*	1975. K. Mwauluka *et al. Phytochemistry* **14**, 1657.
Stachydrine (L-proline betaine) 	*Medicago sativa*	1974. J. K. Sethi and D. P. Carew. *Phytochemistry* **13**, 321.
Homostachydrine (N,N'-dimethylpipecolic acid)	*Medicago sativa*	1974. J. K. Sethi and D. P. Carew. *Phytochemistry* **13**, 321.

Compound	Occurrence / Notes	Reference
S-(+)-N^α-Methyltryptophan	Main alkaloid in *Aotus subglauca*; conversion of 4-γ,γ-dimethyl-allyltryptophan to clavicipitic acid	1971. S. R. Johns *et al. Aust. J. Chem.* **24**, 439. 1976. M. Saini *et al. Phytochemistry* **15**, 1497.
N^4-Methylasparagine	Seeds of *Corallocarpus epigaeus*	1965. P. M. Dunnill and F. Fowden. *Phytochemistry* **4**, 933.
1,3-Dimethylhistidine, 1-methylhistidine, and other known free amino acids	Marine algae, *Gracilaria secundata* (red) and *Phyllospora comosa* (brown)	1970. J. C. Madgwick *et al. Arch. Biochem. Biophys.* **141**, 766.
L(+)-N-(3-Amino-3-carboxypropyl)-β-carboxypyridinium betaine (nicotianine)	Tobacco (*Nicotiana*) leaves	1968. M. Noguchi *et al. Arch. Biochem. Biophys.* **125**, 1017; *Phytochemistry* **7**, 1861.

i. Sulfur- and selenium-containing amino acids and their derivatives[e]

Compound	Occurrence / Notes	Reference
Methylmethionine sulfonium $$\left[\; \begin{array}{c} H_2C \\ \; \\ H_2C \end{array} \!\! \underset{+}{>} SCH_2CH_2CH(NH_2)COOH \right] OH^- $$	Precursor of dimethylsulfide in *Theobroma cacao*	1976. A. S. Lopez and V. C. Quesnei. *J. Sci. Food Agric.* **27**, 85.
Thiothreonine (α-Amino-β-thiobutyric acid)	In extracts of pea (*Pisum*) seedlings exposed to $H_2{}^{35}S$ Biosynthesis from O-phosphohomoserine	1973. J. Schnyder and K. H. Erismann. *Experientia* **29**, 232. 1975. J. Schnyder *et al. Biochem. Physiol. Pflanz.* **167**, 605.
N-Methylmethionine sulfoxide	Red algae	1974. K. Miyazawa and K. Ito. *Nippon Suisan Gakkaishi* (*Bull. Jpn. Soc. Sci. Fish.*) **40**, 655.
N-Acetyl-L-djenkolic acid	*Acacia farnesiana*	1962. R. Gmelin *et al. Phytochemistry* **1**, 233.

(continued)

APPENDIX II (*Continued*)

Name and structure	Occurrence, recognition, etc.	References
CH_3 | C=O | NH H | | HOOCCCH$_2$SCH$_2$SCH$_2$CCOO$^-$ | | H NH$_3$$^+$	*Acacia: Gummiferae* series only	1968. A. S. Seneviratne and L. Fowden. *Phytochemistry* **7**, 1039.
S-Allylmercapto-L-cysteine	Garlic (*Allium sativum*)	1964. M. Sugii *et al. Chem. Pharm. Bull.* **12**, 145, 1114.
L-Homomethionine (L-5-methylthionorvaline)	Cabbage (*Brassica oleracea* var. *capitata*)	1964. M. Sugii *et al. Chem. Pharm. Bull.* **12**, 1115.
	Isolation and biosynthesis	1970. Y. Suketa *et al. Chem. Pharm. Bull.* **18**, 249.
N-Malonyl-D-methionine	*Nicotiana rustica*	1968. D. Keglevic *et al. Arch. Biochem. Biophys.* **124**, 443.
S-(2-Hydroxy-2-carboxyethanethiomethyl)-L-cysteine	*Acacia georginae* seeds	1972. K. Ito and L. Fowden. *Phytochemistry* **11**, 2541.
N-Methyl-L-methionine-S-sulfoxide and N-methylmethionine sulfonium sulfoxide	Red algae	1974. K. Miyazawa and K. Ito. *Nippon Suisan Gakkaishi (Bull. Jpn. Soc. Sci. Fish.)* **40**, 655.

Compound		References
Cycloalliin (5-methyl-1,4-thiazane-3-carboxylic acid oxide) 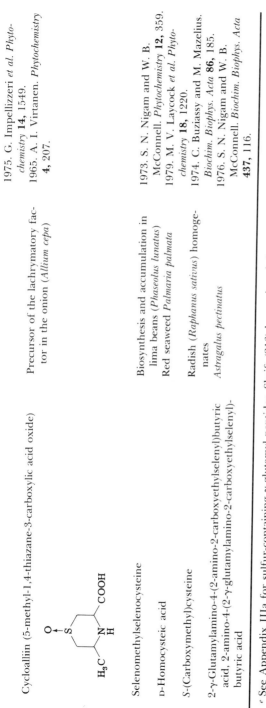	Precursor of the lachrymatory factor in the onion (*Allium cepa*)	1975. G. Impellizzeri et al. *Phytochemistry* **14**, 1549. 1965. A. I. Virtanen. *Phytochemistry* **4**, 207.
Selenomethylselenocysteine	Biosynthesis and accumulation in lima beans (*Phaseolus lunatus*)	1973. S. N. Nigam and W. B. McConnell. *Phytochemistry* **12**, 359.
D-Homocysteic acid	Red seaweed *Palmaria palmata*	1979. M. V. Laycock et al. *Phytochemistry* **18**, 1220.
S-(Carboxymethyl)cysteine	Radish (*Raphanus sativus*) homogenates	1974. C. Buziassy and M. Mazelius. *Biochim. Biophys. Acta* **86**, 185.
2-γ-Glutamylamino-4-(2-amino-2-carboxyethylselenyl)butyric acid, 2-amino-4-(2-γ-glutamylamino-2-carboxyethylselenyl)-butyric acid	*Astragalus pectinatus*	1976. S. N. Nigam and W. B. McConnell. *Biochim. Biophys. Acta* **437**, 116.

e See Appendix IIIa for sulfur-containing γ-glutamyl peptides. Shrift (216) has reviewed the metabolism of selenium by plants and animals,

APPENDIX III

Substituents on Glutamyl, Carbamoyl, and Guanidino Groups

Name and substituent group	Occurrence, recognition, etc.	References
a. γ-Glutamyl compounds (listed by the group attached to the γ-carboxyl of glutamic acid)[a]		
Homoglutathione (cysteinyl-β-alanine)	Phaseolus vulgaris seedlings	1963. P. R. Carnegie. Biochem. J. **89**, 459.
D-Alanine	Pea (Pisum) seeds	1973. M. Fukuda et al. Phytochemistry **12**, 2593.
Valylglycine	Juncus	1958. A. I. Virtanen and T. Ettala. Acta Chem. Scand. **12**, 78.
L-Arginine and S-(2-carboxy-N-propyl)-L-cysteine	Allium cepa	1970. E. J. Matikkala and A. I. Virtanen. Suom. Kemistil. B **43**, 435.
γ-Aminobutyric acid	Lunaria annua	1965. P. O. Larsen. Acta Chem. Scand. **19**, 1071.
L-α-(Methylenecyclopropyl)glycine	Seeds of Billia hippocastanum and fruits of Acer pseudoplatanus; Billia seeds also contained peptides with aspartate, asparagine glutamate, valine, alanine, and threonine	1972. L. Fowden et al. Phytochemistry **11**, 3521.
2-Methylenecycloheptene-1,3-diglycine	Mushroom (Lactarius helvus)	1964. E. Honkanen et al. Acta Chem. Scand. **18**, 1320.
β-Phenyl-β-alanine	Vigna (Phaseolus) angularis (Azuki bean)	1966. M. Koyama and Y. Obata. Agric. Biol. Chem. **30**, 472.
Ethylamine (theanine) and methylamide	Tea [Camellia (Thea) sinensis] leaves; N-methyl carbon of γ-glutamyl-methylamide enters purine RNA nucleotides and caffeine	1966. S. Konishi and E. Takahashi. Plant Cell Physiol. **7**, 171. 1972. S. Konishi et al. Plant Cell Physiol. **13**, 365, 695.

232

		Reference
Tyramine	[^{14}C]Methylamine yields γ-glutamylmethylamide, threo-bromine, caffeine, and CO_2 in *Camellia* (*Thea*) *sinensis*	1973. T. Suzuki. *Biochem. J.* **132**, 753.
Albizziine	*Tephrosia noctiflora*	1980. P. Forgacs *et al. Phytochemistry* **19**, 1225.
Asparagine Glutamic acid Djenkolic acid sulfoxide Aspartic acid	*Acacia georginae* seeds	1972. K. Ito and L. Fowden *Phytochemistry* **11**, 2541.
N-Amino-D-proline (linatine) Vitamin B_6 antagonist	Linseed (*Linum usitatissimum*)	1967. H. J. Kosterman *et al. Biochemistry* **6**, 170.
S-(1-Propenyl)-L-cysteine and its sulfoxide	Trans form in leaves of *Santalum album*	1974. R. Kuttan *et al. Biochemistry* **13**, 4394.
S-Allylmercapto-L-cysteine	Garlic (*Allium sativum*)	1964. M. Sugii *et al. Chem. Pharm. Bull.* **12**, 1114, 1115.
Cystine (N,N'-bis derivative)		1969. Cited by R. L. M. Synge. *Annu. Rev. Plant Physiol.* **19**, 117.
3,3'-(2-Methylethylene-1,2-dithio)dialanine and the N,N'-bis derivative	Seeds of *Allium schoenoprasum*	1964. E. J. Matikkala and A. I. Virtanen. *Acta Chem. Scand.* **18**, 2009.
Numerous γ-glutamylpeptides, e.g., N,N'-bis-(γ-glu)-L-cystine; N,N'-bis-(γ-glu)-3-3'-(2-methylethylene-1,2-dithio)dialanine; γ-glu-S-(propenyl-1-yl)-L-cysteine; γ-glu-S-(propenyl-1-yl)cysteinyl-S-(propen-1-yl)cysteine sulfoxide; and others	Bulbs and seeds of *Allium* spp.	1966. A. I. Virtanen. *Bot. Mag.* **79**, 506. 1966. E. J. Matikkala and A. I. Virtanen. *Suom. Kemistil. B* **39**, 201.

[a] Earlier reports on naturally occurring peptides have been reviewed by Waley (275). Selenopeptides are cited in Appendix IIi.

(continued)

233

Name and substituent group	Occurrence, recognition, etc.	References
	Seeds of *Fagus sylvatica*	1974. I. Kristensen *et al. Phytochemistry* **13**, 2803.
3-1-Uracil)-L-alanine and L-phenylalanyl-3-(1-uracil)-L-alanine	*Fagus sylvatica*	1974. I. Kristensen and P. O. Larsen. *Phytochemistry* **13**, 2799.
L-Pipecolic acid	*Gleditsia caspica*	1974. G. Dardenne *et al. Phytochemistry* **13**, 1515.
Lentinic acid	*Lentinus edodes*	1976. K. Yasumoto *et al. Nippon Nogei Kagaku Kaishi* **50**, 563.
$\underset{\mid}{\overset{NH_2}{MeSO_4CH_2(SOCH_2)_3CHCOOH}}$ (Me = methyl)		1976. G. Höfle *et al. Tetrahedron Lett.*, p. 3129.

b. Selected references to the occurrence of urea and carbamoyl derivatives in plants

Urea $\overset{O}{\underset{\parallel}{NH_2CNH_2}}$	Urea was earlier regarded as a specifically animal product, and in plants. asparagine was considered to be its metabolic counterpart	
	Traces of urea in higher plants: *Brassica, Cichorium, Daucus,* etc.	1912. R. Fosse. *C. R. Hebd. Seances Acad. Sci. Ser. D* **155**, 851; **156**, 1938.
	Urea in ferns and horsetails (*Equisetum*)	1912. H. Weyland. *Jahrb. Wiss. Bot.* **51**, 1.
	Isolation and crystallization of urease	1926. J. B. Sumner. *J. Biol. Chem.* **69**, 435.
	Free urea in plants	1932. G. Klein and K. Taubóck. *Biochem. Z.* **251**, 10.
	Urea converted to guanidine by *Aspergillus niger*	1931. N. N. Ivanov and A. N. Ivetisova. *Biochem. Z.* **231**, 67.
	Urea or ornithine cycle in liver	1932. H. A. Krebs and K. Henseleit. *Z. Physiol. Chem.* **210**, 33.

234

Allophanic acid

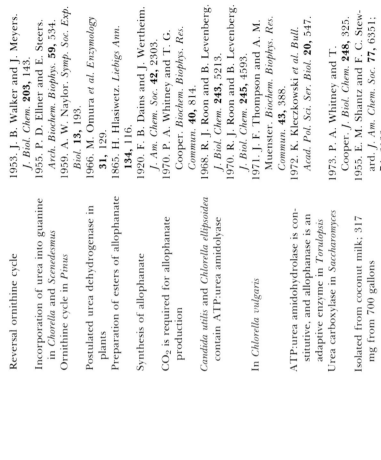

$$NH_2\overset{O}{\overset{\|}{C}}NH\overset{O}{\overset{\|}{C}}{-}OH$$

Diphenylurea

Reversal ornithine cycle	1953. J. B. Walker and J. Meyers. *J. Biol. Chem.* **203**, 143.
Incorporation of urea into guanine in *Chorella* and *Scenedesmus*	1955. P. D. Ellner and E. Steers. *Arch. Biochem. Biophys.* **59**, 534.
Ornithine cycle in *Pinus*	1959. A. W. Naylor. *Symp. Soc. Exp. Biol.* **13**, 193.
Postulated urea dehydrogenase in plants	1966. M. Omura et al. *Enzymology* **31**, 129.
Preparation of esters of allophanate	1865. H. Hlasiwetz. *Liebigs Ann.* **134**, 116.
Synthesis of allophanate	1920. F. B. Dains and J. Wertheim. *J. Am. Chem. Soc.* **42**, 2303.
CO_2 is required for allophanate production	1970. P. A. Whitney and T. G. Cooper. *Biochem. Biophys. Res. Commun.* **40**, 814.
Candida utilis and *Chlorella ellipsoidea* contain ATP:urea amidolyase	1968. R. J. Roon and B. Levenberg. *J. Biol. Chem.* **243**, 5213.
	1970. R. J. Roon and B. Levenberg. *J. Biol. Chem.* **245**, 4593.
In *Chlorella vulgaris*	1971. J. F. Thompson and A. M. Muenster. *Biochem. Biophys. Res. Commun.* **43**, 388.
ATP:urea amidohydrolase is constitutive, and allophanase is an adaptive enzyme in *Torulopsis*	1972. K. Kleczkowski et al. *Bull. Acad. Pol. Sci. Ser. Biol.* **20**, 547.
Urea carboxylase in *Saccharomyces*	1973. P. A. Whitney and T. Cooper. *J. Biol. Chem.* **248**, 325.
Isolated from coconut milk; 317 mg from 700 gallons	1955. E. M. Shantz and F. C. Steward. *J. Am. Chem. Soc.* **77**, 6351; **74**, 6133.

(continued)

235

Name and substituent group	Occurrence, recognition, etc.	References
Hydantoin H_2N — (N, H) — COOH structure (hydantoic acid; N-carbamoylglycine)	Etiolated sugar beet (*Beta vulgaris*) sprouts Tree buds and white shoots of sugar beets Urea condenses with C_2 fragment to hydantoin and hydantoic acid in *Candida flareri*	1896. E. O. Von Lippmann. *Ber. Dtsch. Chem. Ges.* **29**, 2645. 1950. E. Ware. *Chem. Rev.* **46**, 403. 1963. A. R. Cook and D. Boulter. *Biochem. J.* **88**, 69.
Carbamoyl phospate $NH_2C\!-\!O\!\sim\!PO_3H_2$ (with O double bond)	Synthesis and donor in enzymatic citrulline synthesis Carbamoyl transfer to amino acids in rat liver Precursor for ureido carbon in biotin Decomposition products in water	1955. M. E. Jones *et al. J. Am. Chem. Soc.* **77**, 819. 1956. J. Lowenstein and P. Cohen. *Biochem. J.* **63**, 11. 1963. A. Lezius *et al. Biochem. Z.* **336**, 510. 1964. C. M. Allen, Jr. and M. E. Jones. *Biochemistry* **3**, 1238.
Ureidoglycine H_2N $O\!=\!C$ COOH $HH\!-\!CHNH_2$	Hydrolytic degradation of allantoin by *Streptococcus* and *Arthrobacter* spp.	1963. G. D. Vogels. "On the Microbial Metabolism of Allantoin."
Carbamoylaspartic acid (ureidosuccinic acid) HOOC H_2N — (N, H) — COOH structure	Enzymatic synthesis of pyrimidines Possible intermediate in pyrimidine biosynthesis in plants Early product of [^{14}C]urea assimilation in *Picea*	1954. I. Lieberman and A. Kornberg. *J. Biol. Chem.* **207**, 911. 1955. I. Lieberman and A. Kornberg. *J. Biol. Chem.* **212**, 909. 1961. J. Buchowicz *et al. Acta Biochim. Pol.* **8**, 377. 1973. D. J. Durzan. *Can. J. Bot.* **51**, 1197.

Compound	Source/Description	Reference
L(+)-Citrulline $$NH_2CNH(CH_2)_3CH(NH_2)COOH$$ (with O double-bonded to C)	Watermelon [*Citrullus lanatus* (*C. vulgaris*)]	1930. M. Wada. *Biochem. Z.* **224**, 420.
	Reported in edestin, gluten, and mucin	1933. M. Wada. *Biochem. Z.* **227**, 1.
	Free in *Alnus* root and root nodules	1939. W. R. Fearon. *Biochem. J.* **33**, 902.
	Sap of several trees	1952. J. K. Miettinen and A. I. Virtanen. *Physiol. Plant.* **5**, 540.
	Sap of several trees	1954. G. Reuter and H. Wolffgang. *Flora (Jena)* **142**, 146.
	Porcupine quills and rabbit fur	1957. E. G. Bollard. *Aust. J. Biol. Sci.* **10**, 292.
		1962. G. E. Rogers. *Nature (London)* **194**, 1149.
Homocitrulline	Claimed on crown gall	1964. A. Ménagé and G. Morel. *C. R. Hebd. Seances Acad. Sci. Ser. D* **259**, 4795.
4-Hydroxycitrulline	Suspected in *Vicia faba* seeds	1964. A. Bell and A. S. L. Tirimanna. *Biochem. J.* **91**, 356.
L-Citrullinyl-L-arginine	Red algae	1974. K. Miyazawa and K. Ito. *Nippon Suisan Gakkaishi (Bull. Jpn. Soc. Sci. Fish.)* **40**, 815.
β-Ureidopropionic acid (carbamoyl-β-alanine) COOH | CH$_2$ | CH$_2$ | O NH ‖ | H$_2$N—C—NH	Enzymatic conversions of dihydrouracil and β-ureidopropionic acid	1955. S. Grisolia and D. P. Wallach. *Biochim. Biophys. Acta* **18**, 449.
	Synthesis from carbamoyl phosphate	1958. J. Caravaca and S. Grisolia. *J. Biol. Chem.* **231**, 357.
	Pyrimidine precursor in plants	1963. J. Buchowicz et al. *Acta Biochim. Pol.* **10**, 157.
	Product of uracil degradation in *Pinus banksiana*	1973. D. J. Durzan et al. *Can. J. For. Res.* **3**, 209.

(*continued*)

Name and substituent group	Occurrence, recognition, etc.	References
L-Amino-3-(1-hydroxyureido)propionic acid	Alkali hydrolysis product of quisqualic acid	1975. T. Takemoto et al. Yakugaku Zasshi (J. Pharm. Soc. Jpn.) **95**, 326, 448.

L(−)-α-Amino-β-ureidopropionic acid (albizziine)
 	
HOOC—CHCH$_2$NHCNH$_2$
    | ∥
  NH$_2$ O
	Albizia species and synthesis	1958. R. Gmelin et al. Z. Naturforsch. **13**, 252.
		1959. R. Gmelin. Z. Physiol. Chem. **314**, 28.
		1959. A. Kjaer et al. Experientia **15**, 253.
		1959. A. Kjaer et al. Acta Chem. Scand. **13**, 1565.

β-Ureidoisobutyric acid
   COOH
   |
   CHCH$_3$
   |
O CH$_2$
∥ |
H$_2$N—C —NH
	Formed during incubation of tissue slices with pyrimidines
	
	Thymine is degraded to β-aminoisobutyric acid in germinating seedlings
	
	Product of thymine degradation in *Pinus banksiana*
		1953. R. M. Fink et al. J. Biol. Chem. **201**, 349.
		1961. W. R. Evans and B. Axelrod. Plant Physiol. **36**, 9.
		
		1975. J. Pitel and D. J. Durzan. Can. J. Bot. **53**, 673.

Allantoic acid H$_2$N HN$_2$ | | O=C COOH C=O | | | HN—CH——NH	Prepared from allantoin First report in nature (*Phaseolus vulgaris*) Leaves of *Acer pseudoplatanus* Allantoin and allantoin products of [8-^{14}C]guanine degradation in *Pinus banksiana*	1848. A. Schlieper. Liebigs Ann. **67**, 214, 231. 1928. R. Fosse. Bull. Soc. Chim. Biol. **10**, 301. 1928. R. Fosse and A. Hieulle. Bull. Soc. Chim. Biol. **10**, 308. 1970. M. Pandita and D. J. Durzan. Proc. Can. Soc. Plant Physiol. **10**, 62.

Allantoin

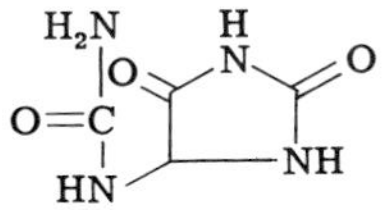

H_2N ... structure (with $O=C$, NH, HN, $C=O$ ring)

Prepared from uric acid; name

First report in plants in shoots of *Platanus orientalis*

Acer pseudoplatanus and *Aesculus hippocastanum*

Allantoin not formed from urea but from glycine

Derived from arginine in potassium-deficient barley (*Hordeum vulgare*)

Intermediate in nicotine biosynthesis

1838. F. Wöhler. *Liebigs Ann.* **26**, 241, 244.

1881. E. Schulze and J. Barbieri. *Ber.* **14**, 1602.

1885. E. Schulze and E. Bosshard. *Z. Physiol. Chem.* **9**, 420.

1959. R. M. Krupka and G. H. N. Towers. *Can. J. Bot.* **37**, 539.

1964. T. A. Smith and J. L. Garraway. *Phytochemistry* **3**, 23.

1965. T. A. Smith. *Phytochemistry* **4**, 599.

1966. D. Yoshida and T. Mitake. *Plant Cell Physiol.* **7**, 301.

N-Carbamoylputrescine

NH_2
|
$C=O$
|
NH
|
$(CH_2)_4$
|
NH_2

Gongrine

$$NH_2CNHCNH(CH_2)_3COOH$$
(with NH and O)

Red alga *Gymnogongrus flabelliformis*

1965. K. Ito and Y. Hashimoto. *Agric. Biol. Chem.* **29**, 832.

Gigartinine

$$NH_2CNHCNH(CH_2)_3CHCOOH$$
(with NH, O, and NH_2)

Red alga *Gymnogongrus flabelliformis*

1966. K. Ito and Y. Hashimoto. *Nippon Suisan Gakkaishi* (*Bull. Jpn. Soc. Sci. Fish.*) **32**, 274; *Nature* (*London*) **211**, 417.

(continued)

239

APPENDIX III (Continued)

Name and substituent group	Occurrence, recognition, etc.	References
N-[(9-β-D-Ribofuranosylpurin-6-yl)carbamoyl]threonine and N-(purin-6-ylcarbamoyl)threonine, or N-(nebularin-6-yl-carbamoyl)threonine 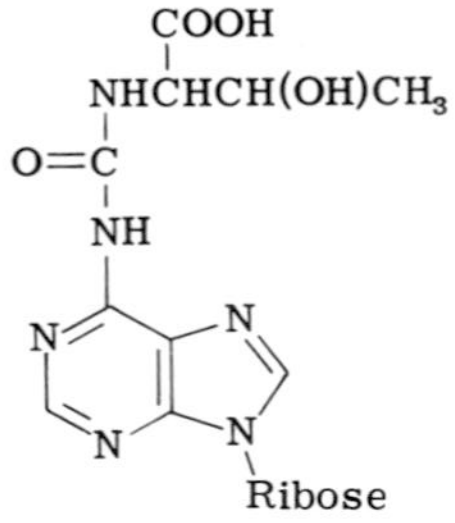	Yeast tRNA tRNA species with codons starting with A	1969. G. B. Chheda et al. Biochemistry **8**, 3278. 1969. M. P. Schweizer et al. Biochemistry **8**, 3283. 1969. H. Ishikura et al. Biochem. Biophys. Res. Commun. **37**, 990. 1972. D. M. Powers and A. Peterkofsky. J. Biol. Chem. **247**, 6394.

COOH
|
NHCHCH(OH)CH₃

O=C   NH

(purine ring with Ribose)

| N-[(9-β-D-Ribofuranosylpurin-6-yl)-N-methylcarbamoyl]-threonine 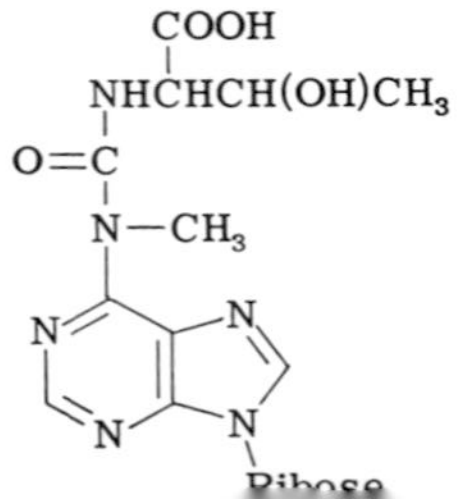 | Escherichia coli tRNA^Thr | 1972. F. Kimura-Harada et al. Biochemistry **11**, 3910. |

COOH
|
NHCHCH(OH)CH₃

O=C   N—CH₃

(purine ring with Ribose)

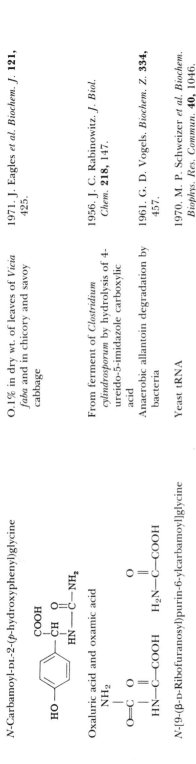

N-Carbamoyl-DL-2-(*p*-hydroxyphenyl)glycine

0.1% in dry wt. of leaves of *Vicia faba* and in chicory and savoy cabbage

1971. J. Eagles *et al. Biochem. J.* **121**, 425.

Oxaluric acid and oxamic acid

From ferment of *Clostridium cylindrosporum* by hydrolysis of 4-ureido-5-imidazole carboxylic acid

1956. J. C. Rabinowitz. *J. Biol. Chem.* **218**, 147.

Anaerobic allantoin degradation by bacteria

1961. G. D. Vogels. *Biochem. Z.* **334**, 457.

N-[9-(β-D-Ribofuranosyl)purin-6-ylcarbamoyl]glycine

Yeast tRNA

1970. M. P. Schweizer *et al. Biochem. Biophys. Res. Commun.* **40**, 1046.

Mass spectra

1972. S. M. Hecht and J. J. McDonald. *Anal. Biochem.* **47**, 157.

c. Free guanidine and amidino compounds in plants[b,c]

Guanidine

Seedling of *Vicia sativa* and *Pisum sativum*

1891. E. Schulze. *Z. Physiol. Chem.* **15**, 140.

1893. E. Schulze. *Z. Physiol. Chem.* **17**, 193.

[b] Apart from enduracididine, divicin, vicin, and lathyrine, other substances containing guanidine in a ring (e.g., guanine, pteridines, etc.) are not included.

[c] Many examples of other substances in this category that occur naturally but mainly in microorganisms and animals are cited by Van Thoai (266).

(*continued*)

APPENDIX III (Continued)

Name and substituent group	Occurrence, recognition, etc.	References
	Sugar beets (*Beta vulgaris*)	1892. E. Schulze. *Ber. Dtsch. Chem. Ges.* **25**, 658, 661.
		1896. E. O. von Lippmann. *Ber. Dtsch. Chem. Ges.* **29**, 2645.
	Maize (*Zea mays*) seedlings	1915. E. Winterstein and F. Wünsch. *Z. Physiol. Chem.* **95**, 310.
	Occurrence not confirmed in *Vicia*	1922. A. Kiesel. *Z. Physiol. Chem.* **118**, 267.
	Rye (*Secale cereale*)	1924. A. Kiesel. *Z. Physiol. Chem.* **135**, 61.
	Guanidine in higher plants	1940. E. Mueller and K. Armbrust. *Z. Physiol. Chem.* **263**, 41.
	Traces in concentrated dialysates of different plants	1954. M. Mourgue and R. Dokhan. *C. R. Hebd. Seances Acad. Sci. Ser. D* **239**, 1518; *C. R. Seances Soc. Biol. Ses Fil.* **148**, 1434.
	Product of arginine catabolism in cell suspensions of sugar cane	1969. A. Maretzki *et al.* *Phytochemistry* **8**, 811.
	Precursor of galegine in *Galega officinalis*	1967. G. Reuter and A. Barthel. *Pharmazie* **22**, 261.
	In congocidine from *Streptomyces*	1967. M. Julia and N. Preau-Joseph. *Bull. Soc. Chim. Fr.* **11**, 4348.
	Precursor of dimethylallylguanidine	1974. J. Steiniger and G. Reuter. *Biochem. Physiol. Pflanz.* **166**, 275.
	Potato, cowpea, and other plants (e.g., clover, wheat, and rye)	1911. M. X. Sullivan. *J. Am. Chem. Soc.* **33**, 2035.
		1912. E. C. Shorey. *J. Am. Chem. Soc.* **34**, 99.

Glycocyamine (guanidinoacetic acid)

$$NH_2C(=NH)NHCH_2COOH$$

Creatine (methylguanidinoacetic acid)

$$NH_2C(=NH)N(CH_3)CH_2COOH$$

Creatinine (anhydride) 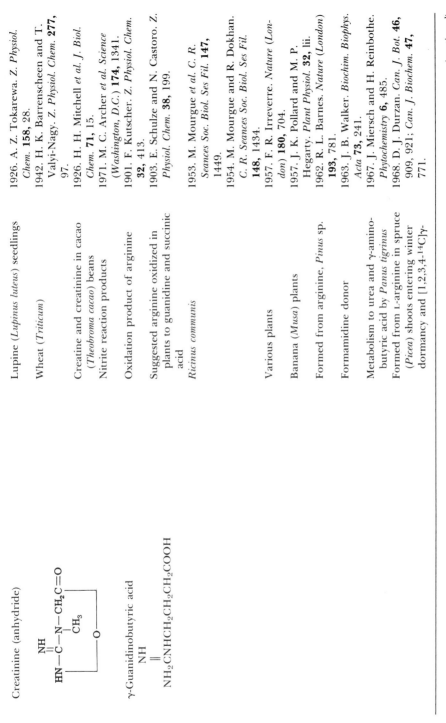	Lupine (*Lupinus luteus*) seedlings	1926. A. Z. Tokarewa. *Z. Physiol. Chem.* **158**, 28.
	Wheat (*Triticum*)	1942. H. K. Barrenscheen and T. Valyi-Nagy. *Z. Physiol. Chem.* **277**, 97.
	Creatine and creatinine in cacao (*Theobroma cacao*) beans	1926. H. H. Mitchell *et al. J. Biol. Chem.* **71**, 15.
	Nitrite reaction products	1971. M. C. Archer *et al. Science* (*Washington, D.C.*) **174**, 1341.
	Oxidation product of arginine	1901. F. Kutscher. *Z. Physiol. Chem.* **32**, 413.
	Suggested arginine oxidized in plants to guanidine and succinic acid	1903. E. Schulze and N. Castoro. *Z. Physiol. Chem.* **38**, 199.
γ-Guanidinobutyric acid	*Ricinus communis*	1953. M. Mourgue *et al. C. R. Seances Soc. Biol. Ses Fil.* **147**, 1449.
	Various plants	1954. M. Mourgue and R. Dokhan. *C. R. Seances Soc. Biol. Ses Fil.* **148**, 1434.
		1957. F. R. Irreverre. *Nature* (*London*) **180**, 704.
	Banana (*Musa*) plants	1957. J. K. Pollard and M. P. Hegarty. *Plant Physiol.* **32**, lii.
	Formed from arginine, *Pinus* sp.	1962. R. L. Barnes. *Nature* (*London*) **193**, 781.
	Formamidine donor	1963. J. B. Walker. *Biochim. Biophys. Acta* **73**, 241.
	Metabolism to urea and γ-amino-butyric acid by *Panus tigrinus*	1967. J. Miersch and H. Reinbothe. *Phytochemistry* **6**, 485.
	Formed from L-arginine in spruce (*Picea*) shoots entering winter dormancy and [1,2,3,4-^{14}C]γ-	1968. D. J. Durzan. *Can. J. Bot.* **46**, 909, 921; *Can. J. Biochem.* **47**, 771.

(*continued*)

Name and substituent group	Occurrence, recognition, etc.	References
	guanidinobutyrate conversion to γ-aminobutyric acid in buds in spring	
	Arctium lappa	1975. Y. Yamada et al. Phyto-chemistry 14, 582.
4-Guanidinobutanol NH ‖ NH₂CNH(CH₂)₃CH₂OH	Leonurus sibiricus	1971. G. Reuter and H. J. Diehl. Pharmazie 26, 777.
L-Arginine (α-amino-δ-guanidinovaleric acid) NH NH₂ ‖ │ NH₂CNHCH₂CH₂CH₂CHCOOH	Lupinus luteus	1886. E. Schulze and E. Steiger. Chem. Ber. 19, 1177; Z. Physiol. Chem. 11, 43.
	Protein hydrolysates of horn	1894. S. G. Hedin. Z. Physiol. Chem. 20, 186
	Picea, Pinus, and Abies seeds	1896. E. Schulze. Z. Physiol. Chem. 24, 276; 25, 360.
	Color reaction with α-naphthol	1925. S. Sakaguchi. J. Biochem. 5, 25.
	Ornithine or urea cycle	1932. H. A. Krebs and K. Hense-leit. Z. Physiol. Chem. 32, 413.
	Ornithine cycle in plants (Pinus sp.)	1959. A. W. Naylor. Symp. Soc. Exp. Biol. 13, 19.
	Precursor of numerous monosubstituted guanidines in Picea	1969. D. J. Durzan. Can. J. Bot. 46, 909.
		1969. D. J. Durzan. Can. J. Biochem. 47, 771.
Agmatine (4-aminobutylguanidine) NH ‖ NH₂CNHCH₂CH₂CH₂CH₂NH₂	Ergot (Claviceps purpurea)	1910. R. Engeland and F. Kutscher. Zentralbl. Physiol. 24, 589; Chem. Zentralbl. 11, 1762.

Ragweed (*Ambrosia*) pollen (protein extract)	1919. F. W. Heyl. *J. Am. Chem. Soc.* **41**, 670.
Rye (*Secale cereale*)	1924. A. Kiesel. *Z. Physiol. Chem.* **135**, 61.
Potassium deficient in barley (*Hordeum vulgare*) plants	1965. C. Hackett *et al. Ann. Bot.* N S **29**, 331.
Ricinus communis	1953. M. Mourgue *et al. C. R. Seances Soc. Biol. Ses Fil.* **147**, 1449.
Arginine decarboxylase in barley	1963. T. A. Smith. *Phytochemistry* **2**, 241.
Intermediate in putrescine biosynthesis (*Hordeum vulgare*)	1965. T. A. Smith. *Phytochemistry* **4**, 599.
Degradation to putrescine and urea (*Escherichia coli*)	1966. D. R. Morris and A. B. Pardee. *J. Biol. Chem.* **241**, 3129.
Intermediate in nicotine biosynthesis in tobacco (*Nicotiana*) plants	1966. D. Yoshida and T. Mitake. *Plant Cell Physiol.* **7**, 301.
Metabolism to arcaine, urea, and putrescine (*Panus tigrinus*)	1971. A. Boldt *et al. Phytochemistry* **10**, 731.
Lathyrus sativus seedlings	1973. S. Ramakrishna and P. R. Adiga. *Phytochemistry* **12**, 2691.
Homoagmatine	
p-Coumaroylagmatine	
Barley (*Hordeum vulgare*)	1965. A. Stoessl. *Phytochemistry* **4**, 977.
Antifungal factors in barley	1966. A. Stoessl. *Adv. Chem. Ser.*, No. 53, 80; *Tetrahedron Lett.* p. 2287, p. 2849.
Phenylpropanoid derivatives of agmatine and their glucosides (hordatine A and B)	1963. K. Koshimizu *et al. Can. J. Bot.* **41**, 744.
Isolation and synthesis	1966. A. Stoessl. *Tetrahedron Lett.* **21**, 2287; **25**, 2849.

(*continued*)

Name and substituent group	Occurrence, recognition, etc.	References
erythro-γ-Hydroxy-L-arginine $$\underset{\substack{\|\| \\ NH}}{NH_2C}NHCH_2CH\underset{\substack{\| \\ OH}}{C}H\underset{\substack{\| \\ NH_2}}{C}H_2CHCOOH$$	*Polychera rufescens, erythro* and L-configuration	1959. Y. Fugita. *Bull. Chem. Soc. Jpn.* **32**, 43.
		1960. Y. Fugita. *Bull. Chem. Soc. Jpn.* **33**, 1379.
	Lathyrus sp. (lactone)	1964. E. A. Bell and A. S. L. Tiri-manna. *Biochem. J.* **91**, 35.
	Lactone from *Vicia sativa*	1964. G. E. Hirst and R. G. Foster. *Biochem. J.* **91**, 361.
L-Homoarginine $$\underset{\substack{\|\| \\ NH}}{NH_2C}\underset{\substack{\| \\ NH}}{}NH(CH_2)_4CHCOOH$$	Seeds of *Lathyrus* spp.	1962. E. A. Bell. *Nature (London)* **196**, 1978.
		1962. E. A. Bell. *Biochem. J.* **85**, 91.
		1963. S. L. N. Rao *et al. Biochemistry* **2**, 298.
	Lathyrus sp.	1963. E. A. Bell. *Nature (London)* **199**, 70.
	Metabolism in crown-gall tumors	1966. A. Petit and G. Morel. *C. R. Seances Soc. Biol. Ses Fil.* **160**, 1806.
γ-Hydroxyhomoarginine $$\underset{\substack{\|\| \\ NH}}{NH_2C}NHCH_2CH_2\underset{\substack{\| \\ OH}}{C}HCH_2\underset{\substack{\| \\ NH_2}}{C}HCOOH$$	*Lathyrus tingitanus*	1964. E. A. Bell. *Biochem. J.* **91**, 356.
	Suggested that homoarginine is directly hydroxylated and prop-erties of lactone	1964. G. E. Hirst and R. G. Foster. *Biochem. J.* **91**, 361.
	Threo isomer, synthesis and isola-tion	1965. Y. Fujita *et al. J. Am. Chem. Soc.* **87**, 2030.
	In pea seedlings	1976. T. A. Smith and G. R. Best. *Phytochemistry* **15**, 1565.

α-Keto-δ-guanidinovaleric acid

$$NH_2CNHCH_2CH_2CH_2CCOOH$$
(=NH on the first C, =O on the last C)

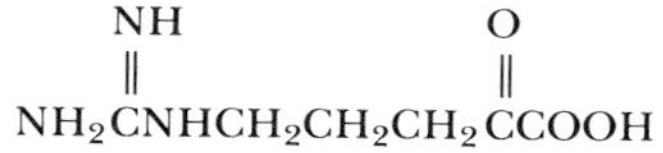

Phlox decussata	1964. G. Brandner and A. I. Virtanen. Acta Chem. Scand. **18**, 574.
Picea glauca	1966. D. J. Durzan and R. G. Richardson. Can. J. Biochem. **44**, 141.
Germinating seedlings of jack pine (Pinus banksiana)	1971. P. K. Ramaiah et al. Can. J. Bot. **49**, 2151.
Intermediate in Pseudomonas putida	1971. D. L. Miller and V. W. Rodwell. J. Biol. Chem. **246**, 5053.

Arginic acid (α-hydroxy-δ-guanidinovaleric acid)

$$NH_2CNHCH_2CH_2CH_2CHCOOH$$
(=NH on the first C, OH on the last C)

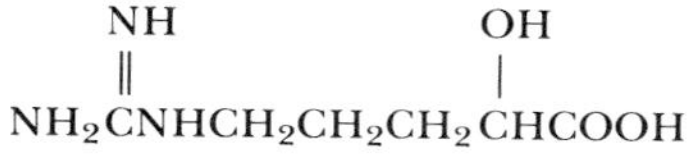

| Suggested intermediate in tissue cultures of Jerusalem artichoke (Helianthus tuberosus) | 1958. G. Morel and H. Duranton. Bull. Soc. Chim. Biol. **40**, 2155. |

N^G-Methyl and N^G,N^G-dimethylarginine

Seeds of broad bean (Vicia faba)	1976. T. Kasai et al. Agric. Biol. Chem. **40**, 2449.
Methylamine breakdown product in acid hydrolysates	1980. W. K. Paik and S. Kim. Biochem. Biophys. Res. Commun. **97**, 8.
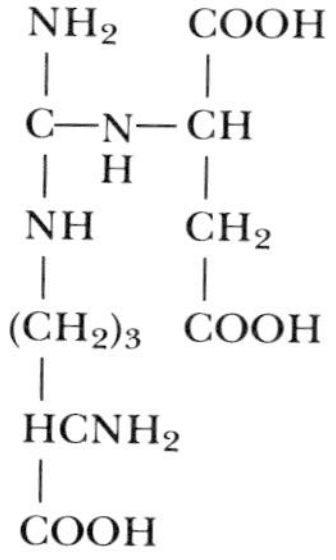 Extracts of pea seedlings	1952. D. C. Davidson and W. H. Elliott. Nature (London) **169**, 313.
Chlorella	1952. J. B. Walker. Proc. Natl. Acad. Sci. USA **38**, 561.
Anhydrides	1953. S. Ratner et al. J. Biol. Chem. **204**, 95.

Argininosuccinic acid

$$
\begin{array}{ccc}
NH_2 & & COOH \\
| & & | \\
C & -N- & CH \\
\| & H & | \\
NH & & CH_2 \\
| & & | \\
(CH_2)_3 & & COOH \\
| & & \\
HCNH_2 & & \\
| & & \\
COOH & &
\end{array}
$$

Separation of spontaneous conversion products	1960. R. G. Westall. Biochem. J. **77**, 135.
	1970. C. R. Lee and R. J. Pollitt. Tetrahedron **26**, 3113.
	1966. S. Ratner and M. Kunkemueller. Biochemistry **5**, 1821 (cf. J. Biol. Chem. **204**, 95).
New derivatives in urine	1972. C. R. Lee and R. J. Pollitt. Biochem. J. **126**, 79.

(continued)

247

APPENDIX III (Continued)

Name and substituent group	Occurrence, recognition, etc.	References
Galegine (isoamylideneguanidine) $\underset{\|\|}{NH}$ $CH_3C=CHCH_2NHCNH_2$ $\|$ CH_3	Seeds of *Galega officinalis* Leaves	1914. G. Tanret. *Bull. Soc. Chim. Fr.* **15**, 613. *C. R. Hebd. Seances Acad. Sci. Ser. D* **158**, 1182, 1426. 1923. G. Barger and F. D. White. *Biochem. J.* **17**, 827. 1925. H. Müller. *Z. Biol. (Munich)* **83**, 239.
4-Hydroxygalegine $\underset{\|\|}{NH}$ $HOCH_2C=CHCH_2NHCNH_2$ $\|$ CH_3	*Galega officinalis* Isolation; hydroxyleucenine may be a precursor for 4-hydroxy-galegine Postulates transamidination of arginine with isopentenylamine Seeds of *Galega officinalis* Biosynthesis from arginine	1961. K. Pufahl and K. Schreiber. *Experientia* **17**, 302. 1963. N. V. Thoai and G. Desvages. *Bull. Soc. Chim. Biol.* 45, 413. 1964. G. Reuter. *Flora (Jena)* **154**, 136. 1964. K. Schreiber *et al. Liebigs Ann. Chem.* **671**, 147. 1968. A. Barthel and G. Reuter. *Pharmazie* **23**, 26.
Sphaerophysin	Isolation from *Sphaerophysa salsula* as isopropylvinylagmatine *Smirnowia turkestana*	1944. M. M. Rubinschtein and G. P. Menschikow. *J. Gen. Chem. USSR (Engl. Transl. of Zh. Obshch. Khim.)* **14**, 161, 172. 1947. A. A. Rjabinin. *J. Gen. Chem. USSR (Engl. Transl. of Zh. Obshch. Khim.)* **17**, 2265.
Identifications as 4-[3-methylbut-2-enylamino-(1)]butyl-1-guanidine	Isolation from *Sphaerophysa*	1957. A. J. Birch *et al. J. Chem. Soc.,* p. 410. 1968. I. Krone and G. Reuter. *Dtsch. Pharm. Ges.* **301**, 64.

Smirnovin	*Smirnovia turkestana*	1947. A. A. Rjabinin. *J. Gen. Chem. USSR (Engl. Transl. of Zh. Obshch. Khim.)* **17**, 2265; **27**, 221 (1954).

$$\underset{\underset{CH_3}{|}}{CH_3C}=CHCH_2\underset{\underset{R}{|}}{N}CH_2NH(CH_2)_4NH\overset{\overset{NH}{||}}{C}NH_2$$

R = acetyl
Smirnovinin
R = malonyl

1951. A. A. Rjabinin and E. M. Iljina. *Dokl. Nauk. Acad. Sci. USSR* **76**, 851.

1954. A. A. Rjabinin and E. M. Iljina. *J. Gen. Chem. USSR (Engl. Transl. of Zh. Obshch. Khim.)* **27**, 221.

N^1,N^2,N^3-triisopentenylguanidine

$$Me_2C=CHCH_2NH\overset{\overset{=NCH_2CH=CMe_2}{}}{C}NHCH_2CHCMe_2$$

Bark and leaves of Alchornea javanensis (Euphorbiaceae)

1970. N. K. Hart *et al. Aust. J. Chem.* **23**, 1679.

N^1,N^2-diisopentenylguanidine

Pterogynine

1969. R. A. Corral *et al. Experientia* **25**, 1020.

Biosynthesis of isoprenoid guanidines

1974. G. Reuter and J. Steiniger. *Pharmazie* **29**, 72!

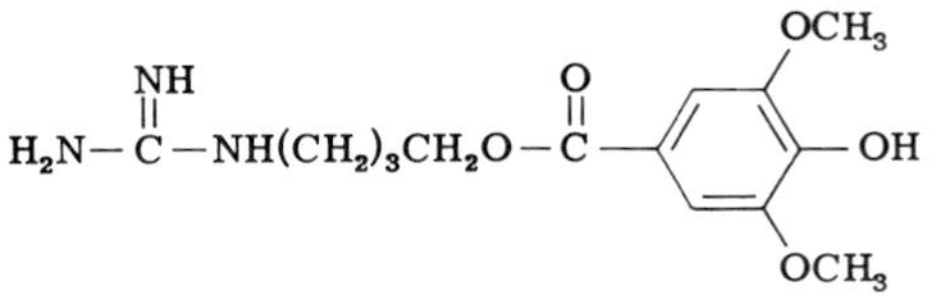

1962. Y. Hayashi. *Yakugaku Zasshi (J. Pharm. Soc. Jpn.)* **82**, 1020.

Leonurin

$$H_2N-\overset{\overset{NH}{||}}{C}-NH(CH_2)_3CH_2O-\overset{\overset{O}{||}}{C}-\text{(aromatic ring: 3,5-OCH}_3\text{, 4-OH)}$$

Leonurus sibiricus

1968. Y. Kishi *et al. Tetrahedron Lett.*, p. 637.

1971. G. Reuter and H. J. Diehl. *Pharmazie* **26**, 777.

(continued)

Name and substituent group	Occurrence, recognition, etc.	References

D-Octopine [N-α-(1-carboxyethyl)]arginine
 	
NH
‖
NH₂CNH(CH₂)₃CHCOOH
    |
   NH
    |
  CH₃CHCOOH
	Scallop and octopus muscle
	
	Pyruvate condensation
	
	Crown gall
	
	Octopine dehydrogenase from muscles of *Pecten maximus*
	Stereochemistry of hydrogen transfer
	Two dehydrogenases in crown gall
	
	One enzyme makes all 4 N^2-(1-carboxyethyl)amino acid derivatives in crown-gall tumors; purification of octopine synthase
		1927. K. Morizawa. *Acta School Med. Univ. Kyoto* **9**, 285.
		1937. S. Akasi. *J. Biochem.* **25**, 261, 281, 291.
		1939. F. Knoop and C. Martius. *Z. Physiol. Chem.* **258**, 238.
		1964. A. Ménagé and G. Morel. *C. R. Hebd. Seances Acad. Sci. Ser. D* **259**, 4795; **261**, 2001.
		1972. A. Olomucki. *Eur. J. Biochem.* **28**, 261.
		1973. J. F. Biellman *et al. FEBS Lett.* **32**, 254.
		1977. P. R. Birnberg *et al. Phytochemistry* **16**, 647.
		1980. E. Hack and J. D. Kemp. *Plant Physiol.* **65**, 949.

Nopaline, N^2-(1,3-dicarboxypropyl)-L-arginine
 	
NH
‖
NH₂CNH(CH₂)₃CHCOOH
    |
   NH
    |
  HOOC(CH₂)₂CHCOOH
	Structure and occurrence in crown galls
	Role of dehydrogenases and bacteria in tumor formation
		1969. A. Goldman *et al. C. R. Seances Acad. Sci. Ser. D* **268**, 852.
		1970. A. Petit *et al. Physiol. Veg.* **8**, 205.

N^α-(2-Hydroxysuccinyl) and N^α-(2-carboxymethyl-2-hydroxysuccinyl)arginine	Tubers of *Smilax china*, seeds of *Vicia faba*, and bulbs of *Lilium maximowiczii*	1983. T. Kasai *et al. Phytochemistry* **22**, 147.
Canavanine  $$\begin{array}{c} NH_2 \\ NH \quad | \\ \parallel \\ NH_2CNHOCH_2CH_2CHCOOH \end{array}$$	Seeds of *Canavalia ensiformis*	1929. M. Kitagawa and T. Tomiyama. *J. Biochem.* **11**, 265. 1933. M. Kitagawa and S. Monobe. *Biochemistry* **18**, 333.
	Cleavage by *Streptococcus* to homoserine and guanidine *Colutea arborescens*	1955. M. Kihara *et al. J. Biol. Chem.* **217**, 497. 1955. W. R. Fearon and E. A. Bell. *Biochem. J.* **59**, 221.
	Cleavage by *Streptococcus faecalis* to homoserine and hydroxyguanidine Various legumes	1958. G. D. Kalyankar *et al. J. Biol. Chem.* **223**, 1175. 1960. E. A. Bell. *Biochem. J.* **75**, 618.
Canavaninosuccinic acid 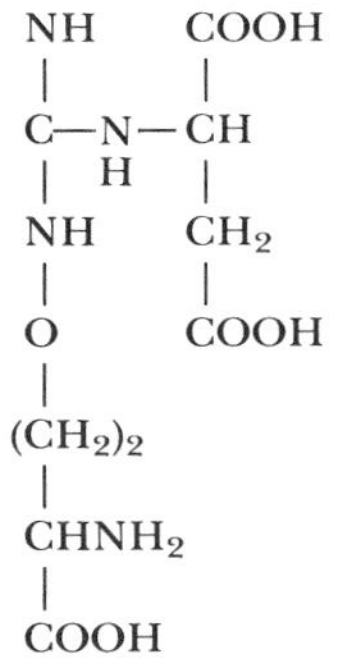 $$\begin{array}{ll} NH & COOH \\ | & | \\ C-N-CH \\ \quad H \\ | & | \\ NH & CH_2 \\ | & | \\ O & COOH \\ | \\ (CH_2)_2 \\ | \\ CHNH_2 \\ | \\ COOH \end{array}$$	Various legumes	1955. J. B. Walker. *J. Biol. Chem.* **204**, 119. 1960. E. A. Bell. *Biochem. J.* **75**, 618.

(continued)

251

APPENDIX III (Continued)

Name and substituent group	Occurrence, recognition, etc.	References
Desaminocanavine NH=CNHOCH₂CH₂CHCOOH $\quad\lfloor$ ──NH ┘ $NH=CNHOCH_2CH_2CHCOOH$ with NH bridge	Prepared from canavanine	1937. M. Kitagawa and J. Tsuka-moto. *J. Biochem.* **26**, 373.
2-Aminoimidazole	Seeds of *Mundulea sericea* 4,5-Dihydroderivative in chaksine from *Cassia abrus*	1977. L. E. Fellows *et al. Phytochemistry* **16**, 1399. 1958. K. Weisner *et al. J. Am. Chem. Soc.* **80**, 1521.
3-[2-Amino-2-imidazolin-(4S)-yl]alanine (enduracididine)	Seeds of *Lonchocarpus sericeus*	1977. L. E. Fellows and E. A. Bell. *Phytochemistry* **16**, 1957.
2-[2-Amino-2-imidazolin-(4S)-yl]acetic acid		

252

Divicin (2,4-diamino-5,6-dioxypyrimidine) and vicin

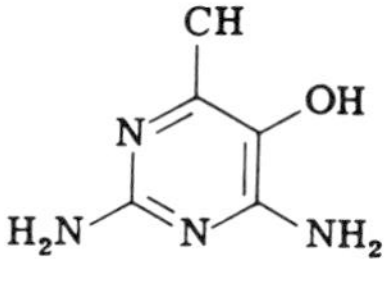

divicin

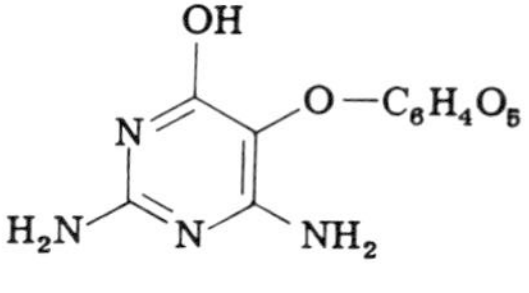

vicin

Lathyrine (tingitanine) (2-Aminopyrimidin-4-ylalanine)

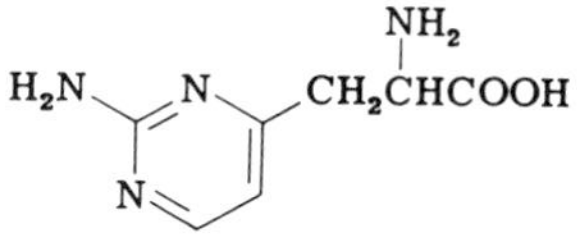

Gongrine (γ-guanidoureidobutyric acid)

$$NH_2CNHCNHCH_2CH_2COOH$$
(NH, O above)

Gigartinine N^5-(amidinocarbamoyl)-l-ornithine

$$NH_2CNHCNHCH(CH_2)_3CHCOOH$$
(NH, O, NH_2 above)

Aglycone of 5-β-D-glucoside in *Vicia sativa* and *V. faba*	1881. H. Ritthausen. *J. Prakt. Chem.* **24**, 202.
	1884. H. Ritthausen. *J. Prakt. Chem.* **29**, 359.
Vicin from *V. sativa*	1910–11. E. Schulze. *Z. Physiol. Chem.* **70**, 143; **71**, 33.
Constitution	1953. A. Bendich and G. C. Clement. *Biochim. Biophys. Acta* **12**, 46.
Synthesis of divicin	1956. J. Davoll and D. H. Laney. *J. Chem. Soc.*, p. 2124.
Lathyrus tingitanus	1961. E. Nowacki and J. Przyblyska. *Bull. Acad. Pol. Ser. Sci. Biol.* **9**, 279.
	1962. E. A. Bell and R. G. Foster. *Nature (London)* **194**, 91.
	1964. A. E. Bell. *Nature (London)* **203**, 378.
Biosynthesis from homoarginine and 4-hydroxyhomoarginine	1965. A. E. Bell and J. Przybylska. *Biochem. J.* **94**, 35.
Suggested biosynthesis from 4-oxo-homoarginine	1973. R. C. Hider and D. I. John. *Phytochemistry* **12**, 119.
Biosynthesis from orotate and serine	1977. E. G. Brown and N. F. Al-Baldawi. *Biochem. J.* **164**, 589.
Red alga *Gymnogongrus flabelliformis*	1965. K. Ito and Y. Hashimoto. *Agric. Biol. Chem.* **29**, 832.
Gymnogongrus flabelliformis	1966. K. Ito and Y. Hashimoto. *Nature (London)* **211**, 417; *Nippon Suisan Gakkaishi (Bull. Jpn. Soc. Sci. Fish.)* **32**, 274.

(continued)

253

APPENDIX III (Continued)

Name and substituent group	Occurrence, recognition, etc.	References
Indospicine (L-2-amino-6-amidinohexanoic acid) $H_2NCCH_2CH_2CH_2CH_2CHNH_2$ $\parallel$ $\mid$ NH $COOH$	Seasonal variation in *Chondrus crispus* Hepatotoxic amino acid from *Indigofera spicata* Effect on incorporation into protein and tRNA of rat liver Arginine antagonist	1977. M. V. Haycock and J. S. Craigie. *Can. J. Biochem.* **55**, 27. 1968. M. P. Hegarty and A. W. Pound. *Nature (London)* **217**, 354. 1970. N. P. Madsen *et al. Biochem. Pharmacol.* **19**, 853. 1972. T. Leisinger *et al. Biochim. Biophys. Acta* **262**, 214.

Arcaine
 	
   NH NH
   $\parallel$ $\parallel$
$NH_2CNH(CH_2)_4NH\,CNH_2$
	First isolation (mussel *Arca noae*)
	
	Claimed in marine algae
	
	Isolation from *Panus tigrinus*
	
	Biosynthesis from agmatine in fungus *Panus tigrinus*
		1931. F. Kutscher *et al. Z. Physiol. Chem.* **199**, 273.
		1962. E. J. Lewis and E. A. Gonsalves. *Ann. Bot. (London)* **26**, 301.
		1968. J. Miersch. *Naturwissenschaften* **35**, 493.
		1971. A. Boldt *et al. Phytochemistry* **10**, 731.

Argininylglutamine	Green alga *Cladophora*	1959. S. Makisumi. *J. Biochem.* **46**, 63.
γ-L-Glutamyl-L-arginine	*Allium cepa*	1970. E. J. Matikkala and A. I. Virtanen. *Suom. Kemistil. B* **43**, 435.
Various arginine peptides	*Chlorella* (e.g., Arg, Arg, Arg, Arg, Glu, Arg, Arg, Arg, etc.) *Chlorella pyrenoidosa* (^{14}C-labeled) Seasonal levels of citrullinyl-L-arginine	1965. T. Kanzawa *et al. Plant Cell Physiol.* **6**, 631. 1972. Z. Nejedly and K. Hybs. *J. Labelled Compd.* **8**, 183. 1977. M. V. Haycock and J. S. Craigie. *Can. J. Biochem.* **55**, 27.

Polypeptides of aspartic acid and arginine
 	
    Arg $\left(\begin{array}{c} \text{Arg} \\ \mid \\ \text{Asp}_n \end{array} \right)$ Arg
    $\mid$ $\mid$
^+H_3N-Asp-   - Asp · COO^-
	Cyanophycin granules from blue-green alga *Anabaena cylindrica*	1971. R. D. Simon. *Proc. Natl. Acad. Sci. USA* **68**, 265.
		1976. R. D. Simon and P. Weathers. *Biochim. Biophys. Acta* **420**, 165.

Stizolamine (1-methyl-3-guanidino-6-hydroxymethylpyrazin-2-one)   Seeds of *Mucuna* (*Stizolobium*) *hassjoo*   1976. T. Yoshida. *Phytochemistry* **15**, 1723.

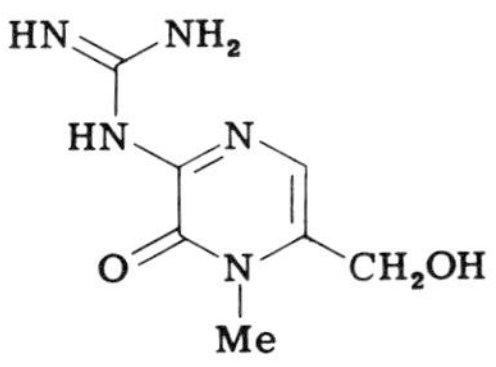

HN NH₂
N   N
HN   O   Me
CH₂OH

d. Selected properties of monosubstituted guanidines

Observation	Guanidine compound	References
Growth inhibition of plant tissue		
Tobacco	Arginine with a methylated guanidino N	1975. E. Tyihák et al. *Experientia* **31**, 818.
		1975. E. Tyihák et al. *Acta Agron. Acad. Sci. Hung.* **24**, 315.
Jack pine (*Pinus banksiana*)	γ-Guanidinobutyric acid	1976. D. J. Durzan and V. Chalupa. *Can. J. Bot.* **54**, 483.
Occurrence in plant tumors	Octopine and nopaline	1964. A. Ménagé and G. Morel. *Physiol. Vég.* **2**, 1; *C. R. Hebd. Seances Acad. Sci. Ser. D* **259**, 4795.
In normal plants and in crown gall	Octopine and nopaline	1974. R. Johnson et al. *Proc. Natl. Acad. Sci. USA* **71**, 536.
Possible occurrence in human tissue	Octopine and nopaline	1964. M. B. Lipsett. *Ann. Intern. Med.* **61**, 733.
Inactivation of trypsin	Binding of amidines and guanidines to active center	1965. M. Mares-Guia and E. Shaw. *J. Biol. Chem.* **240**, 1579.
	Alkyguanidines (e.g., methylguanidine and *n*-butylguanidine)	1969. T. Ingami and H. Hatano. *J. Biol. Chem.* **244**, 1176.

(continued)

APPENDIX III (*Continued*)

Observation	Guanidine compound	References
	4-Guanidinobenzoic acid and 4'-nitrobenzyl ester	1968. H. Mix et al. Physiol. Chem. **349**, 1237.
Inhibition of urease	Guanidinated lima bean inhibitor	1973. R. F. Steiner et al. FEBS Lett. **38**, 106.
	Guanidine derivatives	1974. P. Mildner and B. Mihanovic. Croat. Chem. Acta **46**, 79.
Inhibition of N-methyltransferase	Benzamidines, phenylacetamidines, benzylguanidines, phenylethylguanidines	1975. R. W. Fuller et al. J. Med. Chem. **18**, 304.
Effect on cholinesterase	Various antihypertensive guanidine derivatives	1973. P. Juul. Acta Pharmacol. Toxicol. **32**, 500.
Inhibition of protein kinase	Guanidinium analogs interact with adenosine 3',5'-monophosphate-dependent protein kinases	1977. R. Roskoski, Jr. and J. J. Witt. Fed. Proc. Fed. Am. Soc. Exp. Biol. **36**, 689. 1980. J. J. Witt and R. Roskoski, Jr. Arch. Biochem. Biophys. **201**, 36.
Protease inhibitor inactivating virus	p-Nitrophenyl-p-guanidinobenzoate	1977. M. Bracha et al. Virology **77**, 45.
Effect on urate oxidase	p-Nitrophenyl-p-guanidinobenzoate	1969. K. W. Bently and R. Truscoe. Enzymologia **37**, 285.
Inhibition of phenylalanyl-tRNA synthetase	Aromatic guanidines	1975. P. V. Danenberg and D. V. Santi. J. Med. Chem. **18**, 528.
Herbicide	2-Phenyl-1,1,3,3-tetramethylguanidine	1962. E. Kuehle and L. Eue. German Patent No. 1,089,210.
Dodine applied against scabs on fruit trees	N-Dodecylguanidine (dodine)	1959. G. Lamb. U.S. Patent No. 2,867,562.
	Dodine for apple (*Malus*) scab	1968. D. Woodcock. Chem. Br. **4**, 394.
Effect on swelling of mitochondria	Guanidine derivatives	1970. C. Bhuvaneswaran and K. Dakshinamurti. Biochemistry **9**, 5070.
Respiration of mung bean [*Vigna radiata* (*Phaseolus aureus*)] mitochondria	Guanidine derivatives	1970. S. B. Wilson and W. D. Bonner. Plant Physiol. **46**, 21.
Inhibition of NAD^+ reduction in photosynthetic bacteria	Phenethylbiguanide	1972. D. L. Keister and N. J. Minton. Arch. Biochem. Biophys. **151**, 549.

Effect	Compound	Reference
Reversal of guanidine-derivative inhibition of mitochondrial function by free fatty acids	Guanidine, galegine, bimethylguanide, decamethylene, and diguanide	1968. J. Davidoff. *Clin. Invest.* **47**, 2344.
Effect on citric acid intermediates and NADH	Mechanism of action of guanidine compounds	1965. A. L. A. Boura and A. F. Green. *Annu. Rev. Pharm.* **5**, 183. 1972. F. Davidoff. *Bioenergetics* **3**, 481.
Inhibits energy transfer in mitochondria	Guanidine, synthalin A (decamethylene diguanidine) Phenylethylbiguanide	1955. G. Hollunger. *Acta Pharmacol. Toxicol.* **11** (Suppl. 1), 7. 1963. B. Chance and G. Hollunger. *J. Biol. Chem.* **238**, 432. 1963. B. C. Pressman. *J. Biol. Chem.* **238**, 401.
Inhibition of Pasteur effect in brain slices	Monosubstituted guanidines	1939. F. Dickens. *Biochem. J.* **33**, 2017. 1950. H. McIlwain. *Biochem. J.* **46**, 612.
Inhibits oxygen consumption of brain slices		1966. P. T. Lascelles and W. H. Taylor. *Clin. Sci.* **31**, 403.
Irritability increase in *Nitella* and animals	Guanidine, methylguanidine, and dimethylguanidine	1942. W. J. V. Osterhout. *J. Gen. Physiol.* **26**, 65.
Effect on Na^+, K^+-ATPase activity	Various guanidines	1976. M. Matsumoto and A. Mori. *J. Neurochem.* **27**, 635.
Effect on Na^+ and water uptake	Tetrodotoxin	1973. K. Okamoto and J. H. Quastel. *Proc. R. Soc. London Ser. B* **184**, 83.
Effect on K^+ and $^{86}Rb^+$ transport and cell permeability	Octylguanidine and decamethylene biguanidine (synthalin) Alkylguanidines and barley (*Hordeum vulgare*) roots Permeability in *Allium* epidermal cells	1973. A. Pena. *FEBS Lett.* **34**, 117. 1975. B. Gómez-Lepe and E. J. Avila. *Plant Physiol.* **56**, 540. 1979. B. Gómez-Lepe et al. *Plant Physiol.* **64**, 131.
Blockage of nerve conduction by interference with sodium conductance	Tetrodotoxin and various guanidine esters	1968. B. K. Ranney et al. *Arch. Int. Pharmacodyn. Ther.* **175**, 193.
Interaction with biogenic amines	Chlordimeform (formamidine derivative)	1974. C. O. Knowles and S. A. Aziz. *ACS Symp. Ser.* No. 2, 92–99.

(continued)

APPENDIX III (Continued)

Observation	Guanidine compound	References
Inhibits growth of soybean (*Glycine max*) cotyledon	Arginine–indoleacetic acid conjugate	1977. C.-S. Feung *et al. Plant Physiol.* **59**, 91.
Mutagen	N-Methyl-N′-nitro-N-nitrosoguanidine in bacteria	1960. W. W. Kilgore *et al. Proc. Soc. Exp. Biol. N.Y.* **105**, 469. 1960. J. D. Mandell and J. Greenberg. *Biochem. Biophys. Res. Commun.* **3**, 575; **18**, 788 (1965).
Chloroplast mutagenesis	N-Methyl-N′-nitro-N-nitrosoguanidine (MNNG) (Chloroplast mutagenesis)	1965. D. R. McCalla. *Science (Washington, D.C.)* **145**, 497.
Mutagen in *Scenedesmus*	Homologous N-nitrosonitroalkylguanidines	1975. N. N. Nikolov and L. N. Mladenova. *Genetika (Moscow)* **11**, 73.
Mutagen	N-Nitrosoarginine and related guanidines	1973. H. Endo and K. Takahashi. *Biochem. Biophys. Res. Commun.* **52**, 254.
Reaction with DNA	MNNG reacts with DNA and TMV to yield 7-methylguanine	1968. V. M. Craddock. *Biochem. J.* **106**, 921. 1968. B. Singer *et al. Science (Washington, D.C.)* **164**, 1235.
Carcinogenesis	MNNG	1966. R. Schoental. *Nature (London)* **209**, 726. 1971. M. Archer *et al. Science (Washington, D.C.)* **174**, 1341. 1972. S. Mirvish. *Top. Chem. Carcinog. Proc. Int. Symp.*, p. 279.
	Creatine and creatinine reaction with nitrate	
	N-Nitrosation reaction with ureides and guanidines	
Inhibition of DNA synthesis	Various guanidine compounds	1974. G. Ku *et al. Kidney Int.* **6**, 10.
Induction of *rho* mutants in yeast	Guanidine · HCl	1975. M. H. Juliani *et al. Mutat. Res.* **29**, 67.
Analgesic and antiviral activity	N-(ω-Guanidinoalkyl)bornylamines	1969. G. Minardi and P. Schenone. *Farmaco* **23**, 1040.
Inhibits rhinoviruses (common cold)	N-*p*-Chlorophenyl-N′-(*m*-isobutylguanidinophenyl)urea · HCl	1976. T. H. Maugh II. *Science (Washington, D.C.)* **192**, 128.
Inhibition of rhinovirus RNA synthesis		1974. S. I. Koliais *et al. J. Gen. Virol.* **23**, 341.
Inhibition of tobacco mosaic virus RNA synthesis at two stages	Guanidine	1976. W. O. Dawson. *Intervirology* **6**, 83.

Effect / Activity	Compound / Subject	References
Algicides Synergistic effect in fungicides	N-Ethyl-N-2-benothiazolylguanidines Guanidinooctylamine	1968. G. C. Singh. J. Indian Chem. Soc. **45**, 27. 1973. G. L. Hey. U.K. Patent No. 1337454 (12 Nov. 1971).
Protection against lethal doses of X radiation	Mercaptoalkylguanidines (e.g., S-2-amino-ethylisothiouronium bromide)AET	1955. D. G. Doherty and W. T. Burnett, Jr. Proc. Soc. Exp. Biol. **19**, 312. 1957. R. Shapira et al. Radiat. Res. **7**, 22.
Protection of proteins and DNA against ionizing radiation	Bis(2-guanidinoethyl)disulfide (GED)	1966. G. Kollmann and B. Shapiro. Radiat. Res. **27**, 474; **31**, 721 (1967).
Protection of proteins against ionizing radiation	Bis(2-guanidoethyl)disulfide (GED)	1966. G. Kollmann and B. Shapiro. Radiat. Res. **27**, 474.
Protection of GED against radiation damage to DNA	GED	1967. G. Kollmann et al. Radiat. Res. **31**, 721.
	1,1-(Dithioethylene)diguanidines	1966. T. Hino et al. Chem. Pharm. Bull. **14**, 1193.
Botulism	Treatment with guanidine · HCl	1967. M. Cherington and D. W. Ryan. Lancet Dec. 23, p. 1360.
Antimalarial activity	Substituted diguanidines	1946. F. H. S. Curd and F. L. Rose. J. Chem. Soc., p. 729.
	Nitroguanil and other guanidine derivatives	1960. Y. C. Chin et al. Nature (London) **186**, 170.
	Comparative toxicity of Chloroguanide and nitroguanil	1968. D. M. Aviado et al. Toxicol. Appl. Pharmacol. **13**, 228.
Trypanocoidal activity	Synthesis of related guanidines	1968. D. H. Jones and K. R. H. Woolridge. J. Chem. Soc. C, p. 550.
Antibacterial activity	Bisguanidines	1951. A. F. Crowther et al. J. Chem. Soc., p. 1174. 1956. F. L. Rose and G. Swain. J. chem. Soc., p. 4422. 1950. F. H. S. Curd et al. Br. J. Pharmacol. **5**, 438. 1963. A. F. McKay et al. J. Med. Chem. **6**, 587.

(continued)

APPENDIX III (Continued)

Observation	Guanidine compound	References
Animal virus inhibition	Guanidine hydrochloride	1963. I. Tamm and H. J. Eggers. *Science* (*Washington, D.C.*) **142**, 24.
Virus inhibition (tobacco necrosis virus); aim at virus control	Guanidine carbonate	1968. J. P. Varma. *Virology* **36**, 305.
Inhibition of cytopathic effect of virus and infectivity of animal RNA viruses	Guanidine hydrochloride	1965. L. A. Caliguiri *et al. Virology* **27**, 551.
		1965. W. Levinson. *Virology* **27**, 559.
Inhibition of polio virus	Guanidine	1961. W. A. Rightsel *et al. Science* (*Washington, D.C.*) **134**, 558.
		1961. J. L. Melnick *et al. Science* (*Washington, D.C.*) **134**, 557.
Antitubercular activity	Benzothiazolylguanidines	1968. P. N. Bhargava and M. R. Chaurasia. *Curr. Sci.*, No. 12, 347.
	Bisguanidines	1951. B. N. Jayasimha *et al. Curr. Sci.*, No. 20, 158.
	Biguanido derivatives of diaryl sulfones and sulfides	1951. B. N. Jayasimha *et al. Curr. Sci.*, No. 20, 158, 237.
Antitubercular activity against *Mycoplasma tuberculosis in vitro*	Viomycin	1952. T. H. Haskell *et al. J. Am. Chem. Soc.* **74**, 599.

APPENDIX IV

KETO ACIDS THAT MAY OCCUR IN PLANTS AND THEIR AMINO ACID ANALOGS[a]

Keto acid (corresponding amino acid)	Occurrence, recognition, etc.	References
Glyoxylic acid (glycine)	*Vitis vinifera*	1886. H. Brunner and E. Chuard. *Chem. Ber.* **19**, 595.
	Photosynthetic product	1951. M. Calvin *et al. Symp. Soc. Exp. Biol.* **5**, 284.
	Tulipa gesneriana	1954. G. N. H. Towers and F. C. Steward. *J. Am. Chem. Soc.* **76**, 1959.
	Malus sylvestris (*Pyrus malus*)	1957 A. C. Hulme. *Adv. Food Res.* **9**, 297.
	Wheat (*Triticum*)	1958. R. M. Krupka and G. N. H. Towers. *Can. J. Bot.* **36**, 165.
	Spinach (*Spinacia oleracea*)	1962. P. C. Kearney and N. E. Tolbert. *Arch. Biochem. Biophys.* **98**, 164.
Pyruvic acid (alanine)	*Trifolium* and *Medicago*	1939. A. I. Virtanen *et al. Nature* (*London*) **144**, 597.
	Allium cepa	1945. J. P. Bennett. *Soil. Sci.* **60**, 91.
	Allium cepa	1946. E. J. Morgan. *Nature* (*London*) **157**, 512.
	Solanum tuberosum	1955. J. Barber and L. W. Mapson. *Proc. R. Soc. London Ser. B* **143**, 523.
Mesoxalic acid (amino-malonic acid)	*Medicago sativa*	1909. H. von Euler and I. Bolin. *Z. Physiol. Chem.* **61**, 1.
		1956. H. A. Stafford. *Plant Physiol.* **31**, 135.
Hydroxypyruvic acid (serine)	*Asplenium septentrionale*	1954. A. I. Virtanen and M. Alfthan. *Acta Chem. Scand.* **8**, 1720.
		1955. I. C. Gunsalus *et al. Bacteriol. Rev.* **19**, 79.
Tartronic semialdehyde CHOCHOHCOOH (hydroxyaminopropionic acid)	Intermediate in glycolate metabolism in diatoms	1976. J. S. Paul *et al. Arch. Microbiol.* **110**, 247.

[a] Some expected keto acids such as imidazolepyruvic acid (histidine) have not been studied in detail, at least in higher plants. Keto acids can now be produced by immobilized enzyme technology (e.g., 76).

(*continued*)

APPENDIX IV   (Continued)

Keto acid (corresponding amino acid)	Occurrence, recognition, etc.	References
β-Keto-β-hydroxybutyric acid (threonine)	Vaccinium vitis-idaea	1955. A. I. Virtanen and M. Alfthan. Acta Chem. Scand. 9, 188.
Oxaloacetic acid	Tobacco (Nicotiana) leaves	1937. G. W. Pucher et al. J. Biol. Chem. 119, 523.
	Root nodules of legumes	1943. A. I. Virtanen et al. J. Prakt. Chem. 162, 71.
α-Keto-β-hydroxysuccinic acid	Hordeum vulgare (H. sativum)	1948. P. A. Kolesnikov. Dokl. Nauk Acad. Sci. USSR 60, 1205.
		1956. E. Kun and M. G. Hernandez. J. Biol. Chem. 218, 201.
α-Keto-γ-hydroxybutyric acid (homoserine)	Vaccinium vitis-idaea	1955. A. I. Virtanen and M. Alfthan. Acta Chem. Scand. 9, 188.
α-Keto-γ-hydroxyglutaric acid (γ-hydroxyglutamic acid)	Oxalis	1963. R. K. Morton and J. R. E. Wells. Nature (London) 200, 477.
	Phlox decussata	1964. G. Brandner and A. I. Virtanen. Acta Chem. Scand. 18, 574.
Acetoacetic acid (β-amino-butyric acid)	Flax (Linum usitatissimum) Isoprene biosynthesis and fat metabolism	1954. J. A. Johnston et al. Proc. Natl. Acad. Sci. USA 40, 1031.
Succinic semialdehyde (γ-aminobutyric acid)	Precursor of glutamic acid and glutamine	1956. F. C. Steward and J. K. Pollard. In "Inorganic Nitrogen Metabolism" (W. D. McElroy and B. Glass, eds.), pp. 377–407. Johns Hopkins Univ. Press, Baltimore.
	Pisum sativum	1953. J. K. Miettinen and A. I. Virtanen. Acta Chem. Scand. 7, 1243.
	Hordeum vulgare (H. sativum)	1959. W. L. Ketovich and E. Galas. Dokl. Nauk Acad. Sci. USSR 124, 217.
Aspartic-β-semialdehyde (α,γ-diaminobutyric acid)	Precursor of homoserine and azetidine-2-carboxylic acid	1955. S. Black and N. Wright. J. Biol. Chem. 213, 39.
Diketosuccinic acid	Higher plants	1954. H. Stafford et al. J. Biol. Chem. 207, 621.
α-Ketoglutaric acid (glutamic acid)	Medicago sativa and Trifolium pratense	1939. A. I. Virtanen et al. Nature (London) 144, 597.

APPENDIX IV (*Continued*)

Keto acid (corresponding amino acid)	Occurrence, recognition, etc.	References
	Tulipa gesneriana	1943. A. I. Virtanen *et al. J. Prakt. Chem.* **162**, 71. 1954. G. H. N. Towers and F. C. Steward. *J. Am. Chem. Soc.* **76**, 1959.
α-Ketoglutaric semi-aldehyde (from hydroxy-proline)	*Pseudomonas*	1964. R. M. M. Singh and E. Adams. *Science* (*Washington, D.C.*) **144**, 67.
	Malus sylvestris (*Pyrus malus*)	1957. A. C. Hulme. *Adv. Food. Res.* **9**, 297.
	Wheat (*Triticum*)	1958. R. M. Krupka and G. H. N. Towers. *Can. J. Bot.* **36**, 165.
α-Ketoisovaleric acid (valine)	*Escherichia coli*	1951. H. E. Umbarger and B. Magasanik. *J. Biol. Chem.* **189**, 287.
	Wheat	1959. W. L. Ketovitch and Z. S. Kagan. *Biochem. USSR* **24**, 717. 1962. T. Satyanarayana and A. N. Radhakrish-nan. *Biochim. Biophys. Acta* **56**, 197.
	Wheat	1966. Z. S. Kagan *et al. Enzymology* **30**, 343.
δ-Aminolevulinic acid	Intermediate in porphyrin biosynthesis	1953. D. Shemin and C. S. Russell. *J. Am. Chem. Soc.* **75**, 4873.
	Rhodopseudomonas sphaeroides	1958. G. Kikuchi *et al. J. Biol. Chem.* **233**, 1214.
α-Amino-β-ketoadipic acid	Intermediate in porphyrin biosynthesis	1953. D. Shemin and C. S. Russell. *J. Am. Chem. Soc.* **75**, 4873.
γ-Methylene-α-ketoglutaric acid (γ-methylene-glutamic acid)	*Tulipa gesneriana*	1954. G. H. N. Towers and F. C. Steward. *J. Am. Chem. Soc.* **76**, 1959.
	Arachis hypogaea	1955. L. Fowden and J. A. Webb. *Biochem. J.* **59**, 228.
γ-Methyl-α-ketoglutaric acid (γ-methylglutamic acid)	*Phyllitis scolopendrium*	1955. A. I. Virtanen and A. M. Berg. *Acta Chem. Scand.* **8**, 1085. 1956. M. E. Wickson and G. H. N. Towers. *Can. J. Biochem. Physiol.* **34**, 502.

(*continued*)

Keto acid (corresponding amino acid)	Occurrence, recognition, etc.	References
γ-Hydroxy-γ-methyl-α-ketoglutaric acid (pyruvic aldol) (γ-hydroxy-γ-methylglutamic acid)	*Phyllitis scolopendrium*	1955. A. I. Virtanen and A. M. Berg. *Acta Chem. Scand.* **8**, 1085.
	Adiantum pedatum; inhibits Krebs cycle	1955. N. Grobbelaar *et al. Nature (London)* **175**, 703.
		1955. F. C. Steward *et al. Am. J. Bot.* **42**, 946.
α-Keto-β-methylvaleric acid (isoleucine)	*Proteus vulgaris*; L-amino acid oxidase	1944. P. K. Stumpf and D. E. Green. *J. Biol. Chem.* **153**, 387.
	Neurospora crassa	1955. R. P. Wagner and A. Bergquist. *J. Biol. Chem.* **216**, 251.
	Wheat (*Triticum*)	1962. W. L. Kretovitch *et al. Biochem. USSR (Engl. Transl.)* **27**, 181.
α-Ketoisocaproic acid (leucine)	*Proteus vulgaris*; L-amino acid oxidase	1944. P. K. Stumpf and D. E. Green. *J. Biol. Chem.* **153**, 387.
	Prunus avium	1954. P. H. Abelson. *J. Biol. Chem.* **206**, 335.
		1955. C. Mentzer and L. Cronenberger. *Bull. Soc. Chim. Biol.* **37**, 371.
α-Ketoadipic acid (α-aminoadipic acid)	*Pisum sativum*; intermediate in lysine biosynthesis	1954. A. I. Virtanen and M. Alfthan. *Acta Chem. Scand.* **8**, 1720.
α-Ketoglutaramate (glutamine)	*Pinus pinea*	1963. Y. Guitton. *C. R. Hebd. Seances Acad. Sci. Ser. D.* **257**, 506.
	Conversion to succinamate and succinamide in *Picea*	1973. D. J. Durzan. *Can. J. Bot.* **51**, 359.
α-Ketosuccinamic acid (asparagine)	*Pinus pinea*	1963. Y. Guitton. *C. R. Hebd. Seances Acad. Sci. Ser. D* **257**, 506.
	Transaminase in soybean (*Glycine max*)	1977. J. G. Streeter. *Plant Physiol.* **60**, 235.
	2-Hydroxysuccinamic acid in peas (*Pisum sativum*)	1977. N. D. H. Lloyd and K. N. Joy. *Biochem. Biophys. Res. Commun.* **81**, 186.
Oxalosuccinic acid (α-aminotricarbalyllic acid)	Krebs cycle intermediate	1948. S. Ochoa. *J. Biol. Chem.* **174**, 115.
α-Keto-ε-aminocaproic acid (lysine)	Precursor of pipecolic acid	1955. R. S. Schweet *et al. In* "Amino Acid Metabolism" (B. Glass, ed.), pp. 496–506. Johns Hopkins Univ. Press, Baltimore.

Keto acid (corresponding amino acid)	Occurrence, recognition, etc.	References
Dimethylpyruvic acid (dimethylalanine)	*Prunus avium*	1955. C. Mentzer and L. Cronenberger. *Bull. Soc. Chim. Biol.* **37**, 371.
	Tulipa gesneriana	1954. G. H. N. Towers and F. C. Steward. *J. Am. Chem. Soc.* **76**, 1959.
β-Indolepyruvic acid (tryptophan)	*Proteus vulgaris*; L-amino acid oxidase	1942. H. Waelsh and H. K. Miller. *J. Biol. Chem.* **145**, 1.
		1944. P. K. Stumpf and D. E. Green. *J. Biol. Chem.* **153**, 387.
	Proteus vulgaris	1944. P. K. Stumpf and D. E. Green. *J. Biol. Chem.* **153**, 387.
α-Keto-γ-methylthiobutyric acid (methionine)	Mung bean [*Vigna radiata* (*Phaseolus aureus*)]	1954. D. G. Wilson *et al.* *J. Biol. Chem.* **208**, 863.
α-Keto-δ-guanidinovaleric acid (arginine)	*Proteus vulgaris*; L-amino acid oxidase	1944. P. K. Stumpf and D. E. Green. *J. Biol. Chem.* **153**, 387.
	Insects	1951. G. Ehrensvärd *et al.* *J. Biol. Chem.* **189**, 93.
	Phlox decussata	1964. G. Brandner and A. I. Virtanen. *Acta Chem. Scand.* **18**, 574.
	Picea glauca	1966. D. J. Durzan and R. G. Richardson. *Can. J. Biochem.* **44**, 141.
	Cyclic forms of keto analogs of arginine, homoarginine, and homocitrulline	1978. A. J. L. Cooper and A. Meister. *J. Biol. Chem.* **253**, 5407.
α-Ketopimelic acid (γ-aminopimelic acid)	*Asplenium septentrionale*	1954. A. I. Virtanen and M. Alfthan. *Acta Chem. Scand.* **8**, 1720.
α-Keto-γ-hydroxypimelic acid (γ-hydroxyaminopimelic acid)	*Asplenium septentrionale*	1954. A. I. Virtanen and M. Alfthan. *Acta Chem. Scand.* **8**, 1720.
Phenylpyruvic acid (phenylalanine)	*Escherichia coli*	1947. S. Simmonds *et al.* *J. Biol. Chem.* **169**, 91.
	Salvia splendens	1959. D. R. McCalla and A. C. Neish. *Can. J. Biochem. Physiol.* **37**, 531, 537.
p-Hydroxyphenylpyruvic acid (tyrosine)	*Escherichia coli*	1947. S. Simmonds *et al.* *J. Biol. Chem.* **169**, 91.
	Salvia splendens	1959. D. R. McCalla and A. C. Neish. *Can. J. Biochem. Physiol.* **37**, 531, 537.

CHAPTER THREE

Protein Metabolism[1]

R. C. HUFFAKER

[1]Abbreviations used in this chapter: BA, benzyladenine; CAM, chloramphenicol; CHI, cyclohexamide; CTN, N-carbobenzoxy-L-tyrosine-p-nitrobenzyl ester; ER, endoplasmic reticulum, FMN, $FMNH_2$, flavin mononucleotide (oxidized, reduced); GA, gibberellic acid; GDP, guanosine diphosphate; GlcNac, N-acetylglucosamine; NAD, NADH, nicotinamide-adenine dinucleotide (oxidized, reduced); NR, nitrate reductase; PAL, phenylalanine ammonia-lyase; P_{fr}, P_r, phytochrome; P_i, inorganic phosphate; PP, pyrophosphate; RNA, ribonucleic acid; RuBPcase, ribulosebisphosphate carboxylase; UDP, UMP, uridine diphosphate, monophosphate.

Plant Physiology
A Treatise
Vol. VIII: Nitrogen Metabolism

I. Introduction

Proteins (the enzyme complement) not only cause the metabolism of almost all compounds in the plant but are themselves metabolized. Protein metabolism can be thought of as both synthesis and degradation, or can simply be termed *turnover*. This chapter uses the terms metabolism and turnover interchangeably. The overall picture emerging from research is that protein metabolism is a highly regulated process that permits a plant to modify the concentrations of key enzymes as it develops and adapts to environmental stresses (135, 231). This function is critical in enabling the plant to compete in its environment. Protein represents a major storage sink for nitrogen and carbon skeletons, both of which can be recalled when the plant is faced with limitations in external nitrogen supply or light or with increased sink demand. Under unfavorable conditions protein metabolism is probably the only available means of providing amino acids for either modifying the protein complement or completing development.

Protein metabolism is crucial to the maturation and reproduction of plants. Limitations in nitrogen supply consistently place severe limitations on the photosynthetic productivity of plants. This is especially true when soil nitrogen becomes depleted and the plants mature and produce seeds, which require nitrogen. Final yields then often depend on the ability of the plants to metabolize the leaf protein and translocate the reduced nitrogen into the seeds, where it is further metabolized. Crop varieties vary considerably in ability to accomplish these alternative means of obtaining nitrogen.

Since Vickery *et al.* (273) obtained evidence that plant proteins are highly dynamic, being both synthesized and degraded, information on their synthesis has increased greatly. Their degradation, in contrast, has received much less attention, especially in plant systems. As a result, progress is just beginning toward understanding the regulation and physiological significance of protein metabolism. Are these processes partially regulated by separation into specific cellular compartments? Are degradative enzymes always present and functioning in the cytoplasm? Are proteins constantly broken down according to the ease with which they react with the available proteinases? Each scheme seems viable. Organelle-bound protein, for example, chloroplastic proteins, seem protected, whereas many cytoplasmic proteins show high turnover. Little information is available on the reasons that proteins vary in half-lives.

Protein degradation is subjected to different types of regulation. Catabolism generally increases when plants are placed under "hard times." Degradation rates seem strongly influenced by primary, secondary, and tertiary structures of proteins, as well as by their intracellular location.

Protein catabolism is important in destroying abnormal proteins arising from mutations, chemical modification, spontaneous denaturation,
or cellular wear and tear. That destruction helps explain why the intracellular protein composition remains highly constant over the life of an
organism even though the proteins are highly labile (97).

This chapter deals primarily with the metabolism of individual proteins. It is difficult to assess the significance of results from studies dealing with turnover of total protein. Information on the metabolism of
individual proteins has increased markedly during the last few years
because of the development of relatively rapid procedures for protein
purification.

II. Problems of Methodology

A. POOLS AND PRECURSORS

Although it is not difficult to determine qualitatively that protein is
being metabolized, it is difficult to determine accurately the rates of
synthesis and degradation that regulate the steady-state concentration of
proteins. Estimates of the rates of protein synthesis and degradation rely
strongly on isotopic methods, which label tissue proteins with radioactive
amino acids and then follow the decrease in trichloroacetic acid–precipitable radioactivity. Such methods can be fraught with error and artifact
(132). Some of the important problems involved are dealt with briefly
herein.

In a series of extensive studies Steward and Bidwell and Hellebust and
Bidwell showed evidence for different pools of amino acids in various
plants (35, 36, 117–119, 249, 250). The presence of different precursor
pools of amino acids in plant cells (173) represents a major difficulty and
the largest error involved in determining protein turnover rates. Immediate precursor pools may not be in rapid equilibrium with pools of
stored amino acids, making it difficult to determine the specific radioactivity of the amino acid precursor pool from which a given protein may
be synthesized. Oaks and Bidwell (194) have described the problem in an
excellent review. Instances have occurred in which the linear increase of
^{14}C-labeled amino acid incorporation into protein produce preceded
saturation of the extractable precursor (192, 250). Also, the specific
activity of the product can be greater than that of the total precursor (44,
192).

The lag time before the increase of labeled precursor into protein

becomes linear is correlated with the size of the precursor pool (44, 192).
By measuring such lag periods, Oaks (192) showed two leucine pools in
maize (*Zea mays*) roots: one metabolically active and one apparently a
storage pool. Two similar pools of leucine and valine were found in
soybean (*Glycine max*) hypocotyl, one closely and one remotely related to
protein synthesis (126). The leucine precursor pool turned over 20 times
as fast as the valine pool. The immediate precursor pools of valine in
soybean hypocotyl (126) and leucine in maize root tips (192) expanded
with an increasing concentration of amino acid.

Turnover of the precursor pools is also affected by the rate of amino
acid catabolism. In barley (*Hordeum vulgare*) embryos, more CO_2 was
generated from aspartate and glutamine than from other amino acids
(153). In carrot (*Daucus carota* var. *sativus*) explants, more CO_2 was respired from [1-^{14}C]glutamic acid than from [3-^{14}C]glutamic acid (36).
The latter resulted in a greater distribution of label into other amino
acids (36). Amino acids derived from Krebs-cycle intermediates are most
likely to be labeled from sucrose or glutamine and to lose their radioactivity rapidly (249). Hence it is important to consider which carbon of the
amino acid precursor is labeled and how much of the radioactive carbon
is reworked to other amino acids. Refixation of labeled CO_2 from amino
acids can also affect the specific activity of the precursor pool. Plants can
readily refix some of the CO_2 released through respiration (13, 113,
140). Bidwell (35) reported that $^{14}CO_2$ labeled all the protein amino
acids in wheat (*Triticum*) leaves. Hellebust and Bidwell (118) compared
carbon incorporation into protein coming directly from photosynthate
with that from soluble amino acid pools in wheat leaves. Newly assimilated CO_2 accounted for more than one-half of the label in serine and
glycine incroporated into protein. The glutamate, aspartate, and alanine
incorporated into protein came more from soluble pools formed by
other pathways.

It is important to consider where the labeled amino acid is located in
the metabolic "family" (114).[2] Glutamate, for example, is readily metabolized and, as the head of the family, can be reworked to label arginine,
ornithine, proline, and hydroxyproline.[3] Amino acids at the end of the
family sequence, such as proline and arginine, do not readily contribute
carbon to other amino acids (2, 3). Bidwell (35) found that [1-^{14}C]gluta-
mate also labeled aspartate in wheat leaves. Aspartate can also be reworked to members of its family: threonine, isoleucine, lysine, and

---

[2]For a description of the amino acid families, see metabolic charts 9–14, Chapter 2.
(Eds.)
[3]Chart 11, Chapter 2. (Eds.)

methionine (91).[4] Nair and Vining (187) reported direct conversion of phenylalanine to tyrosine in spinach (*Spinacia oleracea*) leaves.

End-product repression in biosynthetic pathways can also occur when sugars and amino acids are given as carbon sources for protein synthesis (153, 191, 192, 194). This, in turn, could affect the specific activity of the amino acid precursor pools.

Since amino acids can come through different precursor pools with varying turnover rates, the rate of protein turnover found can vary with the amino acid chosen for study. Poole *et al.* (205) found the following half-lives for rat liver catalase: $t_{1/2} = 3.7$, 2.5, and 1.8 days with labeled leucine, arginine, and aminolevulinic acid, respectively.

Another factor that has been ignored in many cases is that the uptake of an amino acid precursor can be inhibited by other amino acids in the medium (192). Different plant cells take up amino acids at differing rates (59, 199). This characteristic can affect the comparative specific activities of precursor pools, especially in short-term experiments. Any changes in the size of the pool feeding protein synthesis could influence the values obtained for turnover (230).

Reutilization of amino acids from hydrolyzed protein is a serious problem in determining protein turnover. Reutilization causes actual rates of degradation to exceed estimates, the error increasing with proteins having high rates of degradation. Zielke and Filner (289) found that reutilization of amino acids prevented an accurate determination of nitrate reductase turnover. Zucker (291) reported a similar problem with phenylalanine ammonia-lyase in *Xanthium* discs. Such errors can be modified by removal of the amino acid from the site of protein synthesis. Bidwell *et al.* (36) found that a part of the amino acids released from protein is removed by respiration and can account for 25% of the total CO_2 released in carrot explants. In contrast, very small amounts of the amino acids released from protein were respired in wheat leaves (117). The amino acids released from protein in carrot explants went into pools not closely associated with protein synthesis (36, 249). Another way of dealing with the problem of reutilization of amino acids is to use a tritium exchange method (see following discussion).

B. METHODS FOR STUDYING PROTEIN METABOLISM

It is difficult to obtain accurate estimates of protein turnover when synthesis and degradation occur simultaneously. Accurate estimates of

[4]Chart 12, Chapter 2. (Eds.)

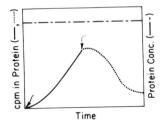

FIG. 1. Idealized curve showing the incorporation of ^{14}C into and the loss of ^{14}C from protein from a pulse application of a ^{14}C-labeled amino acid. The ^{14}C-labeled amino acid is administered to the plant material as indicated (—). The ^{14}C-labeled amino acid is removed and chased with a ^{12}C-labeled amino acid (·······).

turnover require rate constants for both synthesis (k_s) and degradation (k_d). The following equation can be used to describe turnover (132, 234):

$$dp/dt = k_s - k_d P$$

where P is the concentration of protein [g(tissue)$^{-1}$], k_s the rate constant for synthesis, usually zero order (g/g tissue × time) (234), and k_d the rate constant for degradation, usually first order (time^{-1}) (234). At steady state, when $dp/dt = 0$, $k_s = k_d P$, and $P = k_s/k_d$, the turnover rate can be determined by finding either k_s or k_d along with P.

1. Pulse Procedure

This method, which is used extensively, involves a single application of the labeled precursor to the tissue. The exogenous labeled precursor is then removed, unlabeled precursor is added, and the loss of radioactivity from the tissue protein is studied. At a steady-state level of protein, $k_s = k_d P$, the loss of label from the protein is an apparent first-order reaction, allowing an estimate of k_d (Fig. 1).

A problem with this method is that the measured half-life may vary with the dosage time of the labeled precursor. Proteins having a high turnover rate can be highly labeled from a short-term pulse; conversely, proteins with slow turnover may be less labeled, resulting in inaccurate determinations of turnover in the latter. As just mentioned, an apparent increase in half-life may be seen after longer exposure to the labeled precursor.

2. Kinetic Analysis

Kinetic analysis can be incorporated into many of the approaches currently used in studying protein turnover. By following a time course for amino acid uptake and incorporation into protein, Oaks (191) showed that equilibration times and relative sizes of precursor pools could be estimated. Furthermore, when incorporation of labeled precursor into protein was linear, the precursor pool was saturated, and the

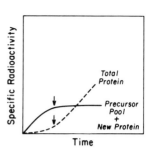

FIG. 2. Idealized curve showing the equilibration of the specific radioactivities (cpm pmol^{-1} amino acid or cpm mg^{-1} protein) of the amino acid precursor pool, new protein, and total protein to an application of a given concentration of a ^{14}C-labeled amino acid to the plant tissue. Arrows denote the time to equilibrium of each.

newly labeled protein had the same specific activity as the precursor pool. At this time the specific activity of the total protein was shown to increase linearly (Fig. 2). The use of increasing concentrations of labeled precursor in such an approach shows when the precursor pool is swamped, that is, when the specific activity of the pool is near that of the label. It has been shown that incorporation of precursor can be quite constant at several concentrations of the label (126, 191, 230). One must still remain aware of the dangers of reutilization of released amino acids and of swamping, that is, inhibition of enzymes, amino acid antagonisms, and disruption of metabolic pathways.

3. Inhibitors of Protein Synthesis

The rate of degradation of protein could theoretically be determined by following the loss of a protein after its synthesis has been stopped by an inhibitor. This approach has proved feasible with some proteins, but not with others. Glasziou et al. (95) used the method successfully in estimating the turnover rate of invertase in sugar cane (*Saccharum officinale*). Others have done similar studies with proteins in animal systems (208, 222). In contrast, inhibitors of protein synthesis have slowed losses of several proteins, including nitrate reductase (263), phenylalanine ammonia-lyase (78), thymidine kinase (128, 129), UDPgalactose polysaccharide transferase (253), and fatty acid synthetase enzymes (283).

4. Constant Infusion

The method of constant infusion involves a continuous supply of the precursor at a constant specific activity. Rate of decay is determined from the time course of the approach of the specific activity of the protein to that of the labeled precursor (234). A major limitation of this method is that problems of amino acid reutilization become important as the specific activity of the protein increases with time.

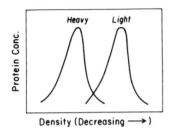

FIG. 3. Idealized curve showing distribution profiles of heavy and light protein from plant tissue treated to media containing heavily and normally labeled isotopes (e.g., D_2O and H_2O).

5. Dual Labeling

An example of the technique of dual labeling includes the incorporation of a ^{14}C-labeled amino acid into the tissue protein; after a period of decay, a second application of the same amino acid is made, but ^{3}H labeled. The [3]H : [14]C ratios are then followed with time. Proteins with high [3]H : [14]C ratios are then degraded faster than those with lower ratios (10).

6. Density Labeling

Density labeling is a technique whereby plant tissue is placed in a medium containing a heavy isotope such as D_2O, $H_2^{18}O$, or [^{15}N]nitrate. The heavy label is absorbed and incorporated into proteins undergoing synthesis (87, 130). The plant tissues are harvested with time and the proteins isolated and analyzed by isopyknic centrifugation in cesium chloride or potassium bromide gradients. The proteins sediment in the centrifugal force field until reaching equilibrium in the, for example, cesium chloride density gradient. The synthesized proteins can thus be separated by their greater buoyant density (Fig. 3).

To determine whether proteins are undergoing turnover, plant tissue is transferred from a heavily labeled medium into a lightly labeled medium. The plant tissue is harvested with time, and the proteins are isolated and subjected to isopyknic centrifugation. The bandwidth at half-peak height has been used to determine whether degradation and synthesis have occurred simultaneously (100, 212). If preexisting heavy-labeled protein is not degraded, the peak widths should widen with time as lighter-labeled protein is synthesized and added to the protein pool. If degradation occurs simultaneously, bandwidth should remain more constant.

In addition, ^{14}C label can be introduced into the heavy protein by administering a ^{14}C-labeled amino acid to the organism at the end of the period in the heavy-labeled medium. The organism is then transferred to a growth medium lacking the heavy label and the ^{14}C-labeled amino acid. The heavy and light proteins are then separated by isopycnic centrifugation. If protein degradation occurs, ^{14}C label is lost from the heavy-protein peak. If amino acid recycling occurs, ^{14}C label appears in the light-protein peak with time. If no protein turnover occurs, ^{14}C label remains in the heavy peak. The method requires a good separation of heavy and light proteins. Boudet *et al.* (40) found potassium bromide gradients give a much better separation than does cesium chloride.

This method also has problems. Boudet *et al.* (40) found the density-labeling method with D_2O to give a shorter half-life for the *Lemna* proteins than occurs with the Trewavas method (determination of amino acyl-tRNA pools to estimate specific activity of precursor pools) and the ^{3}H-exchange method (134). The difference was ascribed to a toxic effect of D_2O on amino acid uptake, protein synthesis, respiration, and photosynthesis. Reducing the D_2O level from 80 to 50% markedly reduced toxicity. The method also lacks the precision of the ^{3}H-labeling method (64).

7. Labeling with ^3H_2O

During the period in which the organism is permitted to grow in the presence of ^3H_2O, a rapid transaminase reaction in the tissue places ^3H at the C-2 position of the amino acid. Humphrey and Davies (64, 134) showed that ^3H at C-2 is stabilized when the amino acid is bound into protein. After the ^3H_2O is replaced with H_2O in the growth medium, if the protein is degraded the transaminase reaction replaces the ^3H with H at the C-2 position of the free amino acid. As a result, even if the same amino acid is reincorporated into protein, no label is found. This method would then give a much truer estimate of protein half-life by greatly decreasing the problems of amino acid reutilization and of precursor pool sizes.

8. Labeling with [^{14}C]Acetic Anhydride

Because of the ready entry and movement of acetic anhydride within cells, labeling with [^{14}C]acetic anhydride has been proposed as a rapid means of radiolabeling proteins to follow their turnover (224).

9. Kinetic Analysis of Enzyme Activities

The method of kinetic analysis of enzyme activities relies on a determination of enzymatic activity as an indication of the concentration of the protein involved. As many factors can affect enzymatic activity with-

out changing the concentration of the enzyme protein, the results are
readily confounded. Joy (151, 152) showed that glutamate dehydrogenase, rather than being induced, changed from EDTA-sensitive
to an EDTA-insensitive state. The activities of several other enzymes
may be decreased by inhibitors binding to the proteins, rather than by
protein degradation (28, 135, 145, 291).

Definitive results require isolation of the protein being investigated.
Because protein purification procedures are often time consuming and
insufficiently quantiative, often resulting in poor yields, many investigators are using antibodies that are highly specific and that permit rapid
quantitative assay of the antigen protein. The use of antibodies enables
detection of very small amounts of antigen.

III. Metabolism of Total Protein

A. Measurements of Protein Turnover

Many previous studies estimating protein turnover in plants suffered
from a lack of recognition of the presence and correct size of precursor
amino acid pools or the reutilization of amino acids released from protein (132). As Boudet *et al.* (40) pointed out, the turnover rate found can
vary with the experimental procedure used. Even with perfect methodology, it would be difficult to assess what a quantitative number might
mean for the turnover rate of total protein. Such a study is inherently
difficult to interpret because of the great number of proteins present in
a cell, the percentage of the total protein they constitute, and the wide
variation in their turnover rates. An example of some of the difficulties
involved is found in protein turnover in green plant leaves. RuBPcase
constitutes up to 50% of the total soluble protein in many species [65% in
some alfalfa (*Medicago sativa*) varieties]. This portion turns over very
slowly in plant leaves until senescence, after which it is rapidly degraded.
In contrast, nitrate reductase constitutes a very small percentage of the
total soluble protein, yet is turned over rapidly. The large percentage of
slowly turned-over protein can confound assessments of the turnover of
total protein. The overall turnover rate is not an average of the turnover
rates of all individual proteins in the sample. For such studies to become
more representative, turnover rates must be worked out for individual
proteins. This section deals with only a select few examples of studies of

total protein turnover in which the investigators recognized the above problems and tried to deal with them.

Davies's laboratory has done a series of excellent studies using a combination of many of the methods described above to determine protein turnover and amino acid recycling in *Lemna minor*. Boudet *et al.* (40) and Davies and Humphrey (64) subjected *Lemna minor* fronds to density labeling by growing them on D_2O culture medium for 7 days. [^{14}C]Leucine was then added. After 8 hr, the fronds were removed from the labeled leucine, washed, placed back in the D_2O growth medium for 40 hr, and then washed and transferred to a deuterium-free growth medium. At various intervals, protein was extracted and the ^{14}C distribution in heavy and light protein was determined after isopycnic centrifugation. Amino acid recycling was estimated by the transfer of radioactivity from the heavy to the light fraction (64). Their calculation was based on the assumption that the protein precursor pool of amino acids is small. Protein half-lives of about 4 days were reported (40).

Davies and Humphrey (64) also used [3]H_2O labeling to estimate protein turnover with and without recycling of amino acid label. *Lemna minor* fronds were floated for 48 hr on a growth medium containing [3]H_2O and ^{14}C-labeled amino acid. The fronds were then washed and transferred to unlabeled growth medium, and the ^{3}H and ^{14}C contents of the protein were followed with time. The results of this study showed that the percentage of recycling varied from 29 to 50%, depending on which amino acid had furnished the label. Glutamate showed the least recycling (29%), as it is more involved in amino acid metabolism. In contrast, leucine and isoleucine were recycled 50% because of lesser involvement in metabolism. Thus the loss of ^{3}H from the proteins would estimate the protein half-life, whereas the ^{14}C count would estimate amino acid recycling.

The [3]H_2O-labeling method identified two groups of proteins: one with a half-life of 3 days, and one with a half-life of 7 days (134).

Trewavas's (266, 267) earlier observation that protein turnover increased when *Lemna minor* was grown under limiting nutritional conditions was confirmed by Davies and Humphrey (64). Optimally growing fronds had protein half-lives of 143 hr, whereas fronds growing under limiting nutrition had a protein half-life of 73 hr. *Schizosaccharomyces pombe* had a protein half-life of 24 hr in full medium and of 6 hr in limiting medium. A faster protein turnover in organisms reacting to limiting growth conditions is consistent with a strategy to change protein constituents in order to meet environmental changes.

Trewavas (266, 267) studied protein synthesis and degradation in

Lemna minor, with a combination of procedures such as constant infusion, pulse labeling, dual labeling, and kinetic analysis. He isolated the amino acyl-tRNA fraction to estimate the specific activity of the amino acid precursor pools. Low levels of amino acids attached to tRNA were determined by growing *Lemna* to isotopic equilibrium on $^{35}SO_4$. The plants were then transferred to [^{3}H]methylmethionine. Isolation of the protein methionine from the exponentially growing fronds and determination of the ^{3}H : ^{35}S ratio gave a direct measure of the protein-specific radioactivity. The specific radioactivity of the precursor methionine pool was estimated by isolation of the methionyl-acyl-tRNA.

In a constant-infusion study, the specific radioactivity of methionyl-tRNA, the specific radioactivity of the protein, and the concentration of protein were followed with time. Rates of synthesis and degradation were determined when the specific radioactivity of the methionyl-tRNA had reached a constant value. At that time, the incorporation of label into protein and the protein concentration were increasing exponentially.

In another method, the protein was pulse-labeled and then chased with unlabeled precursor until the specific activity of the methionyl-tRNA was essentially zero. Estimates of rates of synthesis and degradation were in good agreement with the results of the experiment described above. An intrinsic problem is that the methionyl-tRNA is relatively unstable, and losses may occur during extraction.

From those experiments, the rate constant in culture under a good nutritional regime was 0.46 day^{-1} for protein synthesis and 0.09 day^{-1} for degradation. Under limiting conditions, the rate of synthesis decreased (0.26 day^{-1}), whereas the rate of degradation increased (0.135 day^{-1}).

In an earlier study, Hellebust and Bidwell (119) devised a method similar to constant infusion to help determine the specific activity of the precursor pool. Wheat seedlings were allowed to carry out photosynthesis with $^{14}CO_2$ for about 12 hr each day for 6 days. The specific activities of some amino acids and proteins were followed. The object was to allow all the possible precursors of protein synthesis to approach the same specific activity and to follow turnover when incorporation of label into the proteins was increasing linearly. After the first day the specific activities of free amino acids and sugars remained quite constant throughout the 6-day period. As a result, the incorporation of ^{14}C into protein was nearly linear during the experiment. The turnover rates obtained were 0.4–0.5% per hour for rapidly growing secondary wheat leaves and 0.2–0.3% per hour for nongrowing primary leaves. These results contrast with results of Davies and Humphrey (64), who found that turnover was greater in nongrowing but nitrogen-deficient *Lemna* fronds.

Humphrey and Davies (134) have used equations described by Hellebust and Bidwell (118), Kemp and Sutton (158), and Trewavas (267) to determine protein turnover and found that Hellebust and Bidwell's equations (118) yield turnover values higher than those of the others.

Holleman and Key (126) estimated protein turnover in soybean hypocotyl with kinetic methods similar to those of Oaks (191). These workers determined the rate of transport of leucine and valine into the tissues, their contents in the protein, and the steady-state level of the protein. The sizes of precursor pools for the two amino acids were estimated by multiplying the output rate (in micromoles per minute) of the precursor pools by the lag time until detection of linear incorporation of ^{14}C into protein. This method permitted calculation of the relative turnover rates of the precursor pools of leucine and valine. The lecuine pool turned over 20 times as fast as the valine pool. The calculated turnover of protein was 2.5% per hour.

To determine protein turnover in tobacco (*Nicotiana*) callus, Kemp and Sutton (158) employed a combination of methods: constant influsion, a kinetic analysis of absorption of labeled leucine into the amino acid pools, and incorporation of precursor into protein. After the first half-hour of the time course, the specific activity of leucine in the free amino acid pool remained constant, and its incorporation into protein was found to be linear. The soluble leucine pool thus appeared to be the precursor pool. By determining the difference between the rates of protein synthesis and protein accumulation, a turnover rate of 1.1% per hour was estimated.

The sequential separation of major RuBPcase synthesis from degradation confounds interpretation of turnover rates obtained for total leaf protein, as exemplified in the results of Mizrahi *et al.* (183). These workers pulse-labeled detached *Tropaeolum majus* leaves with $^{14}CO_2$ for 1 hr, transferred the leaves to $^{12}CO_2$ and to kinetin or water control, and then traced the radioactivity in the amino acids and protein. The specific radioactivity of the leaf proteins increased continuously in both the control and hormone-treated plants as the leaves senesced. Mizrahi *et al.* concluded that the proteins labeled during the experiment degraded at a slower rate than the unlabeled, previously synthesized proteins. A possible explanation is that RuBPcase protein may be synthesized much more slowly than other proteins in the mature green leaves and that it may be degraded more rapidly in senescing leaves, resulting in the greater specific radioactivity of total protein.

B. Protein as a Source of Respiratory Substrate

It has been proposed that protein metabolism can strongly contribute to the CO_2 released during respiration. Gregory and Sen (103) suggested that protein was the source of a significant part of the CO_2 produced, since the respiratory rate of barley leaves was not correlated with sugar content. They proposed that proteins were being replaced in a protein cycle. Bidwell et al. (36) proposed that protein degradation contributed about 7% of the CO_2 produced in slow-growing carrot cultures and about 27% of that in actively growing cultures. Davies and Humphrey (64) found that 50% of the amino acids released by protein degradation were reincorporated into protein (method described above) in actively growing Lemna. Hence calculations showed that protein metabolism could account for 2% of the respiratory CO_2. Under conditions of nitrogen starvation, protein metabolism could account for 3–4% of the CO_2 released by respiration.[5]

C. Protein Metabolism: A Regulatory Feature of Senescence

As the plant matures it must mobilize much of its stored nitrogen products in its shoot portion and translocate them to the seeds for reproduction to take place. In most instances, this occurs at a time when external nitrogen is largely limiting. Evidence is mounting that senescence is a highly regulated process and that protein metabolism has a key role in the metabolic shifts and translocation of metabolites.

It appears that during the early phases of plant senescence a stepdown occurs in the rate of certain metabolic pathways, e.g., glycolysis (135), some caused by protein degradation (132), and others by possible reaction with inhibitors developing during the same period (231). Humphrey et al. (135) followed the activities of nine enzymes as Lemna minor adapted from high to low nitrogen in the growth medium. Nitrate reductase went to zero activity, while other glycolytic and tricarboxylic acid cycle enzymes

[5]See Chapter 2, Section III,B,4, for further discussion of this question. (Eds.)

reached a lower steady-state rate. Hedley and Stoddart (115) and Thomas (257) showed that the plastid enzyme alanine aminotransferase decreased early in the senescence of *Lolium temulentum*.

Chloroplasts suffer major degradation early in senescence. The intracellular location of the hydrolases involved in the turnover of chloroplastic and cytoplasmic protein has just begun to emerge. These concepts are dealt with in greater detail in the following section.

IV. Metabolism of Specific Proteins

A. RIBULOSEBISPHOSPHATE CARBOXYLASE: SYNTHESIS

RuBPcase is a multifunctional plant protein with a central position in photosynthesis and photorespiration. It has both carboxylase and oxygenase activities, responsible for photosynthetic CO_2 fixation and for producing the first product for photorespiration (phosphoglycolate). In addition to its catalytic properties, it serves as a major leaf storage protein. It constitutes up to 50% of the total soluble protein in many plants (up to 65% in some alfalfa varieties) and shows many properties characteristic of storage proteins. It is assembled and sequestered in a discrete organelle, the chloroplast, wherein it is protected from the proteolytic enzymes in the cytoplasm. After its synthesis and assembly, little turnover is detected until circumstances require its mobilization. RuBPcase can be mobilized during senescence or when the plant requires its reserves to offset deficits of either nitrogen or carbohydrates. As such, the RuBPcase concentration in a leaf is quite responsive to environmental stresses.

Perturbing the environment allows an approach to the study of the metabolism of RuBPcase. In dark-grown barley plants, about one-half of the maximal concentration of RuBPcase is synthesized. When the plants are placed in light, synthesis continues to a final concentration, quite constant (131). During the greening period, radioactive amino acids can be incorporated into the protein to follow its synthesis and degradation.

1. Biosynthesis of RuBPcase

The synthesis of RuBPcase requires both chloroplast and nuclear genomes. There is now strong evidence that the small subunit is synthesized in the ctyoplasm on 80-S polyribosomes and the large subunit in the chloroplast on 70-S polyribosomes. The small subunit then somehow traverses the chloroplastic membrane and is assembled with the large subunit into the native enzyme.

2. Biosynthesis of Large and Small Subunits

a. In Situ.   Antibiotics that show preferential inhibition of protein synthesis driven by 70- and 80-S ribosomes have provided valuable circumstantial evidence concerning the intracellular location of the synthesis of the subunits. Chloramphenicol (CAM) inhibits protein synthesis on 70-S polyribosomes, whereas cycloheximide (CHI) can inhibit protein synthesis on both 80- and 70-S ribosomes, but preferentially on 80-S ribosomes. Criddle *et al.* (61) showed that CAM specifically inhibited synthesis of the large subunit, whereas CHI preferentially inhibited synthesis of the small subunit in greening barley leaves.

b. In Vitro.   Several laboratories have now reported successful synthesis of the large and small subunits by *in vitro* protein-synthesizing systems. Blair and Ellis (37) showed that isolated chloroplasts incorporated labeled amino acids into the large subunit of RuBPcase. In work by Hartley *et al.* (110), the products of a heterologous system containing pea (*Pisum*) chloroplast RNA and *Escherichia coli* extract were compared with the products of protein synthesis in isolated pea chloroplasts. Co-electrophoresis of products of the *in vitro* and *in vivo* system showed a slight mobility difference between the presumptive large subunits. Further chymotryptic analysis, however, showed that the products were identical. Alscher *et al.* (5) showed that the large subunit was a major product of an *in vitro* protein-synthesizing system consisting of polyribosomes isolated from greening barley leaves and soluble components isolated from *E. coli*. The peptide synthesized was shown by sucrose density-gradient sedimentation to be largely associated with 70-S ribosomes. The synthesis of the large subunit by an *in vitro* heterologous protein-synthesizing system has been compared with *in vivo* synthesis of the large subunit in several laboratories (5, 110). The antibody-precipitable ^{14}C-labeled large subunit from the heterologous system using mRNA from pea chloroplasts (110) and barley polyribosomes (5) was combined with the ^{3}H-labeled large subunit synthesized *in vivo* in pea and barley leaves. The mixture was then cleaved by cyanogen bromide, and the peptides were solubilized in sodium dodecyl sulfate (SDS) and separated by electrophoresis on polyacrylamide gels. Cyanogen bromide fingerprints from the *in vitro* and *in vivo* systems were identical.

Roy *et al.* (227, 228), using wheat leaf and pea leaf polyribosomes, identified the small subunit as a product of an *in vitro* protein synthesis system. The small subunit was identified by precipitation with specific antibody. Tryptic hydrolysis of the precipitated product showed that labeled amino acids were incorporated into the peptides of the small subunit.

Gooding *et al.* (99) identified the large and small subunits as components of peptides associated with 70-S and 80-S ribosomes in greening wheat leaves. The 70- and 80-S ribosomes were separated with minimum cross-contamination by sucrose density-gradient centrifugation. The ribosomes were then reacted with [³H]puromycin to release the bound peptides (226). The results showed that the large subunit was associated with 70-S ribosomes and the small subunit with 80-S ribosomes.

3. Assembly of Subunits into Native Enzyme

Evidence for partially assembled enzyme molecules was found by Smith *et al.* (245). A time course was determined for (1) incorporation of labeled amino acids into the two subunits, (2) incorporation of labeled amino acids into the mature enzyme, (3) the increase in enzymatic activity, and (4) the increase in concentration of RuBPcase protein. Large amounts of amino acids were incorporated into both subunits and into the mature enzyme several hours before an increase in activity was detected. These events were finally followed, several hours later, by an increase in RuBPcase protein accompanying the increased activity. It was hypothesized that early-synthesized subunits might be used to complete partially assembled molecules in RuBPcase. J. W. Bradbeer (personal communication) observed a similar response in bean leaves.

The transport of the small subunit from the cytoplasm into the chloroplast and its subsequent assembly with the large subunit into the native enzyme appear to be a complicated process. Evidence was previously found for the existence of a possible precursor of the small subunit. Roy *et al.* (227) found a 20,000-dalton protein of unknown function made on wheat 80-S ribosomes that reacted with small subunit antibodies. The molecular weight of the small subunit in the native enzyme is ~16,000 (157). Roy *et al.* (225) later showed the existence of pools of both large and small subunits of RuBPcase in pea leaves.

Using polyadenylated mRNA from *Chlamydomonas* translated by a cell-free wheat-germ system, Dobberstein *et al.* (73) showed *in vitro* synthesis of a putative precursor of the small subunit. The precursor was about 3500 daltons larger than the authentic small subunit and was immunoprecipitated by specific antibody against the small subunit. Completion of nascent chains on free polysomes yielded the small subunit in-

stead of the precursor. This result was attributed to the presence of a
highly specific protease associated with the postribosomal supernatant
and with the polysomes. This association with the polyribosomes might
be attributable to nonspecific binding during cell fractionation, hence
might have no physiological significance (73). This protease cleaved the
putative precursor into two products, one identical to the small subunit,
and a smaller fragment representing the remainder of the precursor.
The precursor made in the heterologous system was processed by
adding back some of the cell-free supernatant containing the protease.
The intracellular location of the protease is not yet known, but could be
in the chloroplast envelope, as proposed by Highfield and Ellis (122).

Highfield and Ellis (122) reported *in vitro* synthesis of a 20,000-dalton
polypeptide that was precipitable with anti-RuBPcase antibody. It was
the product of pea polysomes and a wheat-germ protein-synthesizing
system. The product, termed P 20, was 6000 daltons larger than the
authentic small subunit from purified carboxylase. Highfield and Ellis
(122) showed that intact chloroplasts processed the P 20 product when
the products of the *in vitro* protein-synthesizing system were incubated
with intact chloroplasts. The incubation mixture was treated with SDS
and subjected to disc gel electrophoresis. The results showed that P 20
decreased, whereas the authentic small subunit increased. The processing did not require concomitant protein synthesis, as the former occurred after the latter had ceased. Cycloheximide and chloramphenicol
also had no effect on the processing. Because processing was maximum
with intact chloroplasts and less with chloroplastic fragments, these
workers suggested that the processing protease is in the chloroplast
envelope.

Highfield and Ellis found evidence for the transport of the P 20 product into chloroplasts during incubation. They showed that authentic
small subunits appearing in the chloroplasts from the P 20 precursor
were resistant to trypsin digestion but were hydrolyzed when SDS was
added to break the chloroplasts.

The signal hypothesis was recently advanced to describe polypeptide
transport across ER membranes (38). Accordingly, polypeptides des-
tined to be secreted or packaged are synthesized on mRNAs containing a
set of codons at the 5′ terminal coding for a unique sequence of amino
acid residues on the amino terminal of the nascent chain. As this signal
sequence emerges, it permits attachment of the ribosome to the membrane, directing the growing chain vectorially across the membrane. The
signal portion of the amino acids is apparently removed by a membrane-
associated enzyme during the translation process. This hypothesis then
predicts that membrane-bound polysomes will produce processed poly-

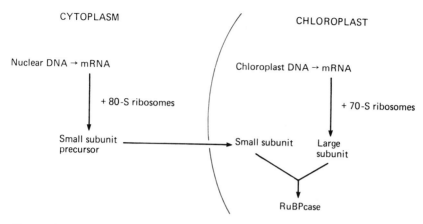

FIG. 4. Scheme for synthesis and assembly of ribulosebisphosphate carboxylase (RuBPcase).

peptides, whereas free polyribosomes will produce unprocessed poly-peptides (38).[6]

Transport of the small subunit across the chloroplastic membrane is not in accord with the signal hypothesis (73, 122). The signal hypothesis requires concomitant protein synthesis by membrane-bound ribosomes to move the growing polypeptide through the membrane. In contrast, the putative precursor of the small subunit was made on apparently free ribosomes of *Chlamydomonas* (73). *In vitro* processing of the precursor occurred after protein synthesis in the absence of ribosomes (73, 122) and in the presence of chloramphenicol or cycloheximide (122).

Ellis and co-workers (37, 122) have proposed an envelope-carrier hy-pothesis according to which P 20 would combine with a specific carrier in the chloroplast envelope. When the processing enzyme removes part of the peptide (rich in acidic amino acids) a conformational change occurs leading to transport and release of the small subunit in the stroma. Figure 4 presents an overall scheme for synthesis of RuBPcase.

B. RIBULOSEBISPHOSPHATE CARBOXYLASE: DEGRADATION AND TURNOVER

The processes of assembly of RuBPcase and its degradation may be separated sequentially, and possibly spatially. Information now available

[6]See Chart 24, Section III,B, Chapter 3. (Eds.)

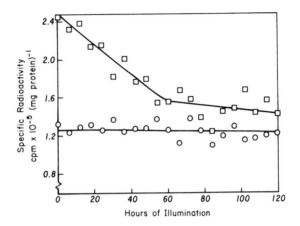

FIG. 5. Time-course curves of specific radioactivity of soluble protein minus RuBPcase protein (□) and of RuDPcase protein (○) (203).

shows turnover characteristics during initial synthesis of RuBPcase, after a steady concentration is attained and during leaf senescence.

1. Turnover during Initial Synthesis

To determine whether RuBPcase turns over during its initial synthesis, leaves were detached from dark-grown barley plants and placed, base down, in light and in the presence of ^{14}C-labeled amino acids (245). After several hours, the labeled amino acids were removed and chased with cold amino acids. No loss of specific-antibody-precipitatable label was detected during the greening period, showing that degradation was absent while the enzyme was undergoing synthesis.

2. Turnover after Steady Concentration Is Attained

After the carboxylase protein reached a steady maximum concentration, essentially no turnover was detected in primary leaves from intact barley plants (203; Fig. 5). This was shown by placing dark-grown plants in light and $^{14}CO_2$ for 6 hr. The labeled CO_2 was then removed and chased with unlabeled CO_2. The specific radioactivity of the carboxylase protein was then followed with time. The concentration of carboxylase protein remained quite constant, as did the specific radioactivity, indicating little turnover. In contrast, soluble protein other than RuBPcase turned over rapidly. Observation of turnover depended on changes in the specific radioactivity of previously highly labeled carboxylase protein. Under this condition it is possible that a low rate of turnover would be undetected. It could well be that a normal, yet slow, turnover of chloroplasts occurs, with the rate species dependent. This concept is amplified below.

Further evidence for slow turnover of carboxylase in barley leaves is that after the carboxylase reached its maximum concentration, no carboxylase messenger activity was detected in isolated polyribosomes in an *in vitro* heterologous protein-synthesizing system (5). During the period of rapid synthesis of the carboxylase, isolated polyribosomes in a heterologous system synthesized the large subunit of the carboxylase. Patterson and Smillie (200) also showed, in wheat leaves, that RNA synthesis decreased and incorporation of labeled amino acids into RuBPcase greatly decreased after the concentration of RuBPcase reached a constant level. Zucker (291) further observed the stability of RuBPcase in green *Xanthium* leaf discs. Little radioactivity was detected in fraction I protein, even though significant amounts were incorporated into phenylalanine ammonia lyase. Likewise, when Kannangara and Woolhouse (154) fed $^{14}CO_2$ to fully expanded *Perilla* leaves, only a small percentage of the activity was in fraction I protein. Brady and Scott (42) showed that incorporation of labeled amino acids into RuBPcase in detached wheat leaves decreased greatly after the maximum content of RuBPcase was reached.

A summary of the evidence, then, is that after RuBPcase reaches its maximum but steady concentration, little turnover or very slow turnover is detected over a number of days. With RuBPcase protected from degradation by remaining inside the chloroplast and the loss of carboxylase often showing characteristics similar to those of chlorophyll, the turnover of carboxylase may represent turnover of the entire chloroplast.

3. RuBPcase and Senescence

When the leaf senesces, as a result of age or limiting environmental factors, RuBPcase protein seems to be the first major protein to be degraded. Earlier workers (74, 155–157) showed that fraction I protein was the main protein constituent lost in senescing tobacco and *Perilla* leaves (154). Fraction I protein can be regarded as crude RuBPcase.

It is well known that light delays the onset of senescence and then slows senescence, once under way. Seasonal variation in irradiance can influence the rate of leaf senescence. The shading of lower leaves by upper leaves in a crop canopy can cause a constant senescing in lower leaves. Therefore senescence can be simulated by detaching the leaves of barley or by placing intact plants in darkness and following the resultant

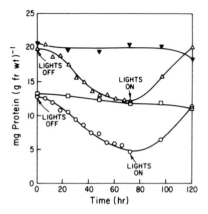

FIG. 6. Effect of light and dark periods on leaf contents of total soluble protein and RuBPcase protein (203). Barley seedlings were grown in light for 6 days, after which light and dark treatments were begun. ▼,△, Total soluble protein; □,○, RuBPcase protein; ▼,□, continuous light; △,○, 72 hr of darkness, then 48 hr of light.

changes in RuBPcase. The loss of RuBPcase showed similar kinetics in both detached and intact leaves of barley, with RuBPcase accounting for 95% of the protein lost over the several-day time course (Fig. 6; 203, 204). Chlorophyll was similarly lost in the senescing leaves. These results suggest that the chloroplast is the major site of degradation early in senescence. When barley plants are placed in darkness, the time course of RuBPcase and carbohydrate loss indicates that RuBPcase is being reutilized as both a carbon and a nitrogen source after carbohydrates have been exhausted (203, 204).

The environmentally induced loss of RuBPcase was totally reversible up to 72 hr in darkness. When the intact plants are returned to light, the leaves again synthesize the full complement of RuBPcase (203). Wittenbach (285) reported a recovery of photosynthesis, chlorophyll, and total protein when wheat leaves were returned to light from 2 days of darkness. After 4 days in dark, recovery did not occur.

4. Simultaneous Synthesis of RuBPcase during Degradation

To determine whether a small amount of synthesis could be detected at a time when RuBPcase was undergoing rapid net degradation, detached barley leaves were placed in tritiated amino acids (203, 204; Fig. 7). Amino acids were incorporated at a low level into RuBPcase protein while it was being rapidly degraded. In comparison, high levels of labeled amino acids were incorporated into the total soluble protein while its concentration remained quite constant, showing a higher rate of turnover.

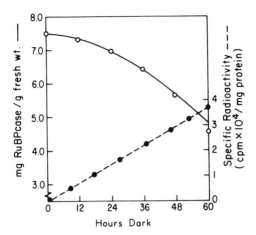

FIG. 7. Simultaneous synthesis and degradation of RuBPcase in detached barley leaves in darkness (203, 204). At the beginning of the experiment detached leaves were placed in the presence of an [³H]amino acid mixture. RuBPcase protein was determined by reaction with specific antibody.

5. Loss of RuBPcase and Proteolytic Activity

In detached leaves of barley, the loss of RuBPcase protein was negatively correlated with an increase in endopeptidase activity (measured against azocasein), while carboxypeptidase activity against N-carbobenzoxy-L-tyrosine-p-nitrophenyl ester (CTN) remained constant during the experiment (204). In contrast, endopeptidase activity remained almost constant in leaves of intact plants while RuBPcase was being degraded (130a).

The increased proteolytic activity in senescing detached leaves may be, at least partially, a function of wounding. Several newly formed proteinases (130a) were found only in the area near the base where the leaves were cut during detachment (B. L. Miller and R. C. Huffaker, unpublished information).

6. Intracellular Location of RuBPcase Degradation

Several lines of evidence indicate that the initial reactions leading to RuBPcase degradation occur in the cytoplasm. In detached barley leaves, CHI totally prevented the loss of RuBPcase protein and chlorophyll and prevented the appearance of increased proteolytic activity, whereas chloramphenicol had no effect on either (Fig. 8; 204). In de-

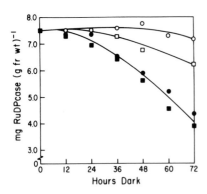

FIG. 8. Inhibition of degradation of RuBPcase by cycloheximide (CHI) and kinetin and chloramphenicol (CAM) in detached barley leaves in darkness (204). Concentrations in micromoles per milliliter: CHI, 5; kinetin, 10; CAM, 50. Inhibitors were added at the start of the experiment at leaf detachment. ○, CHI; ●, CAM; □ kinetin; ■, controls.

tached leaves, CHI prevented the appearance of new proteinase (130a), detectable by isoelectric focusing. During isoelectric focusing a pH gradient is formed where ampholites with varying isoelectric points are placed in an electric field. When added to such a system, proteins migrate until they reach the pH equal to their isoelectric point. Isoelectric focusing can be carried out in solution in a column, in a polyacrylamide disc or slab gel, or in a slab gel consisting of Sephadex.

Choe and Thimann (54) showed that isolated chloroplasts lost chlorophyll and protein much more slowly than did oat (Avena) leaves. Hence, it appears that the new proteolytic activity was developed on cytoplasmic 80-S polyribosomes. Proteolytic activity was actually lost in the cycloheximide treatment, indicating that some of the proteinases themselves can be turned over. Thomas (256), Martin and Thimann (176), and Peterson and Huffaker (204) showed, respectively, that CHI, although not CAM, prevented the loss of chlorophyll in forage grass, oat leaves, and barley leaves. This finding further indicates that the hydrolytic processes are probably initiated and driven by components of the cytoplasm.

Although the initiatory reactions leading to senscence may begin in the cytoplasm, the chloroplasts are the major location for loss of proteins. Using very sensitive methods for gel electrophoresis, good evidence has been obtained for the presence of endoproteinases in chloroplasts (293). Nothing is yet known concerning their regulation; however, they appear capable of degrading RuBPcase (A. B. Tang and R. C. Huffaker, unpublished results).

Instances are reported that show simultaneous loss of chlorophyll and RuBPcase protein, for example, in detached and intact leaves of barley (204), or of chlorophyll and total protein (259). Thomas (258) and Hall et al. (105) recently showed, however, that those events can be separated.

Using a mutant of *Festuca pratensis* that loses very little of its chlorophyll throughout senescence, Thomas showed that RuBPcase protein was lost at the same rate in both the mutant and normal type. The normal type loses all its chlorophyll during the same period in which the carboxylase protein is lost. Electron microscopy of mutant chloroplasts indicated that the stroma matrix was destroyed but that thylakoid membranes persisted in a loose, unstacked condition. The activities of photosystems I and II declined at similar rates in both mutant and normal leaves. Polypeptides of chloroplast membranes were separated into about 30 components. Of five major components, two declined similarly in both mutant and normal, whereas the other three were more stable in the mutant. One of the components was tentatively identified as the apoprotein of the light-harvesting chlorophyll–protein complex.

Hall *et al.* (105) showed that chlorophyll was lost faster than RuBPcase in the flag leaf of wheat during senescence. Deviations in the degradation of chlorophyll, RuBPcase, and some membrane proteins from the normal senescing pattern show another degree of complexity of regulation. Indications are that significant manipulation may be possible of the events occurring during senescence or crop maturation, for example identification of soybean variants maintaining a "green-leaf character" further into the maturation period (4).

7. Schemes for RuBPcase Degradation

Several schemes seem possible at present: chloroplasts may be taken up by vacuoles (300), wherein are located the two major endoproteinases of barley (299) and the main proteolytic activity of wheat (300), or RuBPcase may be degraded by proteinases located in chloroplasts (293). If chloroplastic proteinases are responsible for the degradation of RuBPcase, interesting problems of regulation result, for example, How are they activated at a specific time early in senescence by cytoplasmic reactions? Are they sequestered away from their substrates? Must they be activated before they can react with their protein substrates?

8. Chemical Regulants

Like inhibitors of protein synthesis such as cycloheximide, kinetin greatly retarded the loss of both RuBPcase and chlorophyll as well as inhibiting the increase in proteolytic activity in detached barley leaves (130a, 204). Kinetin is well known for its retardation of senescence in many plants. Kinetin could decrease the loss of RuBPcase either by increasing the rate of synthesis or by decreasing the rate of degradation.

Current evidence favors the latter possibility. Tavares and Kende (255) prelabeled the protein of senescing corn-leaf discs, replaced the ^{14}C label with ^{12}C-labeled amino acid, and treated with benzyladenine. Protein from the BA-treated leaves lost label much more slowly than did the control. Also, the specific activity of the bulk protein did not change, indicating no differences in rates of synthesis. Kinetin prevented increased proteolytic activity (204) and the formation of a second proteinase on 80-S polyribosomes in detached barley leaves (130a). Balz (24) and Atkin and Srivastava (17) showed that cytokinins prevented the increased proteolytic and RNase activities in detached senescing leaves. Thomas (256) reported that kinetin retarded the loss of alanine aminotransferase in senescing *Lolium temulentum* leaves.

C. Relationship of Ribulosebisphosphate Carboxylase Metabolism to Biomass Production

Maintenance of RuBPcase through the growing season is critical for biomass production. Photosynthate furnishes the sink requirements of the developing seed for reduced carbon compounds. It also supplies energy for both nitrogen fixation and nitrate assimilation in order to meet the nitrogen sink requirements of the developing seed. When the supply of reduced nitrogen does not meet the sink requirement of the seed, the leaf protein is degraded to meet the nitrogen requirement. Available information shows that RuBPcase is among the first leaf proteins to be lost at that time. As RuBPcase protein is degraded, photosynthetic CO_2 fixation capacity can be lost. That has been termed a self-destruct phenomenon in soybeans (244).

Maintaining a flow of reduced nitrogen through the plant during the growing season is critical to protecting RuBPcase from being degraded. Crop varieties showing a character of maintaining photosynthetic competence further into seed maturation have a potential for higher yields. Helsel and Frye (294) reported selection of an oat variety that retained chlorophyll longer during seed fill and subsequently had higher yields. Boyd and Walker (41) reported heritable variation in chlorophyll content and stability in the flat leaves of wheat varieties. In crop plants utilizing nitrate as the principal source of nitrogen, plants should be selected that can assimilate nitrate more efficiently in terms of acquiring it and translocating the reduced products to the seed. Wheat varieties vary considerably in ability to assimilate nitrate as well as translocate the reduced products from the straw to the developing seeds (221). Leaf applications of reduced nitrogen during seed fill will very likely become important in the future.

Sinclair and deWit (243, 244) analyzed the photosynthate and nitrogen requirements of various seed crops and found soybean unique among the 24 species studied. Soybean must assimilate large quantities of nitrogen to supply the high protein needs of the seeds. Yet it produces biomass at one of the lowest rates. As the seed develops, the photosynthate supply might not be sufficient to meet the energy needs of both the seed and nodules. As the nodules receive less photosynthate, nitrogen fixation decreases and does not supply sufficient reduced nitrogen. Reduced nitrogen from protein degradation could then be translocated from the leaves. The resulting loss in photosynthetic CO_2-fixation capacity would not sustain both the seed and nodule requirements for photosynthate. This self-destructive characteristic could limit the length of the seed development period and potentially limit total seed production. Mondal et al. (184) showed that when the seed sink of soybeans was removed, the concentration of both RuBPcase and total protein remained higher in the leaves. Hardy et al. (107) suggested that nitrogen input during rapid seed development could be increased by extending the exponential phase of nitrogen fixation or, according to Sinclair and deWit (244), by increasing the proportion of photosynthate allocated to support nitrogen fixation. Those objectives might be reached by selecting for variants that maintain more photosynthetically active leaves throughout reproductive growth.

To test the above hypothesis, several thousand plants in a soybean nursery at the University of California, Davis, were examined for variation in senescence characteristics (4). Five plants having completely green leaves with brown mature pods were selected. Seeds from the plant showing the highest nitrogen fixation were used to produce 14 F_3 plants in a growth chamber (Table I). Of those, 5 plants had delayed senescence. They also had much higher levels of chlorophyll, total leaf protein, RuBPcase protein and activity, and nitrogenase activity than did the normally senescing control plants. Seed yield and protein concentration were not decreased in the plants with delayed senescence. F_4 plants produced from F_3 plants having delayed and normal senescence gave similar results. These findings suggest that the maintenance of chlorophyll and RuBPcase is a heritable quality that can be exploited to allow continued nitrogen fixation during seed production. The variant plants with delayed leaf senescence seemed to have a greater capacity to partition photosynthate, proportionally maintaining the requirements of

TABLE I

Physiological Parameters of Six F_3 Soybean (*Glycine max*) Plants[a]

| F_3 plant | Total dry wt. (g) | Pod dry wt. (g) | Pod dry matter (%) | Chlorophyll (mg g^{-1}, fresh wt.) | Leaf protein (g g^{-1}, fresh wt.) | Ribulosebisphosphate carboxylase | | Acetylene reduction (μmol C_2H_4 hr^{-1} g^{-1} nodule, dry wt.) |
						Activity (μmol CO_2 hr^{-1} g^{-1}, fresh wt.)	Protein (mg g^{-1}, fresh wt.)	
1	74.4	30.8	32.8	1.3	11.7	36.1	2.1	1.4
2	67.4	28.0	32.6	1.6	16.1	54.4	3.3	0.7
3	63.5	25.6	31.6	1.4	14.0	45.8	1.9	1.0
4	84.1	27.5	32.1	0.5	6.1	20.6	0.9	0.7
5	60.0	20.9	28.7	0.5	5.1	16.9	0.7	0.1
6	52.4	19.8	29.1	0.4	4.9	12.0	0.5	0.1

[a] Produced from seed of an F_2 soybean plant that exhibited green leaves and acetylene-reducing nodules in the field at a time when pods were mature. F_3 plants were grown in a growth chamber (4).

both the nodule and seed sinks. Indications are that much genetic variability in regulation of senescence is already available in nature. The processes of photosynthesis and nitrogen assimilation seem tightly correlated with the regulation of senescence.

D. NITRATE REDUCTASE

This enzyme catalyzes the first step in the reduction of nitrate to ammonia, converting nitrate to nitrite. Once nitrate is inside the plant cells, evidence indicates that the activity of NR is the rate-limiting step in the conversion. More is probably known about the turnover characteristics of NR and RuBPcase than of any other plant enzymes.

Studies with NR have been attractive because the enzyme can be induced at will by adding nitrate either to intact plants or to tissue sections. Concurrently, the loss of NR can be studied by withholding nitrate from the plant. Once nitrate disappears from the metabolic pool, NR is rapidly lost. Evidence is now very strong that induction of NR by nitrate is caused by *de novo* synthesis (160, 289) and that at least in some plants the enzyme is simultaneously synthesized and degraded (289).

Environmental factors strongly affect the induction and steady-state level of NR. Tracking the changes in NR activity after modifying the environmental conditions under which the plant is grown has given valuable clues as to the metabolism of NR and regulation of its activity. Among these are light and dark periods (13, 31, 104, 167, 264), nitrate concentrations (31, 51, 104, 210), water deficit (133, 185, 239, 240), and temperature (31, 106, 195, 237).

It has long been known that light increases the induction of NR activity. Recent investigations show that light has manifold effects. Light can affect the flux of nitrate moving through the plant by affecting nitrate uptake and translocation (106). The flux can then regulate the induction of NR. Light might also somehow regulate the movement of nitrate from the storage to the metabolic pool (16). Light affects the concentration of leaf polyribosomes (262), which in turn are required for NR induction. *In vitro* translation of polyribosomes isolated from light-grown leaves was also greater than that of polyribosomes isolated from dark-treated plants (262).

Light per se is not required for NR induction. Beevers *et al.* (31) showed that NR was induced in dark-grown radishes (*Raphanus sativus*). Travis *et al.* (262) showed that light-grown (carbohydrate sufficient) corn (*Zea mays*) leaves placed in darkness readily induce NR in the presence of nitrate. Aslam *et al.* (13) showed that this effect was attributable to the carbohydrate level of the leaves. Carbohydrate-deficient barley leaves did not induce NR in darkness but did in light. With glucose in the external medium they readily induced NR in darkness.

Light-induced activation of NR was proposed in 1978 by Jolly and Tolbert (146). They isolated a protein inhibitor of NR from soybean leaves kept in darkness for 54 hr. The purifield inhibitor was inactivated by light but again was fully activated after 24 hr of darkness. Equal amounts of inhibitor were present in both light and dark. When the active inhibitor was added to a suspension of soybean leaf cells, their NR activity was stopped. A protein activator of NR activity was also isolated from the soybean leaves. Jolly and Tolbert suggested that the increases in NR activity in light and decreases in dark are regulated by specific protein inhibitors and activators. This possibility was also suggested by Tischner and Huttermann (261), who showed that the rapid increase in NR activity in *Chlorella* in light was caused by a light-mediated activation rather than by *de novo* synthesis of NR. Because cycloheximide inhibited the activation the effect could have been caused by an inhibition of the synthesis of an activator.

The occurrence or importance of NR inhibitors may vary with plant species. Jolly and Tolbert reported that NR from dark-grown soybean leaves was inactive until the inhibitor was removed, whereas NR from leaves and roots of dark-grown barley plants is active *in vitro* (13, 51, 264). Unpublished results of M. Aslam and R. S. Huffaker showed that the *in vivo* rates of nitrate reduction by both detached barley roots and leaves in light or dark was a function of an energy source such as glucose. Although the reduction rate in darkness was lower than in light, the barley leaves, in darkness, still reduced more than 90% of the nitrate taken up while in the presence of glucose. It is possible, however, that inhibitors and activators of NR may influence NR activity in barley to some degree; however, substrate supply seems to be a more important regulator of the *in vivo* activity.

Because of the importance of nitrate pools in regulating the steady-state concentration of NR, evidence for the occurrence and regulation is dealt with briefly here. Heimer and Filner (116) and Ferrari *et al.* (86) showed the existence of two pools of nitrate in tobacco cells, barley aleurone layers, and corn leaf section. Ferrari *et al.* (86) showed that after external nitrate was removed from the external medium the tissues ceased reduction after ~1 hr, although most of the nitrate previously taken up was still present in the cells. The cessation of nitrate reduction was not the result of enzyme inactivation, as nitrate reduction occurred promptly when the tissues were again exposed to nitrate.

Aslam and Oaks (14) later found evidence for nitrate pools in tips and mature sections of corn roots, and Aslam *et al.* (16) showed that barley

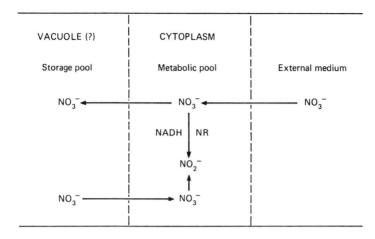

FIG. 9. Relationship between metabolic and storage pools in barley leaves (16).

leaves behaved similarly. The small metabolic pool is most likely located in the cytoplasm, and the large storage pool is probably in the vacuole. There is strong evidence that the cytoplasmic or metabolic pool of nitrate is responsible for the induction of NR (16, 51, 86, 106, 239, 240) where nitrate reduction occurs (14, 16, 51, 106). The size of the metabolic pool in leaf cells is largely a function of the nitrate flux from the roots. Chantarotwong *et al.* (51) showed that, as a result, the rate of nitrate uptake strongly regulated the induction of NR and its subsequent *in vivo* activity. Shaner and Boyer (239, 240) also showed that nitrate flux to the leaves from the roots was much more important than leaf nitrate content in regulating the amount of NR induced. The movement of nitrate from the storage pools to the metabolic pool appears to be regulated differently in different plant tissues. Aslam and Oaks (14) showed that glucose increased the size of the metabolic pool of nitrate in corn roots. On the other hand, Aslam *et al.* (16) showed that glucose had no effect but that light was required to move nitrate from the storage pool to the metabolic pool in green leaves of barley. It was earlier proposed that light was required for the *in vivo* activity of NR in barley leaves, since no nitrate was reduced in leaves in darkness when the external source of nitrate was removed (48, 50). All the same, light could have allowed movement of nitrate from the storage pool into the metabolic pool. Hallmark and Huffaker (106) showed that nitrate reduction could be driven by the storage pool as well as from the flux supplied by uptake in Sudangrass seedlings. Figure 9 summarizes the relationship between nitrate metabolic and storage pools.

1. Product Inhibition of NR Synthesis

It has long been known that end products inhibit both the turnover and activity of NR in bacteria. New information in that area relating to plants has appeared relatively recently. Because the nitrate flux highly regulates NR induction, the effect on nitrate uptake by any assimilation product added to the medium must be accounted for if studies are to be meaningful. Heimer and Filner (116) showed that amino acids can strongly inhibit nitrate uptake and NR activity in cultured tobacco cells.

Radin (215) showed that glycine, glutamine, and asparagine strongly inhibited NR induction in cotton (*Gossypium*) root tips, whereas aspartate and glutamate strongly stimulated induction. Those effects were largely independent of any effects on nitrate uptake. Nitrate uptake was estimated by simultaneously treating the tissue with tungstate, which replaces molybdenum in the NR, resulting in inactivation. The nitrate content was then assayed in those tissues.

Radin (215) detected no regulation by amino acids in green shoot tissue of cotton, indicating basic differences in control between root and leaf NR. Ingle *et al.* (138) showed little effect of amino acids on NR induction in green leaves. The addition of amino acids to *in vitro* reaction systems has no effect on NR activity (138).

The inhibition of NR induction in roots by the above amino acids was alleviated when they were supplied in combination with several other amino acids (216). In some cases the antagonism was the result of competition for uptake. Leucine, isoleucine, and valine did not affect the uptake of asparagine, glutamine, and glycine, but did overcome their inhibition of NR induction: leucine was effective against all three inhibitors, whereas isoleucine and valine were effective against asparagine and glutamine, but not glycine.

Three component activities have been detected in NR: NADH-NR, $FMNH_2$-NR, and NADH-cytochrome c reductase (14, 247, 286). All three activities are induced by nitrate. A marked superinduction of the nitrate-induced NADH-cytochrome c reductase occurred in green leaf tissue in the presence of tungstate (14, 286). When tungstate replaces molybdenum in the complex, an inactive NR results. It was proposed that the molybdenum form of the enzyme might in some way regulate the amount of NR synthesized. When tungstate replaces Mo, more NR may be formed because of effects on either synthesis or degradation (15).

2. Effect of BA and Kinetin on Induction of NR

In the absence of nitrate, BA induced NR synthesis in excised embryos of *Agrostemma githago* (159, 160). NR activity increased very rapidly, being detected within 30 min. Induction was additive when embryos were incubated in the presence of both BA and nitrate. The hormonal induction of NR by BA appeared to be different from that of nitrate. BA ceases to induce NR in the embryos after 64 hr of seed germination, whereas nitrate can induce NR at any time (71). Although NR concentration decreases after removal of BA or nitrate, BA can induce NR a second time, whereas nitrate cannot. Because the inductions by both are additive, the inductions appear to be by different mechanisms. Nitrite reductase was not induced by BA but was induced with nitrate (72). Induction by nitrate and BA also showed differential sensitivity to a variety of metabolic inhibitors (159, 160).

3. Nitrate Reductase Degradation

The steady-state level of NR is modulated by environmental conditions such as nitrate concentration, temperature, light, and moisture. Adverse variations in the above-mentioned conditions can result in rapid loss of NR from plants (31, 106, 133, 239, 240). Inhibitors of protein synthesis such as cycloheximide and actinomycin D inhibited both the appearance and loss of NR in barley seedlings (263). A possible explanation for the retarded loss is that both the synthesis of NR and its degrading system are stopped. The degrading system may then be turned over, accounting for the decreased loss of NR. Another explanation might be that cycloheximide prevented the synthesis of an inhibitor of NR. The half-life for decay for barley leaves was ~12 hr, but was 35 hr in the presence of cycloheximide. Loss of NR in darkness was also greatly decreased by maintaining induced barley plants at low temperature. A half-life of 4 hr was reported for NR in excised corn leaves (237) and for suspension cell cultures of tobacco (289). The mature part of a primary corn root had a faster decay ($t_{0.5}$, 2 hr) than did younger root-tip cells ($t_{0.5}$, 4 hr) (15). $FMNH_2$ and NADH NR activity disappeared at the same rate ($t_{0.5}$, 4 hr), whereas NADH-cytochrome c reductase decayed more slowly ($t_{0.5}$, 7 hr) (15). In the presence of cycloheximide all three activities decayed at similar rates, strongly indicating that NR is turned over in a large number of plants.

The demonstrated presence of inhibitors (146) and activators of NR activity in different plants makes interpretation of turnover studies very difficult when enzymatic activities but not enzyme protein have been followed. Because of this, the elegant experiments of Zielke and Filner (289) become very important. Using triple labeling, they showed that NR is simultaneously synthesized and degraded in tobacco cells (289). Cells were placed in solution containing [^{14}C]arginine to label preexisting protein and in [^{15}N]nitrate to increase their buoyant density. Cells were transferred to a solution containing [^{14}N]nitrate and [^{3}H]arginine, and the changes occurring in the labeled protein were then followed. The effect of the preexisting pools of [^{15}N]amino acids on the density of newly synthesized protein was determined by the ^{3}H label. The change in labels was followed during induction of NR, during steady-state activity, and after a shift to noninducing conditions. During those times the soluble protein was isolated, and isopycnic equilibrium centrifugation was carried out. The results showed that while the enzyme concentration was at a steady state the buoyant density of NR decreased from that of [^{15}N]NR toward that of [^{14}N]NR. Hence both synthesis and degradation were occurring. Preexisting protein labeled with ^{14}C and newly synthesized protein labeled with ^{3}H both showed turnover. It was thus shown that NR was under constant turnover during the induction, the steady state, and the decay phase of noninducing conditions.

An inactivating enzyme of NR isolated from mature roots of 3-day-old corn seedlings (277–279) might account for the degradation of NR, at least in the mature roots of corn. The inactivating enzyme was not detected in root tip and scutella, which have a higher NR activity and exhibit decay. Hence, the inactivating enzyme might not be involved in normal turnover of NR.

In mature roots, inactivation of the NADH-cytochrome c reductase was faster than inactivation of the activities using FMNH$_2$ or NADH. This suggested that the NADH-cytochrome c reductase component might be the primary site for degradation (278, 279).

The addition of amino acids or ammonium had no effect on the rate of loss of NR from corn roots, although amino acids did influence NR induction (193).

E. OTHER PROTEINS

1. Nitrate Transport System

Evidence for the induction of a nitrate transport system in response to nitrate in the substrate medium has been obtained for fungi (98, 236) and higher plants (51, 139, 220). Nitrate transport increases exponentially during the first few hours in nitrate, after which it becomes linear. Inhibitors of protein synthesis such as cycloheximide inhibited the induction of nitrate transport in *Neurospora crassa* (236), *Penicillium* (98), barley seedlings (220), and corn seedlings (139). Casamino acids inhibited formation of the transport system in *Neurospora*, whereas ammonium did not (236).

There is also evidence that the nitrate transport system undergoes turnover. Schloemer and Garrett (236) induced the transport system in *Neurospora* and then applied cycloheximide. The activity of the transport system decreased exponentially, with a half-life of ~3 hr. Goldsmith *et al.* (98) showed that the nitrate transport system in *Penicillium* lost activity faster when ammonium was added to the substrate medium. Using the same method, Rao (219) showed that the activity of the nitrate transport system in barley seedlings decreased after cycloheximide was added to the substrate medium. Those results are inconclusive, however, because cycloheximide can have effects other than inhibiting protein synthesis directly, such as inhibiting energy flow (77). Renosto and Ferrari (223), however, presented evidence that cycloheximide inhibited sulfate transport much later than did dinitrophenol. This finding was taken to indicate that the effect of cycloheximide was to prevent the synthesis of sulfate carriers rather than to inhibit energy flow. It appears that the nitrate transport system or a component thereof is a protein undergoing constant synthesis and degradation.

2. Sulfate Transport System

This system develops by *de novo* synthesis in *Neurospora* (177). The degradation system of the uptake system is not present in dormant conidia, but appears after the onset of germination. The half-life of the sulfate transport system was almost 2 hr. Renosto and Ferrari (223) reported evidence that the sulfate uptake system may be under turnover in both barley roots and potato (*Solanum tuberosum*) tuber tissue. They showed that dinitrophenol (DNP) inhibited sulfate uptake several hours before cycloheximide did. That was interpreted to mean that DNP rapidly inhibited the production of energy required for sulfate uptake, whereas cycloheximide prevented continuing synthesis of the sulfate transport system, permitting net degradation.

3. Phenylalanine Ammonia-lyase

Phenylalanine ammonia-lyase (PAL) catalyzes the conversion of phenylalanine to cinnamic acid and is the first in an enzymatic pathway involved in the synthesis of flavone and flavonol glycosides. Several treatments, including light, induce an increase in PAL activity (290, 291). Several hypotheses have been advanced to explain the increases in PAL activity, that is, that the rate of PAL synthesis was already high in some uninduced plants and that the increase observed was caused by either an increased rate of activation of inactive enzyme from a large precursor pool or a decreased rate of inactivation of active enzyme.

Others proposed that the enzyme was synthesized *de novo*, to account for the activity increase under inducing conditions.

The evidence is now very good that PAL undergoes turnover in several plants. Wellman and Schopfer (280) incubated a suspension cell culture of parsley (*Petroselinum crispum*) in $(^{15}NH_4)_2SO_4$ and $K^{15}NO_3$ during conditions noninducing and inducing for PAL. PAL in the heavy and light fractions was determined by isopycnic centrifugation. Under noninducing conditions (dark), ^{15}N was incorporated into PAL accompanied by band broadening, showing that PAL was being turned over during a 5-hr period. Under inducing conditions (i.e., UV irradiation followed by either red or by far-red irradiation), ^{15}N was incorporated into PAL without significant band broadening over the 5-hr period. These results were taken to mean that phytochrome induced a synthesis of the enzyme greater than that of the constantly occurring degradation. Attridge *et al.* (18) also showed that PAL became density labeled in mustard [*Brassica hirta (Sinapis alba)*] cotyledons, and Attridge and Smith (19) showed the same for *Cucumis sativus* seedlings. Under inducing conditions of either far-red light (18) or blue light (19), PAL was density labeled less than under dark conditions, indicating the activation of a previously inactive pool of PAL.

The synthesis of PAL has been induced both by placing suspension cell cultures of parsley in light and by diluting the cells (218, 238). Evidence for *de novo* synthesis of PAL was obtained using a specific rabbit antibody by following *in vivo* incorporation of radioactive label into a product precipitated by a specific rabbit antibody against the PAL. The *in vitro* synthesis of PAL was accomplished by incubating parsley polyribosomes with a rabbit reticulocyte lysate and immunoprecipitating the product.

In contrast to Wellman and Schopfer (280), Ragg *et al.* (218) showed that polyribosomes isolated from dark-grown parsley cells did not support PAL synthesis. Polyribosomes from light-treated cells, however, synthesized PAL, and the messenger activity of the polysomes increased with the time in light. Tanaka and Uritani (254) showed that PAL appeared as a function of cut injury. PAL protein was not detected in freshly cut leaves but did appear with time after wounding. After a maximum induction, PAL protein most likely decreased as a result of proteolytic activity.

4. Phytochrome Metabolism

Spectrophotometric procedures produced evidence that phytochrome may be both synthesized and degraded (161–164). Immunocytochemical techniques have been used to follow the intracellular location, synthesis, and degradation of phytochrome (206). The protein appears to be in the cytosol, generally diffuse in the P_r form but in discrete regions in the P_{fr} form. It has been speculated that the P_{fr} form may be associated with particulate subcellular fractions that may possibly relate either to its mechanism of action or to its degradation.

Pratt (206) found that phytochrome began to be synthesized in Garry oat and Balboa rye grains about 4 hr after imbibition began. None was detected in seeds maintained at 0°C for 24 hr. Using density-labeling procedures, Quail *et al.* (213) showed that phytochrome was synthesized in pumpkin (*Cucurbita*) hooks grown in darkness and also during recovery in darkness after red irradiation. The rate constant for synthesis was apparently independent of the light treatment.

Quail *et al.* (213, 214) also used density labeling to show that both P_r and P_{fr} are degraded. These investigators found that the apparent k_d of P_{fr} may be up to 100-fold greater than that of P_r. The total phytochrome present, then, is modulated by changing rates of synthesis and degradation of each of the light-absorbing species. Pratt (206) also showed immunochemically that phytochrome is degraded in the oat (*Avena*) and rye (*Secale*) cells.

5. Leucine:tRNA Ligase

The enzymes carrying out the first step in protein synthesis (activation and attachment of amino acids to tRNA) change in activity during the germination of pea (120), bean (*Phaseolus vulgaris*) (7), and wheat (190); during the differentiation of pea (60) and bean roots (125); and during the senescence of soybean cotyledons (34) and tobacco leaves (189). It has been suggested that variation in the activities of these amino acid:tRNA ligases may qualitatively regulate the development of some enzymes (9, 33, 189).

Gore and Wray (100) in 1978 showed that the cellular concentration of leucine:tRNA ligase in tobacco cells was regulated possibly by both synthesis and degradation. When tobacco cells density labeled with D_2O and [^{15}N]nitrate were transferred to light medium, leucine:tRNA ligase decreased in buoyant density with time, showing the incorporation of light isotopes into the protein. Hence, the protein was synthesized *de novo*. Because the bandwidth at half-peak height remained constant, leucine:tRNA ligase was also degraded in the tobacco cells.

V. Metabolism of Seed Protein

Chrispeels (55) has pointed out the necessity of considering the role of cytoplasmic organelles in the metabolism of protein. A working hypothesis is emerging as to how these organelles are intergrated in carrying out the biosynthesis, transport, and secretion of proteins. The rough ER and the Golgi apparatus form a functionally integrated membrane system and carry out the synthesis, processing, and secretion of proteins and other macromolecules. Thus, proteins are synthesized on the polysomes of the rough ER and released into the lumina of the rough ER. The proteins released directly to the cytoplasm may not be processed further; the proteins secreted, however, are apparently initially larger than the mature protein. They are synthesized with a hydrophobic portion of appropriate amino acids and are glycosylated, apparently to facilitate their passage through a membrane. After synthesis, the proteins to be secreted can be transported via the smooth transition elements to the Golgi apparatus. The posttranslational modifications can occur anywhere along the transport route. The proteins accumulate in secretory vesicles that pinch off the maturing face of the dictyosome. The vesicles move to the plasmalemma, where their contents are discharged by reverse pinocytosis.

A. α-Amylase

1. Formation in Seeds

α-Amylase appears to be an example of a protein proceeding through the above-described pathway. This enzyme develops in the aleurone layers, which are living cells having the specialized hydrolytic function of behaving as lysosomes. They produce and release the hydrolytic enzymes required to digest the reserves in the endosperm.

Paleg (198) showed that GA was necessary for the increase in α-amylase activity, and Varner (269) and Varner and Chandra (270) showed evidence for its synthesis. Chrispeels and Varner (58) showed that isolated aleurone layers accounted for all α-amylase activity present in the half-seed when incubation was in the presence of calcium.

GA is produced in the embryo portion of the seed (174, 217) and then translated to the aleurone layer via the vascular system of the nodal region and scutellum (174). Aleurone cells of cereal seeds have become model systems for the study of hormone-induced protein metabolism, because GA is required to initiate the development of hydrolytic enzymes (83).

There is strong evidence that α-amylase is synthesized on rough ER. Studies have shown that proliferation of rough ER is a characteristic response to GA treatments (82, 150, 274) and precedes the synthesis of hydrolytic enzymes. Autoradiography provides evidence that protein synthesis occurs on the rough ER of the GA-treated cells (52, 53, 94). Furthermore, Gibson and Paleg (94) showed that more than one-half the cytoplasmic α-amylase in GA-treated wheat aleurone cells is present in membrane-bound vesicles banding at the same density as ER. Jones and Chen (150) showed that the enzyme in GA-treated aleurone cells of barley is located exclusively around the nucleus in the stacks of rough ER cisternae. Concomitant synthesis of RNA was required for α-amylase synthesis (270). Zwar and Jacobson (292) and Jacobson (142) produced evidence that mRNA is the required species of RNA involved.

2. In Vitro *Synthesis*

Cell-free synthesis of α-amylase has been accomplished using as templates total RNA and poly(A)-RNA isolated from GA_3-treated barley aleurone layers (121). One of the products was identified as α-amylase with a molecular weight of 45,000 by coelectrophoresis with authentic α-amylase. Identification was confirmed by immunoprecipitation with specific rabbit antibody against α-amylase. RNA isolated from aleurone cells not treated with GA did not direct protein synthesis.

Okita and Rappaport (personal communication) have shown evidence for the synthesis of a putative precursor of α-amylase by an *in vitro* protein-synthesizing system from barley aleurone layers. The polypeptide was ~1500 daltons larger than mature α-amylase. In a readout system in which polyribosomes were the template, both the precursor and processed forms were found. Both forms reacted with a specific antibody from rabbits and showed identical fragment patterns after proteolysis. These results indicate that the signal hypothesis (Section IV,A,3) may describe the synthesis and association of α-amylase with the rough ER in barley aleurone cells. Accordingly, α-amylase is synthesized on rough ER processed as it traverses the ER membrane, transported to the ER lumen, glycosylated, and then either transferred to the Golgi, where it accumulates into secretory vesicles, or packaged directly from the rough ER for subsequent discharge across the plasmalemma.

3. Secretion of α-Amylase

Secretion of α-amylase is not yet understood, and different hypotheses (with some controversy) exist regarding the mode of secretion. Jones and Chen (150) propose that secretory organelles are not required for release of amylase from aleurone cells. Others conclude that most of the amylase associated with GA-treated cells was already secreted but not yet released from the tissue. Varner and Meuse (271) nicely separated the processes of α-amylase secretion and transport across the plasma membrane from that of release, the movement of the secreted molecules through the walls into the surrounding medium. Varner and Meuse (271) inactivated most of the α-amylase associated with GA-treated barley aleurone layers by treating them with 1 mM HCl, which did not damage the cells. This finding indicated that the enzyme was probably in the periplasmic space or bound to the wall, but not inside the cell. After first inactivating amylase in the cell walls by treatment with dilute HCl, Firn (88) showed that more than one-half of the α-amylase in GA$_3$-treated tissues was associated with membrane-bound vesicles. Jacobson and Knox (143) showed a similar location of α-amylase. Hence, the ER of the cell may be involved in both the synthesis and secretion of α-amylase. It was shown that the secretion process was energy dependent, whereas release from the cell was diffusion dependent. The process of secretion may be better understood if membrane organelles involved in secretion are not considered as distinct structures, but rather as part of an endomembrane system that has functional continuity. Known as the membrane-flow hypothesis, secretion can then be considered as the transfer of the membrane contents from one part of the system to another.

In addition to α-amylase, several other hydrolytic enzymes are induced by GA in barley aleurone layers. Density-labeling techniques showed ribonuclease (32), laminarase (32), and protease to be synthesized *de novo* (144). Acid phosphatase activity increases in GA-treated half-seeds of barley (141).

Jacobson and Varner (144) showed that the similarity of synthesis of α-amylase and protease in GA-treated aleurone layers shows a similar control. Several lines of evidence indicated that this protease was the one responsible for hydrolysis of endosperm protein. The protease activity was measured against gliadin as substrate, and most of the protease was released into the medium surrounding the aleurone layers.

Little information is available on hydrolysis of the hydrolytic enzymes of the aleurone cells.

B. SEED STORAGE PROTEINS

1. Characteristics

Developing seeds offer intriguing systems for the study of protein synthesis. The endosperm of monocots or the cotyledons of dicots carry out large-scale protein synthesis of apparently only a few specific proteins. During large-scale synthesis the possibility is very good for isolating large amounts of mRNA from the polyribosomes driving such synthesis. These proteins, synthesized in large amounts during seed development, are found primarily in the endosperm of monocots or cotyledons of dicots.

The storage proteins are sequestered in discrete organelles, known as protein bodies (11, 181, 272), that can generally be readily isolated. They have been classified by Osborne (197) on the basis of their solubility in various aqueous solutions. Those that are water soluble are classified as albumins; globulins are water insoluble but soluble in dilute salt; glutelins are insoluble in water and dilute salt, but soluble in weak acid or base; and prolamines are insoluble in all the above but soluble in 70–80% ethanol.

The storage proteins must be characterized to assist in understanding their synthesis and mobilization, as well as determining the number of genes that encode their synthesis. Characterization is a problem because of the difficulty of purifying storage proteins. Millerd (181) reported that many of the reported purifications contain microheterogeneity. A major difficulty has been the development of assay procedures for determining purity during the various steps in isolations. Another complication results from the presence of proteinase activity associated with the protein bodies containing the storage proteins (229), which leads to protein breakdown during the isolation.

Seeds of many species contain only a few major storage proteins (123, 181). In most seeds except those of Gramineae, the principal storage proteins are globulins (11). An exception is oats, the major storage proteins of which are globulins (H. Thomas, personal communication). In cereals except rice (*Oryza sativa*), 40–60% is prolamine and 20–40% glutelins (11). In *Vicia* spp. and *Pisum sativum* the globulins have been further separated into the components legumin and vicilin (63, 197). These, in turn, are high molecular weight proteins comprising several subunits (67).

Albumins constitute up to 40% of the total protein in protein bodies of castor bean (*Ricinus communis*) and are localized in the matrix of the organelle. They have an S value of 25 and are resolved into several proteins of ~12,000 daltons by SDS-acrylamide gel electrophoresis. Evidence is presented that albumins are also important storage proteins. These proteins disappeared rapidly during the first 2 days of germination, providing the first evidence that albumins are storage proteins (287), contrary to the generally held belief that they are mainly enzymes and metabolic proteins. These investigators also found evidence that the castor bean allergens are the albumin storage protein (287, 288), helping explain the wide distribution of allergens in many seeds.

Seed storage proteins have also been examined by sucrose-density centrifugation. Three distinct fractions have resulted: a 2 S, a 7–8 S, and an 11–12 S. In soybean seed proteins the 7- and 12-S fractions accounted for up to 70% of the total seed protein (124). Further examination of the three fractions by acrylamide gel electrophoresis showed that the 2-S fractions gave a diffuse electrophoretic pattern under dissociating conditions. The 7-S component separated into about three components under both nondissociating and dissociating conditions. The 12-S component yielded only one band under nondissociating conditions, but about five or six under dissociating conditions. Similar 8-S and 11-S components have been isolated from *Vigna radiata (Phaseolus aureus)*, the 8-S component showing four electrophoretic bands and the 11-S component three bands (79). Millerd (181) isolated from *Vicia faba* a legumin having three subunits. She has estimated that much of the 11-S component is a single protein. The 7- to 8-S fraction probably contains two or three different proteins

2. Storage Glycoproteins

Glycoproteins have received greater research effort because of recognition of their importance in the biochemistry of the living system. Pea cotyledon storage proteins legumin and vicilin are glycoproteins (25). Glycoproteins are important components of membranes. The glycosyl components (211) are derived from nucleotide sugars. The oligosaccharide chains are apparently assembled by adding the sugar residues sequentially from their nucleotide derivatives (275). The glucosyl components are first transferred to lipid intermediates, forming lipid-linked monosaccharides. These then serve as precursors to form lipid-linked oligosaccharides. The glycosyl moiety is then transferred as a complete unit to the polypeptide (30, 66, 80, 89, 90, 171). The lipid appears to be of the dolichol type (66, 275). These are phosphorylated polyprenols containing 16–22 isoprenoid units and characterized by a saturated α-residue.

The glycoproteins formed contain both mannose and N-acetylglucosamine (GlcNAc) as the carbohydrate moieties. The following illustrates the proposed pathway of glycosylation (81):

$$UDP\text{-}GlcNAc + lipid\text{-}P \rightleftarrows lipid\text{-}PP\text{-}GlcNAc + UMP$$

$$UDP\text{-}GlcNAc + lipid\text{-}PP\text{-}GlcNAc \rightarrow lipid\text{-}PP\text{-}(GlcNAc)_2 + UDP$$

$$GDP\text{-}mannose + lipid\text{-}P \leftrightarrows lipid\text{-}P\text{-}mannose + GDP$$

$$Lipid\text{-}PP\text{-}(GlcNAc)_2 + n \, lipid\text{-}P\text{-}mannose \rightarrow lipid\text{-}PP\text{-}(GlcNAc)_2(mannose)_n + n \, lipid\text{-}P$$

$$Lipid\text{-}PP\text{-}(GlcNAc)_2\text{-}mannose_n + protein \rightarrow glycoprotein + lipid\text{-}PP$$

$$Lipid\text{-}PP \rightarrow lipid\text{-}P + P_i$$

Nagahashi and Beevers (186) showed that the UDP-N-acetyglucosamine *N*-acetylglucosaminyltransferase and GDP-mannose mannosyl-transferase activities that carry out the assembly of lipid-linked sugar intermediates and glycoproteins in pea cotyledons are located mainly in the rough endoplasmic reticulum.

3. In Situ *Synthesis*

Storage proteins increase steadily in the cotyledons during their development (181). Hill and Breidenbach (124) found that the 2-S fraction from soybeans predominated at the early stages of development, but decreased throughout maturation. The 7- and 12-S components were synthesized later than the 2-S fraction, and in larger amounts. In *Vicia faba*, vicilin was detected in the developing seed before legumin, whereas legumin synthesis was greatly increased when cotyledonary cell division was essentially complete (182). The concentration of legumin was finally several times that of vicilin.

At maturity a dry French bean (*Phaseolus vulgaris*) seed contains ~90 mg of protein, 50% of which is globulin. The globulins have been divided into globulin-1 (G-1) and globulin-2 (G-2) fractions, the former requiring high-saline concentrations for solubility, and the latter lower-saline conditions. The G-1 fraction predominates. The major increase in G-1 and G-2 synthesis occurred abruptly at 16 days after flowering, when the cotyledon was 10 mm long (251).

The G-1 fraction contains three subunits: 43,000, 47,000, and 53,000 daltons. Sun *et al.* (251) used polysomes from the growing cotyledons of French bean seeds to synthesize the three G-1 subunits *in vitro*. The products of the *in vitro* reaction were identified quantitatively by acrylamide electrophoresis.

4. *Sites of Synthesis*

Some investigators have not found ribosomes in association with the protein body (22, 272). They propose that the proteins are synthesized elsewhere and transported into the protein body (21). It has been proposed that globulins may be synthesized on membrane ribosomes and then transferred through the ER into the protein bodies (67). Evidence shows that an increase in membrane-bound ribosomes and proliferation of rough ER (43, 196, 201, 202) was correlated with increased protein synthesis in the cotyledons of *Vicia faba* (201, 202). Using a pulse-chase method, Bailey *et al.* (21) showed that unlabeled material moved from the endoplasmic reticulum to the protein bodies. The pathway remains to be worked out, however.

On the other hand, Burr and Burr (47) identified protein bodies in maize endosperm with polyribosomes associated with the exterior surface of their single membranes. The polyribosomes were identified by electron microscopy. The protein bodies were isolated by sucrose density gradients, and the polyribosomes were dissociated by treatment with detergent. When placed in an amino acid–incorporating system, the polyribosomes drove the synthesis of zein. The components required for zein synthesis apparently existed at the protein membrane surface. Hence the protein bodies served as the specific site of synthesis and of deposition of zein.

5. Formation of Protein Bodies

Protein bodies may be formed by subdivision of the main vacuole (43, 196) or by secretion of protein into cisternae of ER (22), or they may originate from Golgi bodies (108). Evidence of protein bodies from ER was taken from pea cotyledon cells. Harris and Boulter (108) reported that most protein bodies of cowpea (*Vigna unguiculata*) seeds originate from Golgi bodies.

6. Corn Storage Protein

Zein, a major storage protein in the endosperm of corn, is localized in protein bodies. At grain maturity it composes up to 60% of the total endosperm protein (147). It has been separated into two major components, termed Z1 and Z2. Mutations at the *opaque*-2 locus and *floury*-2 loci substantially reduce zein concentrations, which are poor in lysine and tryptophan. Thus the proportion of the nutritionally superior nonzein proteins is increased. Not only is total zein reduced in the *opaque*-2 mutant, but the mutant is deficient in the major zein protein, Z1, as well (170).

The ability to isolate undergraded free and membrane-bound polysomes has greatly facilitated studies of the regulation of storage protein biosynthesis (149). *In vitro* translation of the isolated polyribosomes showed that zein biosynthesis was associated with the large-size classes of the membrane-bound fraction. The proportion of these large-size classes increased during kernel development and was highly correlated with *in vivo* synthesis of zein. These classes are associated with the ER, which is forming bodies (148). The *in vitro* translation products were electrophoretically identical to the two *in vivo* zein classes (149).

Jones *et al.* (149) showed that zein synthesis was correlated with the decrease in the large-size classes of polyribosomes in the mutant. As the number of *floury*-2 alleles was increased in the mutant, zein concentration decreased proportionally (147) with both Z1 and Z2 components proportionally decreased. In contrast to findings with the *opaque*-2 mutant, no major alterations were found in polysome size classes. However, a reduction in membrane-bound polysomes correlated with reductions in *in vitro* zein synthesis and *in vivo* zein accumulation.

Zein RNA was isolated from the membrane-bound polysomes of developing maize kernels by oligo(dT)-cellulose chromatography (169). *In vitro* translation of the mRNA yielded both zein components. The development of these *in vitro* zein-synthesizing systems is leading to understanding of the genetic regulation of storage-protein synthesis and the mechanism whereby mutations bring about desired changes.

After this chapter was prepared, many new and exciting discoveries have been reported regarding the molecular biology of seed storage proteins. Please refer to the following reviews by Larkins (295, 296).

C. DEGRADATION OF SEED STORAGE PROTEINS

Ultrastructural studies indicate that the degradation of protein bodies during germination can begin from the inside or at the periphery of the membrane (11). The proteinases responsible for degradation appear to originate internally for the former and externally for the latter. If the proteinases are originally inside the protein body, questions about their regulation become important. The presence of proteinase inhibitors has been invoked as a control mechanism (229). These inhibitors are present while storage proteins are being synthesized in the seed, disappearing during germination.

1. Proteinase Inhibitors

Proteinase inhibitors are plant proteins that inhibit trypsin-like or chymotrypsin-like enzymes from animals or bacteria (229). They are distributed widely in nature, mainly in plants, and are usually small, with molecular weights below 50,000 and some even below 10,000 (229). The inhibitors have been difficult to study since their effects must be measured indirectly. Because many different types can be present in the same tissue, the inhibitors must be highly purified for meaningful interpretation of their effects. Endogenous inhibitors of plant proteinases have been detected in extracts of barley (*Hordeum vulgare*) seeds (165) and seedlings (46), rice (*Oryza sativa*) (127), lettuce (*Lactuca sativa*) (241), sorghum (*Sorghum bicolor*) (93), and mung beans [*Vigna radiata (Phaseolus aureus)*] (26, 27, 56).

A few studies have attempted to correlate increases in proteolytic activity in germinating seeds with decreases in endogenous inhibitors. Several studies showed that increased proteolytic activity did not occur until the inhibitors decreased (26, 165, 242). Shain and Mayer (242) found that trypsin-like enzymatic activity increased during germination, although no $^{35}SO_4$ was incorporated into the enzyme. Concurrently, an endogenous inhibitor of the proteolytic activity disappeared. They proposed that active proteinase is liberated as the inhibitor disappears. Baumgartner and Chrispeels (26) showed the importance of comparing the kinetics of the appearance of the proteinase activity and the disappearance of the inhibitor, as well as the intracellular locations of the two activities. They isolated an endopeptidase and two inhibitors of it from mung bean cotyledons having molecular weights of 12,000 and 2,000. When extracts containing both the endopeptidase and the inhibitors were incubated *in vitro* at 7°C, the inhibitor activity decreased and the endopeptidase activity increased. The two processes differed in kinetics, however. The inhibitor activity was lost faster than the endopeptidase activity increased. Inhibitor decay and increased enzyme activity *in vivo* also differed in kinetics. The inhibitor activity gradually decreased soon after germination until the fifth or sixth day. Endopeptidase activity increased very slowly during the first 3 days and then rapidly during days 4 and 5. These workers also showed that endopeptidase activity (responsible for proteolysis of the protein bodies) was in the protein bodies and that the inhibitor activity was in the cytosol; it therefore seems unlikely that the inhibitors are expressing a regulatory function in the degradation of protein bodies. The degradation of reserve protein seems more dependent on the synthesis of vicilin peptidohydrolase. Chrispeels and Baumgartner (56) showed that the inhibitor does not inhibit the activity of vicilin peptidohydrolase. The endogenous inhibitors may function to protect the cytoplasm against accidental disruption of the protein bodies containing the endopeptidase (26).

2. Inhibitors as Stored Protein

Ryan (229) showed evidence that inhibitors may play a storage role, as they represent a significant percentage of the protein in seeds and tubers.

3. Plant Protection

Ryan (229) showed evidence also that proteinase inhibitors may protect plants against bacterial attack. Many of the bacterial proteinases are trypsin-like or chymotrypsin-like in their specificities and are strongly inhibited by the plant-derived inhibitors. Ryan also reported that inhibitor activity is greatly increased in plants challenged by wounding (102, 276).

VI. Protein Degradation

A. PROTEOLYTIC ENZYMES

Reviews by Dechary (65), Ryan (229), and Ashton (11) describe many of the known proteolytic enzymes in higher plants. It seems unlikely that the numbers and types of proteolytic enzymes in any given plant tissue are completely documented, for comparison of reported results is made difficult by the variety of the isolation procedures and assay substrates used.

Proteolytic enzymes from animal systems have been broadly classed as serine, thiol, acid, and metalloproteases (109). Several plant proteases have been characterized as serine proteases (137, 176, 281, 282), sulfhydryl proteases (8, 142, 168, 179), or acid proteases (6, 93, 179, 248). Most of the carboxypeptidases and aminopeptidases from plants are not metalloproteases. On the other hand, some di- and tripeptidases (12, 45) and endopeptidases (142, 207) have been isolated that require metals for activity. The most extensive studies of proteases have used germinating seeds. Extracts from germinating barley grains show activities against two different carboxypeptidase substrates, Z-Phe-Ala and Z-Phe-Phe; against two aminopeptidase substrates, Phe-β-NAA and Leu-β-NAA; and against two dipeptidase substrates, Leu-Tyr and Ala-Gly (180). Different pH optima indicated the presence of three endopeptidase activities against gelatin (252). Jacobson and Varner (144) also showed the presence of an endopeptidase activity in barley. Germinating corn shows endopeptidase activity (85, 92, 111, 112), carboxypeptidase activity (85), and aminopeptidase activity (85). The aleurone layers of barley show high activity of each of the four classes of peptidases (144, 180).

Much work has been done to relate proteolytic enzymatic activities to the observed mobilization of storage protein during seed germination. During germination, barley aleurone layers showed little change in the carboxy- and aminopeptidase activities, whereas endopeptidase activity increased (180, 252). Carboxypeptidase and endopeptidase activities increased during germination in both barley (180) and corn (85, 92), whereas aminopeptidase activity changed but little. Dipeptidase activity remained high in barley, suggesting that some hydrolysis products are absorbed as peptides and must be further hydrolyzed to amino acids. The starchy endosperm of both barley (180) and corn (85) increased in carboxypeptidase activity during germination. Aminopeptidase activity was not found in barley (180) but was observed in corn, decreasing in activity during germination (85).

Because of difficulties in assay procedures, the activity of a proteinase has seldom actually been tied to the protein it degrades *in vivo*. Progress has been good in showing endopeptidase activity against native substrates (26, 27, 57, 111, 112). Jacobson and Varner (144) showed that GA induced the synthesis of proteases in barley aleurone layers; the proteases were then secreted into the endosperm. Sundblom and Mikola (252) showed these proteases to be endopeptidases. Harvey and Oaks (111, 112) and Fujimaki *et al.* (92) recovered from corn an endopeptidase that was able to degrade the native glutelin and zein. It had a molecular weight of 21,000 and was a sulfhydryl protease. The endopeptidase present was sufficient to account for the *in vivo* degradation of the endosperm protein over the period during which the protein was mobilized. Chrispeels and Boutler (57) correlated the increase in endopeptidase activity with the disappearance of storage proteins in mung bean cotyledons. Baumgartner and Chrispeels (27) later purified the endopeptidase, accounting for more than 95% of the activity. These workers showed that the endopeptidase apparently responsible for hydrolysis of the protein bodies in mung bean is within the protein bodies. This evokes interesting questions as to its regulation. Is it synthesized during formation of the protein bodies? What prevents it from hydrolyzing the storage proteins before germination ensues? Is it inhibited by the protein inhibitors widely found in plant tissues?

1. Roots

Feller *et al.* (85) showed aminopeptidase, carboxypeptidase, and caseolytic (endopeptidase) activities in corn roots.

2. Leaves

The amount of proteinase activity found in leaves of different species depends on the physiological age of the plant and the environmental treatments it has undergone. Leaves can serve as a sink for stored protein, which can later be mobilized as new sinks develop. The proteolytic activities accompanying the mobilization of the leaf protein are now receiving considerable attention. Exo- and endopeptidase activities detected in leaves of intact barley plants before the onset of senescence apparently function in the turnover of cytoplasmic proteins (132).

Soong *et al.* (246) showed that when corn leaves were detached before the onset of senescence, carboxypeptidase, aminopeptidase, and endopeptidase activities remained constant during the next 48 hr of darkness. When the experiment was run with leaves detached after the onset of senescence, aminopeptidase activity against casein increased. Feller *et al.* (85) determined proteolytic activities as foliar senescence symptoms developed in corn leaves in the field. Increasing senescence was paralleled by decreases in exopeptidase activities and increases in endopeptidase activities. Decreases in chlorophyll and protein accompanied those observations. During the latter stages of the grain development, amino- and carboxypeptidase activities decreased and endopeptidase activity increased. The caseolytic activity was highest when protein loss from leaves was greatest. Dalling *et al.* (62) showed a high correlation between protein loss from wheat leaves and acid proteinase activity.

It is not yet known whether the increase of proteolytic activity observed *in vitro* is a requirement for the increased rate of protein loss *in vivo*. Several reports show that protein loss can be just as rapid, even though endopeptidase activity remains constant during senescence. Beevers (29) showed that caseolytic activity was not correlated with protein loss in leaf discs of nasturtium (*Tropaeolum majus*). Soong *et al.* (246) showed that high endopeptidase activity is not a prerequisite for rapid protein degradation in detached corn leaves. Results were similar in detached oat leaves (268). Endopeptidase activity did not increase in intact barley leaves in darkness but greatly increased in detached leaves. Protein was lost at nearly the same rate in both cases (130a). Cyclohex-imide (CHI) prevented the loss of RuBPcase and chlorophyll in detached barley leaves while simultaneously preventing the increase in proteolytic activity (204). CHI stopped the formation in the senescing leaves of a new proteolytic activity that was seen as a second band appearing after isoelectric focusing (130a). The relationship to protein degradation of the appearance of the second band of proteolytic activity is unknown.

3. Purification of Leaf Proteinases

Two endoproteinases, one purified 5800-fold, the other 50-fold from barley leaves, show high activity against RuBPcase, casein, methylated casein, azocasein, hemoglobin, and bovine serum albumin (297). The degradation products of each endoproteinase against RuBPcase were identified (298).

The proteinases have been somewhat difficult to purify, as they appear to hydrolyze themselves during purification. Hence it is important to minimize the time required for purification and to compare the electrophoretic or isoelectrophoretic band patterns of endopeptidase activity at the different steps of purification. Disc gel electrophoresis initially showed single bands of endopeptidase activity from leaves of senescing intact plants and detached leaves of barley (130a). A preparation containing the proteolytic activity from senescing detached leaves was stored for $2\frac{1}{2}$ weeks at $-20°C$ as the ammonium sulfate precipitate. Multiple bands of activity appeared when the preparation was again subjected to electrophoresis, showing the problem of degradation to other active components during storage.

Isoelectric focusing of cell-free extracts showed one band of proteolytic activity from senescing leaves of intact barley plants and two bands from senescing detached leaves. CHI prevented the formation of the second band of proteolytic activity detected in senescing detached leaves. Likewise, kinetin prevented its formation, whereas CAM had no effect (130a).

From senescing oat leaves, Drivdahl and Thimann (75) purified an acid and a neutral protease about 500-fold, using a combination of ammonium sulfate precipitation, affinity chromatography on hemoglobin–Sepharose, and ion-exchange chromatography. The molecular weight of each was estimated at 76,000. The neutral protease was apparently a sulfhydryl enzyme. The corn and (one) barley leaf endopeptidases were also apparently sulfhydryl enzymes (84, 297).

B. Regulation of Protein Degradation

The tertiary structure of a protein seems strongly related to its rate of *in vivo* degradation. Even more important than the amino acid sequence may be the compactness of folding. Proteolytic cleavage is less likely if the bond is deep within the folds or in a rigid region of the chain (260). The flexible exposed regions of the protein molecule allow the attachment to the active site of the protease required for subsequent hydrolysis (188). Hence, the "hinge" region of immunoglobulin (232) and the peptide links between the globular units of plasma albumin (284) are more susceptible to hydrolysis.

Recent studies have attempted to correlate *in vivo* half-life with the susceptibility of a protein to denature, inactivate, or be degraded. The half-lives *in vivo* of some mammalian and bacterial proteins were correlated with their rates of *in vitro* degradation by well-known endopeptidases of different specificities [e.g., trypsin, chymotrypsin, and papain (39, 96)]. Spontaneous denaturation may be rate limiting for degradation of most cell proteins. As the actual free-energy difference between the native conformation and the denatured state is relatively small, denaturation may occur frequently *in vivo* (97). In bacteria and reticulocytes, abnormal proteins tend to aggregate and form precipitates before undergoing rapid hydrolysis. Prouty and Goldberg (209) isolated a rapidly sedimenting fraction of abnormal proteins from *E. coli.* The pellet represented abnormal proteins that had denatured, aggregated, and precipitated. The final phase of the degradation then occurred in the soluble phase of the cell. Ballard *et al.* (23) proposed that in rat liver phosphoenolpyruvate carboxykinase was first denatured, precipitated, and hydrolyzed by limited proteolysis (23). Similarly, Bond (39) showed that the *in vivo* half-life of some liver enzymes correlated with their inactivation at pH 5.0. In these systems, then, the ease of *in vitro* denaturation was correlated with the *in vivo* half-lives of the proteins.

Protein molecular weight seems related to degradation rates in mammalian cells. High molecular weight polypeptides tend to have shorter half-lives than smaller proteins in many kinds of eukaryotic cells (68, 69, 97).

The charge on proteins, independent of their molecular weights, may also relate to their *in vivo* half-lives (70). Proteins with low isoelectric points were degraded faster than were proteins with neutral or basic points. Hence, proteins of low molecular weight and with basic isoelectric points should be the most stable in the living systems (97).

The sensitivity to hydrolysis of various sites on the protein can be modulated by ionic strength or binding of ligands. Such treatments usually produce changes in the relative rates of hydrolysis but have little effect on the products formed (188). The binding of 1 mol of ligand per mole of serum albumin was sufficient to reduce its rate of hydrolysis (175). Binding Ca^{2+} and Mn^{2+} had similar effects (101). Conalbumin was less susceptible to trypsin hydrolysis in the presence of iron (20) or substrate cofactors (265). Homologous haptens reduced the degradation of rabbit antibody (175). Substrate molecules can reduce the rate of hydrolysis of enzyme proteins. Both the *in vivo* and *in vitro* rates of degradation of tryptophan oxygenase are decreased by the presence of tryptophan (235), thymidine retards the degradation of thymidine kinase (166, 172), and iron stabilizes rat liver ferritin (76). It was shown that the effector molecules acted on the protein substrate, not on the proteinase.

C. VACUOLES AS LYSOSOMES

It has been proposed that vacuoles may serve a lysosomal function or may represent a sequestering area for lytic enzymes as a means of protecting the cytoplasmic metabolic protein. Indirect evidence from cytological, ultrastructural, and biochemical studies has indicated a lysosomal function of the vacuole (178). It has been proposed that cytoplasmic materials may be transported into the vacuole, where hydrolysis would occur. The process of autophagy has also been invoked (178). Ultrastructural studies have indicated the presence of remnants of microbodies within vacuoles. A second type of autophagy might result from the trapping of portions of cytoplasm in ER wherein hydrolysis could occur (178). Results from direct analysis were scarce because of problems involved in obtaining unbroken pure vacuoles. Because the commercial enzyme preparations used to prepare protoplasts contain high levels of proteinases, the protoplasts and vacuoles can easily be contaminated with the extracellular proteinases. Careful isolation of leaf vacuoles has shown the presence of proteinases (178, 299, 300). Proteinases are not contained in the vacuoles of all plant cells. In a careful study Butcher et al. (49) found no evidence for proteinases in vacuoles prepared from *Hippeastrum* petals or from tulip petals and leaves. The vacuoles were free of cytoplasm contamination, as determined by marker enzymatic activities. The integrity of the vascuoles was attested to by the following criteria: the anthocyanin was retained in the vacuoles from the flower petals; the vacuoles excluded Evans blue and concentrated neutral red; and the ion content was essentially the same as in the intact protoplasts. High proteolytic activities were found in the cytoplasm. Hence it appears that in some tissues proteolytic activity is present in the cell cytoplasm external to the vacuole and constantly turning over cytoplasmic proteins (204). The degradation of the specific cytoplasmic proteins, then, is probably dependent on the pH value of the cytoplasm, molecular weight, isoelectric points, and tertiary structure.

VII. Conclusions

Protein metabolism enables the plant to adapt to changing environmental conditions and to complete its life cycle. As the plant matures, it can shift its stored protein reserves from foliar parts into reproduction organs. The process seems to be highly regulated, with chloroplast protein (which in many plants contains more than one-half the total soluble protein) being broken down early. Ribulosebisphosphate carboxylase is a major constituent that is lost during that period. As this loss occurs, photosynthesis decreases and the plant becomes more and more dependent on the metabolism of stored reserves for its life cycle.

The proteinases responsible for protein metabolism in various plant parts are being identified using rapid methods of separation and novel assays for determination of activities. Taking cereals as an example, the major leaf endopeptidases seem to be synthesized in the cytoplasm on 80-S ribosomes and to be located in the vacuoles. As a result, much of the cytoplasmic protein undergoes turnover, depending on susceptibility to attack by the proteinases. Nitrate reductase, for example, seems highly susceptible to constant turnover. Organelle-destined proteins such as the small subunit of RuBPcase are synthesized in the cytoplasm. The way in which a sufficient quantity of the subunits is protected from degradation until traversing the organelle membrane remains an open question. Information is just emerging on the synthesis and proteolytic processing of the subunits as they traverse organelle membranes.

Still open to question is how the major endoproteinases, which are located in the vacuole in some plant species, function in turnover of cytoplasmic and possibly organelle-contained proteins. Also, do the endoproteinases in chloroplasts degrade proteins *in vivo*? Questions concerning their importance and regulation remain to be resolved.

More information is required on individual proteins in order to understand the regulation of protein metabolism. Determination of the interaction of the synthetic and degradative processes, especially in pacemaker enzymes, will yield exciting information as to how protein metabolism is regulated to permit appropriate growth, development, and maturation.

References

1. Abe, M., Arai, S., and Fujimaki, M. (1977). Purification and characterization of a protease occurring in endosperm of germinating corn. *Agric. Biol. Chem.* **41**, 893–899.
2. Abelson, P. H. (1954). Amino acid biosynthesis in *E. coli:* isotopic competition with C^{14}-glucose. *J. Biol. Chem.* **206**, 335–343.
3. Abelson, P. H., and Vogel, H. J. (1955). Amino acid biosynthesis in *Torulopsis utilis* and *Neurospora crassa. J. Biol. Chem.* **213**, 355–364.
4. Abu-Shakra, S. S., Phillips, D. A., and Huffaker, R. C. (1978). Nitrogen fixation and delayed leaf senescence in soybeans. *Science (Washington, D.C.)* **199**, 973–975.
5. Alscher, R., Smith, M. A., Peterson, L. W., Huffaker, R. C., and Criddle, R. S. (1976). *In vitro* synthesis of the large subunit of ribulose diphosphate carboxylase on 70S ribosomes. *Arch. Biochem. Biophys.* **174**, 216–225.
6. Amagase, S. (1972). Digestive enzymes in insectivorous plants. III. Acid proteases in the genus *Nepenthes* and *Drosera peltata. J. Biochem.* **72**, 73–81.
7. Anderson, J. W., and Fouden, L. (1969). A study of the aminoacyl-tRNA synthesis of *Phaseolus vulgaris* in relation to germination. *Plant Physiol.* **44**, 60–68.
8. Anderson, J. W., and Rowan, K. S. (1965). Activity of peptidase in tobacco leaf tissue in relation to senescence. *Biochem. J.* **97**, 741–746.

9. Anderson, M. B., and Cherry, J. H. (1969). Differences in leucyl-transfer RNA's and synthetase in soybean seedlings. *Proc. Natl. Acad. Sci. USA* **62**, 202–209.
10. Arias, I. M., Doyle, D., and Schimke, R. T. (1969). Studies on the synthesis and degradation of proteins of the endoplasmic reticulum of rat liver. *J. Biol. Chem.* **244**, 3303–3315.
11. Ashton, F. M. (1976). Mobilization of storage proteins of seeds. *Annu. Rev. Plant Physiol.* **27**, 95–117.
12. Ashton, F. M., and Dahmen, W. J. (1967). A partial purification and characterization of two amino peptidases from *Cucurbita maxima* cotyledons. *Phytochemistry* **6**, 641–653.
13. Aslam, M., Huffaker, R. C., and Travis, R. C. (1973). The interaction of respiration and photosynthesis in induction of nitrate reductase activity. *Plant Physiol.* **52**, 137–141.
14. Aslam, M., and Oaks, A. (1975). Effect of glucose on the induction of nitrate reductase in corn roots. *Plant Physiol.* **56**, 634–639.
15. Aslam, M., and Oaks, A. (1976). Comparative studies on the induction and inactivation of nitrate reductase in corn roots and leaves. *Plant Physiol.* **57**, 572–576.
16. Aslam, M., Oaks, A., and Huffaker, R. C. (1976). Effect of light and glucose on the induction of nitrate reductase and on the distribution of nitrate in etiolated barley leaves. *Plant Physiol.* **58**, 588–591.
17. Atkin, R. K., and Srivastava, B. I. S. (1969). The changes in soluble protein of excised barley leaves during senescence and kinetin treatment. *Physiol. Plant.* **22**, 742–750.
18. Attridge, T. H., Johnson, C. B., and Smith, H. (1974). Density labeling evidence for the phytochrome-mediated activation of phenylalanine ammonia-lyase in mustard cotyledons. *Biochim. Biophys. Acta* **343**, 440–451.
19. Attridge, T. H., and Smith, H. (1974). Density labeling evidence for the blue-light mediated activation of phenylalanine ammonia lyase in *Cucumis sativus* seedlings. *Biochim. Biophys. Acta* **343**, 452–464.
20. Azari, P. R., and Feeney, R. E. (1961). The resistance of conalbumin and its iron complex to physical and chemical treatments. *Arch. Biochem. Biophys.* **92**, 44–52.
21. Bailey, C. J., Cobb, A., and Boulter, D. (1970). A cotyledon slice system for the electron autoradiographic study of the synthesis and intracellular transport of the seed storage protein of *Vicia faba*. *Planta* **95**, 103–118.
22. Bain, J. M., and Mercer, F. V. (1966). Subcellular organization of the developing cotyledons of *Pisum sativum* L. *Aust. J. Biol. Sci.* **19**, 49–67.
23. Ballard, F. J., Hopgood, M. F., Reskef, L., and Hanson, R. W. (1974). Degradation of phosphoenolpyruvate carboxykinase (guanosine triphosphate) *in vivo* and *in vitro*. *Biochem. J.* **140**, 531–538.
24. Balz, H. P. (1966). Intracellulare Lokalisation und Funktion von hydrolytischen Enzymen bei Tabak. *Planta* **70**, 207–236.
25. Basha, S. M. M., and Beevers, L. (1976). Glycoprotein metabolism in the cotyledons of *Pisum sativum* during development and germination. *Plant Physiol.* **57**, 93–97.
26. Baumgartner, B., and Chrispeels, M. J. (1976). Partial characterization of a protease inhibitor which inhibits the major endopeptidase present in the cotyledons of mungbeans. *Plant Physiol.* **58**, 1–6.
27. Baumgartner, B., and Chrispeels, M. J. (1977). Purification and characterization of vicilin peptidohydrolase, the major endopeptidase in the cotyledons of mungbean seedlings. *Eur. J. Biochem.* **77**, 223–233.
28. Bechet, J., and Wiame, J. M. (1965). Indication of a specific regulatory binding protein for ornithinetranscarbamylase in *Saccharomyces cerevisiae*. *Biochem. Biophys. Res. Commun.* **21**, 226–234.

29. Beevers, L. (1968). Growth regulator control of senescence in leaf discs of nasturtium (*Tropaeolum majus*). *In* "Biochemistry and Physiology of Plant Growth Substances" (F. Wightman, and G. Setterfield, eds.), pp. 1417–1435. Runge Press, Ottawa.
30. Beevers, L., and Mense, R. M. (1977). Glycoprotein biosynthesis in cotyledons of *Pisum sativum* L. Involvement of lipid-linked intermediates. *Plant Physiol.* **60,** 703–708.
31. Beevers, L., Schrader, L. E., Flesher, D., and Hageman, R. H. (1965). The role of light and nitrate in the induction of nitrate reductase in radish cotyledons and maize seedlings. *Plant Physiol.* **40,** 691–698.
32. Bennett, P. A., and Chrispeels, M. J. (1972). *De novo* synthesis of ribonuclease and β-1,3-glucanase by aleurone cells of barley. *Plant Physiol.* **49,** 445–447.
33. Bick, M. D., Liebke, H., Cherry, J. H., and Strehler, B. L. (1970). Changes in leucyl and tryrosyl-tRNA of soybean cotyledons during plant growth. *Biochim. Biophys. Acta* **204,** 175–182.
34. Bick, M. D., and Strehler, B. L. (1971). Leucyl transfer RNA synthetase changes during soybean cotyledon senescence. *Proc. Natl. Acad. Sci. USA* **68,** 224–228.
35. Bidwell, R. G. S. (1963). Pathways leading to the formation of amino acids and amides in leaves. *Can. J. Bot.* **41,** 1623–1637.
36. Bidwell, R. G. S., Barr, R., and Steward, F. C. (1964). Protein synthesis and turnover in cultured plant tissue: sources of carbon for synthesis and the fate of the protein breakdown products. *Nature (London)* **203,** 367–373.
37. Blair, G. E., and Ellis, R. J. (1973). Protein synthesis in chloroplasts. I. Light-driven synthesis of the large subunit of fraction I protein by isolated pea chloroplasts. *Biochim. Biophys. Acta* **319,** 223–234.
38. Blobel, G., and Dobberstein, B. (1975). Transfer of proteins across membranes. I. Presence of proteolytically processed and unprocessed nascent immunoglobulin light chains on membrane-bound ribosomes of murine myeloma. *J. Cell Biol.* **67,** 835–851.
39. Bond, J. S. (1975). Relationship between inactivation of an enzyme by acid or lysosomal extracts and its *in vivo* degradation rates. *Fed. Proc. Fed. Am. Soc. Exp. Biol.* **34,** 651.
40. Boudet, A., Humphrey, T. J., and Davies, D. D. (1975). The measurement of protein turnover by density labeling. *Biochem. J.* **152,** 409–416.
41. Boyd, W. J. R., and Walker, M. G. (1972). Variation in chlorophyll *a* content and stability in wheat flag leaves. *Ann. Bot. (London)* **36,** 87–92.
42. Brady, C. J., and Scott, N. S. (1977). Chloroplast polyribosomes and synthesis of fraction I protein in the developing wheat leaf. *Aust. J. Plant Physiol.* **4,** 327–335.
43. Briarty, L. G., Coult, D. A., and Boulter, D. (1969). Protein bodies of developing seeds of *Vicia faba. J. Exp. Bot.* **20,** 358–372.
44. Britten, R. J., and McClure, F. T. (1962). The amino acid pool in *Excherichia coli. Bacteriol. Rev.* **26,** 292–335.
45. Burger, W. C., Prentice, N., Kastenschmidt, J., and Moeller, M. (1968). Partial purification and characterization of barley peptide hydrolases. *Phytochemistry* **7,** 1261–1270.
46. Burger, W. C., and Siegelman, H. W. (1966). Location of a protease and its inhibitor in the barley kernel. *Physiol. Plant.* **19,** 1089–1093.
47. Burr, B., and Burr, F. A. (1976). Zein synthesis in maize endosperm by polyribosomes attached to protein bodies. *Proc. Natl. Acad. Sci. USA* **73,** 515–519.
48. Burstrom, H. (1943). Photosynthesis and assimilation of nitrate by wheat leaves. *Lantbrukshoegsk. Ann.* **11,** 1–50.
49. Butcher, H. C., Wagner, G. J., and Siegelman, H. W. (1977). Localization of acid

hydrolases in protoplasts. Examination of the proposed lysosomal function of the mature vacuole. *Plant Physiol.* **59**, 1089–1103.

50. Canvin, D. T., and Atkins, C. A. (1974). Nitrate, nitrite and ammonia assimilation by leaves: effect of light, carbon dioxide and oxygen. *Planta* **116**, 207–224.

51. Chantarotwong, W., Huffaker, R. C., Miller, B. L., and Granstedt, R. C. (1976). *In vivo* nitrate reduction in relation to nitrate uptake, nitrate content, and *in vitro* nitrate reductase activity in intact barley seedlings. *Plant Physiol.* **57**, 519–522.

52. Chen, R., and Jones, R. L. (1974). Studies on the release of barley aleurone cell proteins: kinetics and labeling. *Planta* **119**, 193–206.

53. Chen, R., and Jones, R. L. (1974). Studies on the release of barley aleurone cell proteins: autoradiography. *Planta* **119**, 207–220.

54. Choe, H. J., and Thimann, K. V. (1975). The metabolism of oat leaves during senescence. III. The senescence of isolated chloroplasts. *Plant Physiol.* **55**, 828–834.

55. Chrispeels, M. J. (1976). Biosynthesis, intracellular transport, and secretion of extracellular macromolecules. *Annu. Rev. Plant Physiol.* **27**, 19–38.

56. Chrispeels, M. J., and Baumgartner, B. (1978). Trypsin inhibitor in mungbean cotyledons. Purification, characteristics, subcellular localization, and metabolism. *Plant Physiol.* **61**, 617–623.

57. Chrispeels, M. J., and Boulter, D. (1975). Control of storage protein metabolism in the cotyledons of germinating mungbeans: role of endopeptidase. *Plant Physiol.* **55**, 1031–1037.

58. Chrispeels, M. J., and Varner, J. E. (1967). Gibberellic acid–enhanced synthesis and release of α-amylase and ribonuclease by isolated barley aleurone layers. *Plant Physiol* **42**, 398–402.

59. Clowes, F. A. L. (1958). Protein synthesis in root meristems. *J. Exp. Bot.* **9**, 229–238.

60. Cowles, J. R., and Key, J. L. (1973). Changes in certain amino-acyl transfer ribonucleic acid synthetase activities in developing pea roots. *Plant Physiol.* **51**, 22–25.

61. Criddle, R. S., Dau, B., Kleinkopf, G. E., and Huffaker, R. C. (1970). Differential synthesis of ribulose diphosphate carboxylase subunits. *Biochem. Biophys. Res. Commun.* **41**, 621–627.

62. Dalling, M. J., Boland, G., and Wilson, J. H. (1976). Relation between acid proteinase activity and redistribution of nitrogen during grain development in wheat. *Aust. J. Plant Physiol.* **3**, 721–730.

63. Danielsson, C. E. (1949). Seed globulins of the Gramineae and Leguminosae. *Biochem. J.* **44**, 387–400.

64. Davies, D. D., and Humphrey, T. J. (1978). Amino acid recycling in relation to protein turnover. *Plant Physiol.* **61**, 54–58.

65. Dechary, J. M. (1970). Seed proteases and protease inhibitors. *Econ. Bot.* **24**, 113–122.

66. Delmer, D. P., Kulow, C., and Ericson, M. C. (1978). Glycoprotein synthesis in plants. II. Structure of the mannolipid intermediate. *Plant Physiol.* **61**, 25–29.

67. Derbyshire, E., Wright, D. J., and Boulter, D. (1976). Legumin and vicilin, storage proteins of legume seeds. *Phytochemistry* **15**, 3–24.

68. Dice, J. F., Dehlinger, P. J., and Schimke, R. T. (1973). Studies on the correlation between size and relative degradation rate of soluble proteins. *J. Biol. Chem.* **248**, 4220–4228.

69. Dice, J. F., and Goldberg, A. L. (1975). A statistical analysis of the relationship between degradative rates and molecular weights of proteins. *Arch. Biochem. Biophys.* **170**, 213–219.

70. Dice, J. F., and Goldberg, A. L. (1975). Relationship between *in vivo* degradative rates and isoelectric points of proteins. *Proc. Natl. Acad. Sci. USA* **72**, 3893–3897.

71. Dilworth, M. F., and Kende, H. (1974). Comparative studies on nitrate reductase in *Agrostemma githago* induced by nitrate and benzyladenine. *Plant Physiol.* **54**, 821–825.

72. Dilworth, M. F., and Kende, H. (1974). Control of nitrite reductase activity in excised embryos of *Agrostemma githago*. *Plant Physiol.* **54**, 826–828.

73. Dobberstein, B., Blobel, G., and Chua, N.-H. (1977). *In vitro* synthesis and processing of a putative precursor for the small subunit of ribulose-1,5-bisphosphate carboxylase of *Chlamydomonas reinhardtii*. *Proc. Natl. Acad. Sci. USA* **74**, 1082–1085.

74. Dorner, R. W., Kahn, A., and Wildman, S. G. (1957). The proteins of green leaves. VII. Synthesis and decay of the cytoplasmic proteins during the life of the tobacco leaf. *J. Biol. Chem.* **229**, 945–952.

75. Drivdahl, R. H., and Thimann, K. V. (1977). Proteases of senescing oat leaves. I. Purification and general properties. *Plant Physiol.* **59**, 1059–1063.

76. Drysdale, J. W., and Munro, H. N. (1966). Regulation of synthesis and turnover of ferritin in rat liver. *J. Biol. Chem.* **241**, 3630–3637.

77. Ellis, R. J., and McDonald, I. R. (1970). Specificity of cycloheximide in higher plant systems. *Plant Physiol.* **46**, 538–542.

78. Engelsma, G. (1968). Photo induction of phenylalanine deaminase in Gherkin seedlings. III. Effects of excision and irradiation on enzyme development in hypocotyl segments. *Planta* **82**, 355–368.

79. Ericson, M. C., and Chrispeels, M. J. (1973). Isolation and characterization of glucosamine-containing storage glycoproteins from the cotyledons of *Phaseolus aureus*. *Plant Physiol.* **52**, 98–104.

80. Ericson, M. C., and Delmer, D. P. (1977). Glycoprotein synthesis in plants. I. Role of lipid intermediates. *Plant Physiol.* **59**, 341–347.

81. Ericson, M. C., and Delmer, D. P. (1978). Glycoprotein synthesis in plants. III. Interaction between UDP-N-acetylglucosamine and GDP-mannose as substrates. *Plant Physiol.* **61**, 819–823.

82. Evins, W. H., and Varner, J. E. (1971). Hormone controlled synthesis of endoplasmic reticulum in barley aleurone cells. *Proc. Natl. Acad. Sci. USA* **68**, 1631–1633.

83. Evins, W. H., and Varner, J. E. (1972). Hormone control of polyribosome formation in barley aleurone layers. *Plant Physiol.* **49**, 348–352.

84. Feller, U. K., Soong, T.-S. T., and Hageman, R. H. (1977). Leaf proteolytic activities and senescence during grain development of field grown corn (*Zea mays* L.). *Plant Physiol.* **59**, 290–294.

85. Feller, U., Soong, T.-S. T., and Hageman, R. H. (1978). Patterns of proteolytic enzyme activities in different tissues of germinating corn (*Zea mays* L.). *Planta* **140**, 155–162.

86. Ferrari, T. E., Yoder, O. C., and Filner, P. (1973). Anaerobic nitrite production by plant cells and tissues: evidence for two nitrate pools. *Plant Physiol.* **51**, 423–431.

87. Filner, P., Wray, J. L., and Varner, J. E. (1969). Enzyme induction in higher plants. *Science (Washington, D.C.)* **165**, 358–367.

88. Firn, R. D. (1975). On the secretion of α-amylase by barley aleurone layers after incubation in gibberellic acid. *Planta* **125**, 227–233.

89. Forsee, W. T., and Elbein, A. D. (1975). Glycoprotein synthesis in plants. *J. Biol. Chem.* **250**, 9283–9293.

90. Forsee, W. T., Yalkovich, G., and Albein, A. D. (1976). Glycoprotein synthesis in plants. *Arch. Biochem. Biophys.* **174**, 469–479.

91. Fowden, L. (1965). Origins of the amino acids. In "Plant Biochemistry" (J. Bonner, and J. E. Varner, eds.), pp. 361–388. Academic Press, New York.

92. Fujimaki, M., Abe, M., and Arai, S. (1977). Degradation of zein during germination of corn. *Agric. Biol. Chem.* **41**, 887–891.

93. Garg, G. K., and Virupakash, T. K. (1970). Acid protease from germinated sorghum. I. Purification and characterization of the enzyme. *Eur. J. Biochem.* **17**, 4–12.

94. Gibson, R. A., and Paleg, L. G. (1976). Purification of gibberellic acid–induced lysosomes from wheat aleurone cells. *J. Cell Sci.* **22**, 1–13.

95. Glasziou, K. T., Waldron, J. C., and Ball, T. A. (1966). Control of invertase synthetases in sugar cane. Loci of auxin and glucose effects. *Plant Physiol.* **41**, 282–288.

96. Goldberg, A. L. (1972). Correlation between rates of degradation of bacterial proteins *in vivo* and their sensitivity to proteases. *Proc. Natl. Acad. Sci. USA* **69**, 2640–2644.

97. Goldberg, A. L., and St. John, A. C. (1976). Intracellular protein degradation in mammalian and bacterial cells. Part 2. *Annu. Rev. Biochem.* **45**, 747–803.

98. Goldsmith, J., Livoni, J. P., Norberg, C. L., and Segel, I. H. (1973). Regulation of nitrate uptake in *Penicillium chrysogenum* by ammonium ion. *Plant Physiol.* **52**, 363–367.

99. Gooding, L. R., Roy, H., and Jagendorf, A. T. (1973). Immunological identification of nascent subunits in wheat ribulose diphosphate carboxylase on ribosomes of both chloroplast and cytoplasmic origin. *Arch. Biochem. Biophys.* **159**, 324–335.

100. Gore, N. R., and Wray, J. L. (1978). Leucine : tRNA ligase from cultured cells of *Nicotiana tabacum* var. *xanthii. Plant Physiol.* **61**, 20–24.

101. Gorini, L., and Audrain, L. (1952). Action de quelques métaux bivalents sur la sensibilité de la serum albumine à l'action de la trypsine. *Biochim. Biophys. Acta* **9**, 180–192.

102. Green, T. R., and Ryan, C. A. (1973). Wound-induced proteinase inhibitor in tomato leaves. Some effects of light and temperature on the wound responses. *Plant Physiol.* **51**, 19–21.

103. Gregory, F. G., and Sen, P. K. (1937). Physiological studies in plant nutrition. VI. The relation of respiration rate to the carbohydrate and nitrogen metabolism of the barley leaf as determined by nitrogen and potassium deficiency. *Ann. Bot. NS* **1**, 521–561.

104. Hageman, R. H., and Flesher, D. (1960). Nitrate reductase activity in corn seedlings as affected by light and nitrate content of nutrient media. *Plant Physiol.* **35**, 700–708.

105. Hall, N. P., Keys, A. J., and Merrett, M. J. (1978). Ribulose-1,5-diphosphate carboxylase protein during flag leaf senescence. *J. Exp. Bot.* **29**, 31–37.

106. Hallmark, W. B., and Huffaker, R. C. (1978). The influence of ambient nitrate, temperature, and light on nitrate assimilation in Sudangrass seedlings. *Physiol. Plant.* **44**, 147–152.

107. Hardy, R. W. F., Burns, R. C., Herbert, R. R., Holsten, R. D., and Jackson, E. K. (1971). *In* "Biological Nitrogen Fixation in Natural and Agricultural Habitats" (T. A. Lie, and E. G. Mulder, eds.). Nijhoff, The Hague.

108. Harris, N., and Boulter, D. (1976). Protein body formation in cotyledons of developing cowpea (*Vigna unguiculata*) seeds. *Ann. Bot. (London)* **40**, 739–744.

109. Hartley, B. S. (1960). Proteolytic enzymes. *Annu. Rev. Biochem.* **29**, 45–72.

110. Hartley, M. R., Wheeler, A., and Ellis, R. J. (1975). Protein synthesis in chloroplasts. V. Translation of messenger RNA for the large subunit of Fraction I protein in a heterologous cell-free system. *J. Mol. Biol.* **91**, 67–77.

111. Harvey, B. M. R., and Oaks, A. (1974). Characteristics of an acid protease from maize endosperm. *Plant Physiol.* **53**, 449–452.

112. Harvey, B. M. R., and Oaks, A. (1974). The hydrolysis of endosperm protein in *Zea mays. Plant Physiol.* **53**, 453–457.

113. Hatch, M. D., and Slack, C. R. (1970). Photosynthetic CO_2-fixation pathways. *Annu. Rev. Plant Physiol.* **21**, 141–162.
114. Haynes, R., and Feeney, R. E. (1967). Fractionation and properties of trypsin and chymotrypsin inhibitors for lima beans. *J. Biol. Chem.* **242**, 5378–5385.
115. Hedley, C. L., and Stoddart, J. L. (1971). Factors influencing alanine aminotransferase activity in leaves of *Lolium temulentum* L. I. Photoperiodically induced variations. *J. Exp. Bot.* **22**, 239–248.
116. Heimer, Y. H., and Filner, P. (1971). Regulation of the nitrate assimilation pathway in cultured tobacco cells. III. The nitrate uptake system. *Biochim. Biophys. Acta* **230**, 362–372.
117. Hellebust, J. A., and Bidwell, R. G. S. (1963). Protein turnover in wheat and snapdragon leaves, preliminary investigation. *Can. J. Bot.* **41**, 969–983.
118. Hellebust, J. A., and Bidwell, R. G. S. (1963). Sources of carbon for the synthesis of protein amino acids in attached photosynthesizing wheat leaves. *Can. J. Bot.* **41**, 985–994.
119. Hellebust, J. A., and Bidwell, R. G. S. (1964). Protein turnover in attached wheat and tobacco leaves. *Can. J. Bot.* **42**, 1–12.
120. Henshall, J. D., and Goodwin, T. W. (1964). Amino acid–activating enzymes in germinating pea seedlings. *Phytochemistry* **3**, 677–691.
121. Higgins, T. J. V., Zwar, J. A., and Jacobsen, J. V. (1976). Gibberellic acid enhances the level of translatable mRNA for α-amylase in barley aleurone layers. *Nature (London)* **260**, 166–169.
122. Highfield, P. E., and Ellis, R. J. (1978). Synthesis and transport of the small subunit of chloroplast ribulose bisophosphate carboxylase. *Nature (London)* **271**, 420–424.
123. Hill, J. E., and Breidenbach, R. W. (1974). Proteins of soybean seeds. I. Isolation and characterization of the major components. *Plant Physiol.* **53**, 742–746.
124. Hill, J. E., and Breidenbach, R. W. (1974). Proteins of soybean seeds. II. Accumulation of the major protein components during seed development and maturation. *Plant Physiol.* **53**, 747–751.
125. Hinde, R. W., Finch, L. R., and Cory, S. (1966). Amino acid dependent ATP-pyrophosphate exchange in normal and boron deficient bean roots. *Phytochemistry* **5**, 609–618.
126. Holleman, J. M., and Key, J. L. (1967). Inactive and protein precursor pools of amino acids in soybean hypocotyl. *Plant Physiol.* **42**, 29–36.
127. Horiguchi, T., and Kitagishi, I. (1971). Studies on rice seed protease. V. Protease inhibitor in rice seed. *Plant Cell Physiol.* **12**, 907–915.
128. Hotta, Y., and Stern, H. (1963). Molecular facets of mitotic regulation. I. Synthesis of thymidine kinase. *Proc. Natl. Acad. Sci. USA* **49**, 638–654.
129. Hotta, Y., and Stern, H. (1963). Molecular facets of mitotic regulation. II. Factors underlying the removal of thymidine kinase. *Proc. Natl. Acad. Sci. USA* **49**, 861–865.
130. Hu, A. S. L., Bock, R. M., and Halverson, H. O. (1962). Separation of labeled from unlabeled proteins by equilibrium density gradient sedimentation. *Anal. Biochem.* **4**, 489–504.
130a. Huffaker, R. C., and Miller, B. L. (1978). Reutilization of ribulose bisphosphate carboxylase. *In* "Photosynthetic Carbon Assimilation" (H. W. Siegelman, and G. Hind, eds.), pp. 139–152. Plenum, New York.
131. Huffaker, R. C., Obendorf, R. L., Keller, C. N., and Kleinkopf, G. E. (1966). Effects of light intensity on photosynthetic carboxylative phase enzymes and chlorophyll synthesis in greening leaves of *Hordeum vulgare* L. *Plant Physiol.* **41**, 913–918.
132. Huffaker, R. C., and Peterson, L. W. (1974). Protein turnover in plants and possible means of its regulation. *Annu. Rev. Plant Physiol* **25**, 363–392.

133. Huffaker, R. C., Radin, T., Kleinkopf, G. E., and Cox, E. L. (1970). Effects of mild water stress on enzymes of nitrate assimilation and of the carboxylative phase of photosynthesis in barley. *Crop Sci.* **10**, 471–474.
134. Humphrey, T. J., and Davies, D. D. (1975). A new method for the measurement of protein turnover. *Biochem. J.* **148**, 119–127.
135. Humphrey, T. J., Sarawek, S., and Davies, D. D. (1977). The effect of nitrogen deficiency on the growth and metabolism of *Lemna minor* L. *Planta* **137**, 259–264.
136. Hunde, R. W., Finch, L. R., and Cory, S. (1966). Amino acid dependent ATP-pyrophosphate exchange in normal and boron deficient bean roots. *Phytochemistry* **5**, 609–618.
137. Ihle, J. E., and Dure, L. S., III (1972). The developmental biochemistry of cottonseed embryogenesis and germination. II. Catalytic properties of the cotton carboxypeptidase. *J. Biol. Chem.* **247**, 5041–5047.
138. Ingle, J., Joy, K. W., and Hageman, R. H. (1966). The regulation of activity of the enzymes involved in the assimilation of nitrate by higher plants. *Biochem. J.* **100**, 577–588.
139. Jackson, W. A., Flesher, D., and Hageman, R. H. (1973). Nitrate uptake by dark-grown corn seedlings: some characteristics of apparent induction. *Plant Physiol.* **51**, 120–127.
140. Jackson, W. A., and Volk, R. J. (1970). Photorespiration. *Annu. Rev. Plant Physiol.* **21**, 385–432.
141. Jacobson, A., and Corcoran, M. R. (1977). Tannins as gibberellin antagonists in the synthesis of α-amylase and acid phosphatase by barley seeds. *Plant Physiol.* **59**, 129–133.
142. Jacobson, J. V. (1977). Regulation of ribonucleic acid metabolism by plant hormones. *Annu. Rev. Plant Physiol.* **28**, 537–564.
143. Jacobson, J. V., and Knox, R. B. (1973). Cytochemical localization and antigenicity of α-amylase in barley aleurone tissue. *Planta* **112**, 213–224.
144. Jacobson, J. V., and Varner, J. E. (1967). Gibberellic acid-induced synthesis of protease by isolated aleurone layers of barley. *Plant Physiol.* **52**, 1596–1600.
145. Jaynes, T. A., and Nelson, O. E. (1971). An invertase inactivator in maize endosperm and factors affecting inactivation. *Plant Physiol.* **47**, 629–634.
146. Jolly, S. O., and Tolbert, N. E. (1978). NADH-nitrate reductase inhibitor from soybean leaves. *Plant Physiol.* **62**, 197–203.
147. Jones, R. A. (1978). Effects of *floury*-2 locus on zein accumulation and RNA metabolism during maize endosperm development. *Biochem. Genet.* **16**, 27–38.
148. Jones, R. A., Larkins, B. A., and Tsai, C. Y. (1976). Reduced synthesis of zein *in vitro* by a high lysine mutant of maize. *Biochem. Biophys. Res. Commun.* **69**, 404–410.
149. Jones, R. A., Larkins, B. A., and Tsai, C. Y. (1977). Storage protein synthesis in maize. II. Reduced synthesis of a major zein component by the *opaque*-2 mutant of maize. *Plant Physiol.* **59**, 525–529.
150. Jones, R. L., and Chen, R. F. (1976). Immuno-histochemical localization of α-amylase in barley aleurone cells. *J. Cell Sci.* **20**, 183–198.
151. Joy, K. W. (1969). Nitrogen metabolism in *Lemna minor*. II. Enzymes of nitrate assimilation and some aspects of their regulation. *Plant Physiol.* **44**, 849–853.
152. Joy, K. W. (1971). Glutamate dehysrogenase changes in *Lemna* not due to enzyme induction. *Plant Physiol.* **47**, 445–446.
153. Joy, K. W., and Folkes, B. F. (1965). The uptake of amino-acids and their incorporation into the protein of excised barley embryos. *J. Exp. Bot.* **16**, 646–666.
154. Kannangara, C. G., and Woolhouse, H. W. (1968). Changes in the enzyme activity of

soluble protein fractions in the course of foliar senescence in *Perilla frutescens* (L.) Britt. *New Phytol.* **67**, 533–542.

155. Kawashima, N., Imai, A., and Tamaki, E. (1967). Studies on protein metabolism in higher plant leaves. III. Changes in the soluble protein components with leaf growth. *Plant Cell Physiol.* **8**, 447–458.

156. Kawashima, N., and Mitake, T. (1969). Studies on protein metabolism in higher plants. VI. Changes in ribulose diphosphate carboxylase activity and fraction I protein content in tobacco leaves with age. *Agric. Biol. Chem.* **33**, 539–543.

157. Kawashima, N., and Wildman, S. G. (1970). Fraction I protein. *Annu. Rev. Plant Physiol.* **21**, 325–358.

158. Kemp, J. D., and Sutton, D. W. (1971). Protein metabolism in cultured plant tissues. Calculation of an absolute rate of protein synthesis, accumulation and degradation in tobacco callus *in vivo*. *Biochemistry* **10**, 81–88.

159. Kende, H., Hahn, H., and Kays, S. E. (1971). Enhancement of nitrate reductase activity by benzyladenine in *Agrostemma githago*. *Plant Physiol.* **48**, 702–706.

160. Kende, H., and Shen, T. C. (1972). Nitrate reductase in *Agrostemma githago*. Comparison of the inductive effects of nitrate and cytokinin. *Biochim. Biophys. Acta* **286**, 118–125.

161. Kendrick, R. E. (1972). Aspects of phytochrome decay in etiolated seedlings under continuous illumination. *Planta* **102**, 286–293.

162. Kendrick, R. E., and Frankland, B. (1968). Kinetics of phytochrome decay in *Amaranthus* seedlings. *Planta* **82**, 317–320.

163. Kendrick, R. E., and Frankland, B. (1969). The *in vivo* properties of *Amaranthus* phytochrome. *Planta* **86**, 21–32.

164. Kendrick, R. E., and Spruit, C. J. P. (1972). Phytochrome decay in seedlings under continuous incandescent light. *Planta* **107**, 341–350.

165. Kirsi, M., and Mikola, J. (1971). Occurrence of proteolytic inhibitors in various tissues of barley. *Planta* **96**, 281–291.

166. Kit, S., Dubbs, D. R., and Frearson, P. M. (1965). Decline of thymidine kinase activity in stationary phase mouse fibrobolase cells. *J. Biol. Chem.* **240**, 2565–2573.

167. Klepper, L., Flesher, D., and Hageman, R. H. (1971). Generation of reduced nicotinamide adenine dinucleotide for nitrate reduction in green leaves. *Plant Physiol.* **48**, 580–590.

168. Kolehmainen, L., and Mikola, J. (1971). Partial purification and enzymatic properties of an aminopeptidase from barley. *Arch. Biochem. Biophys.* **145**, 633–642.

169. Larkins, B. A., Jones, R. A., and Tsai, C. Y. (1976). Isolation and *in vitro* translation of zein messenger ribonucleic acid. *Biochemistry* **15**, 5506–5511.

170. Lee, K. H., Jones, R. A., Dally, A., and Tsai, C. Y. (1976). Genetic regulation of storage protein content in maize endosperm. *Biochem. Genet.* **14**, 641–650.

171. Lehle, L., Fartaczek, F., Tanner, W., and Kaus, H. (1976). Formation of polyprenol-linked mono- and oligosaccharides in *Phaseolus aureus*. *Arch. Biochem. Biophys.* **175**, 419–426.

172. Littlefield, J. W. (1965). Studies on thymidine kinase in cultured mouse fibroblasts. *Biochim. Biophys. Acta* **95**, 14–22.

173. MacLennan, D. H., Beevers, H., and Harley, J. L. (1963). Compartmentation of acids in plant tissues. *Biochem. J.* **89**, 316–327.

174. MacLeod, A. M., and Palmer, G. H. (1969). Interaction of indoyl acetic acid and gibberellic acid in the synthesis of α-amylase by barley aleurones. *New Phytol.* **68**, 295–304.

175. Markus, G. (1965). Protein substrate conformation and proteolysis. *Proc. Natl. Acad. Sci. USA* **54**, 253–258.

176. Martin, C., and Thimann, K. V. (1972). The role of protein synthesis in the senescence of leaves. I. The formation of protein. *Plant Physiol.* **49**, 64–71.

177. Marzluf, G. A. (1972). Control of the synthesis, activity, and turnover of enzymes of sulfur metabolism in *Neurospora crassa. Arch. Biochem. Biophys.* **150**, 714–724.

178. Matile, P. H. (1976). Vacuoles. In "Plant Biochemistry" (J. Bonner, and J. E. Varner, eds.), pp. 189–224. Academic Press, New York.

179. McDonald, C. E., and Chen, L. L. (1969). Properties of wheat flour proteinases. *Cereal Chem.* **41**, 443–455.

180. Mikola, J., and Kolehmainen, L. (1972). Localization and activity of various peptidases in germinating barley. *Planta* **104**, 167–177.

181. Millerd, A. (1975). Biochemistry of legume seed proteins. *Annu. Rev. Plant Physiol.* **26**, 53–72.

182. Millerd, A., Simon, M., and Stern, H. (1971). Legumin synthesis in developing cotyledons of *Vicia faba* L. *Plant Physiol.* **48**, 419–425.

183. Mizrahi, Y., Amir, J., and Richmond, A. (1970). The mode of action of kinetin in maintaining the protein content of detached *Tropaeolum majus* leaves. *New Phytol.* **69**, 355–361.

184. Mondal, M. H., Brun, W. A., and Brenner, M. L. (1978). Effects of sink removal on photosynthesis and senescence in leaves of soybean (*Glycine max* L.) plants. *Plant Physiol.* **61**, 394–397.

185. Morilla, C. A., Boyer, J. S., and Hageman, R. H. (1973). Nitrate reductase activity and polyribosomal content of corn (*Zea mays* L.) having low leaf water potentials. *Plant Physiol.* **51**, 817–824.

186. Nagahashi, J., and Beevers, L. (1978). Subcellular localization of glycosyl transferases involved in glycoprotein biosynthesis in the cotyledons of *Pisum sativum* L. *Plant Physiol.* **61**, 451–459.

187. Nair, P. M., and Vining, L. C. (1965). Phe hydroxylase from spinach leaves. *Phytochemistry* **4**, 401–411.

188. Naslin, L., Spyridakis, A., and Labeyria, F. (1973). A study of several bands hypersensitive to proteases in a complex flavohemoenzyme. Yeast cytochrome b_2 modification of their reactivity with ligand-induced conformational transitions. *Eur. J. Biochem.* **34**, 268–283.

189. Nathan, I., and Richmond, A. (1974). Leucyl-transfer ribonucleic acid synthetase in senescing tobacco leaves. *Biochem. J.* **140**, 169–173.

190. Norris, R. D., Lea, P. J., and Fouden, L. (1973). Amino acyl-tRNA synthetases in *Triticum aestivum* L. during seed development and germination. *J. Exp. Bot.* **24**, 615–625.

191. Oaks, A. (1965). The soluble leucine pool in maize root tips. *Plant Physiol.* **40**, 142–149.

192. Oaks, A. (1965). The effect of leucine on the biosynthesis of leucine in maize root tip. *Plant Physiol.* **40**, 149–155.

193. Oaks, A., Aslam, M., and Boesel, I. (1977). Ammonium and amino acids as regulators of nitrate reductase in corn roots. *Plant Physiol.* **59**, 391–394.

194. Oaks, A., and Bidwell, R. G. S. (1970). Compartmentation of intermediary metabolites. *Annu. Rev. Plant Physiol.* **21**, 43–66.

195. Onwueme, I. C., Laude, H. M., and Huffaker, R. C. (1971). Nitrate reductase activity in relation to heat stress in barley seedlings. *Crop Sci.* **11**, 195–200.

196. Opik. H. (1968). Development of cotyledon cell structure in ripening *Phaseolus vulgaris* seeds. *J. Exp. Bot.* **19**, 64–67.

197. Osborne, T. B. (1908). Our present knowledge of plant proteins. *Science (Washington, D.C.)* **28**, 417–427.

198. Paleg, L. (1961). Physiological effects of gibberellic acid. III. Observations on its mode of action on barley endosperm. *Plant Physiol.* **36,** 829–837.

199. Pate, J. S., and O'Brien, T. P. (1968). Microautoradiographic study of the incorporation of labeled amino acids into insoluble compounds of the shoot of a higher plant. *Planta* **78,** 60–71.

200. Patterson, B. D., and Smillie, R. M. (1971). Developmental changes in ribosomal ribonucleic acid and fraction I protein in wheat leaves. *Plant Physiol.* **47,** 196–198.

201. Payne, E. S., Boulter, D., Brownrigg, A., Lonsdale, D., Yarwood, A., and Yarwood, J. N. (1971). A polyuridylic acid–directed cell-free system from 60-day-old developing seeds of *Vicia faba. Phytochemistry* **10,** 2293–2298.

202. Payne, P. I., and Boulter, D. (1969). Free and membrane-bound ribosomes of the cotyledons of *Vicia faba* (L.). I. Seed development. *Planta* **84,** 263–271.

203. Peterson, L. W., Kleinkopf, G. E., and Huffaker, R. C. (1973). Evidence for lack of turnover of ribulose 1,5-diphosphate carboxylase in barley leaves. *Plant Physiol.* **51,** 1042–1045.

204. Peterson, L. W., and Huffaker, R. C. (1975). Loss of ribulose 1,5-diphosphate carboxylase and increase in proteolytic activity during senescence of detached primary barley leaves. *Plant Physiol.* **55,** 1009–1015.

205. Poole, B., Leighton, F., and deDuve, C. (1969). The synthesis and turnover of rat liver peroxisomes. II. Turnover of peroxisome proteins. *J. Cell Biol.* **41,** 536–546.

206. Pratt, L. H. (1978). Immunocytochemistry and the visualization of phytochrome. *What's New Plant Physiol.* **9,** 1–3.

207. Prentice, N., Burger, W. C., Moeller, M., and Kastenschmidt, J. (1970). The hydrolysis of alpha naphthyl acetate and 1-leucyl-beta naphthylamide by enzymes from wheat embryo. *Cereal Chem.* **47,** 282–287.

208. Price, V. E., Sterling, W. R., Tarantola, V. A., Hartley, R. W., Jr., and Rechcigl, M., Jr. (1962). The kinetics of catalase synthesis and destruction *in vivo. J. Biol. Chem.* **237,** 3468–3475.

209. Prouty, W. F., and Goldberg, A. L. (1972). Fate of abnormal proteins in *E. coli.* Accumulation in intracellular granules before catabolism. *Nature (London) New Biol.* **240,** 147–150.

210. Purvis, A. C., Peters, D. B., and Hageman, R. H. (1974). Effect of carbon dioxide on nitrate accumulation and nitrate reductase induction in corn seedlings. *Plant Physiol.* **53,** 934–941.

211. Pusztai, A. (1964). Hexoseamines in the seeds of higher plants. *Nature (London)* **201,** 1328–1329.

212. Quail, P. H., and Scandalios, J. G. (1971). Turnover of genetically defined catalase isozymes in maize. *Proc. Natl. Acad. Sci. USA* **68,** 1402–1406.

213. Quail, P. H., Schafer, E., and Marme, D. (1973). Turnover of phytochrome in pumpkin cotyledons. *Plant Physiol.* **52,** 138–131.

214. Quail, P. H., Schafer, E., and Marme, D. (1973). *De novo* synthesis of phytochrome in pumpkin hooks. *Plant Physiol.* **52,** 124–127.

215. Radin, J. (1975). Differential regulation of nitrate reductase induction in roots and shoots of cotton plants. *Plant Physiol.* **55,** 178–182.

216. Radin, J. (1977). Amino acid interactions in the regulation of nitrate reductase induction in cotton root tips. *Plant Physiol.* **60,** 467–469.

217. Radley, M. (1969). The effect of the endosperm on the formation of gibberellin by barley embryos. *Planta* **86,** 218–223.

218. Ragg, H., Schroeder, J., and Hahlbrock, K. (1977). Translation of poly(A)-containing and poly(A)-free messenger RNA for phenylalanine ammonia-lyase, plant-specific protein, in a reticulocyte lysate. *Biochim. Biophys. Acta* **474,** 226–233.

219. Rao, K. P. (1976). "Some Physiological and Biochemical Aspects of Nitrogen Assimilation in Barley and Wheat." Ph.D. Thesis, Univ. of California, Davis.
220. Rao, K. P., and Rains, D. W. (1976). Nitrate absorption by barley. II. Influence of nitrate reductase activity. *Plant Physiol.* **57,** 59–62.
221. Rao, K. P., Rains, D. W., Qualset, C. O., and Huffaker, R. C. (1977). Nitrogen nutrition and grain protein in two spring wheat genotypes differing in nitrate reductase activity. *Crop Sci.* **17,** 283–286.
222. Rechcigl, M., Jr. (1962). *In vivo* turnover and its role in the metabolic regulation of enzyme levels. *Enzymologia* **34,** 23–39.
223. Renosto, F., and Ferrari, G. (1975). Mechanism of sulfate transport by cycloheximide in plant tissues. *Plant Physiol.* **56,** 478–480.
224. Roberts, R. M., and Yuan, B. O.-C. (1975). Radiolabeling of mammalian cells in tissue culture by use of acetic anhydride. *Arch. Biochem. Biophys.* **171,** 226–233.
225. Roy, H., Costa, K. A., and Adari, H. (1978). Free subunits of ribulose-1,5-bisphosphate carboxylase in pea leaves. *Plant Sci. Lett.* **11,** 159–168.
226. Roy, H., Gooding, L. R., and Jagendorf, A. T. (1973). Formation, release, and identification of peptidyl-(^{3}H) puromycins from wheat leaf ribosomes *in vitro*. *Arch. Biochem. Biophys.* **159,** 312–323.
227. Roy, H., Patterson, R., and Jagendorf, A. T. (1976). Identification of the small subunit of ribulose 1,5-bisphosphate carboxylase as a product of wheat leaf cytoplasmic ribosomes. *Arch. Biochem. Biophys.* **172,** 64–73.
228. Roy, H., Terenna, B., and Cheong, L. C. (1977). Synthesis of the small subunit of ribulose-1,5-bisphosphate carboxylase by soluble fraction polyribosomes of pea leaves. *Plant Physiol.* **60,** 532–537.
229. Ryan, C. A. (1973). Proteolytic enzymes and their inhibitors in plants. *Annu. Rev. Plant Physiol.* **24,** 173–196.
230. Sacher, J. A. (1966). Permeability characteristics and amino acid incorporation during senescence (ripening) of banana tissue. *Plant Physiol.* **41,** 701–708.
231. Sarawek, S., and Davies, D. D. (1977). The control of aldolase in *Lemna minor* L. in relation to nitrogen deficiency. *Planta* **137,** 271–277.
232. Sarma, V. R., Silverton, E. W., Davies, D. R., and Terry, W. D. (1971). The three-dimensional structure at 6Å resolution of a human γG1 immunoglobulin molecule. *J. Biol. Chem.* **246,** 3753–3759.
233. Schatz, G., and Maston, T. L. (1974). The biosynthesis of mitochondrial proteins. *Annu. Rev. Biochem.* **43,** 51–87.
234. Shimke, R. T., and Doyle, D. (1970). Control of enzyme levels in animal tissues. *Annu. Rev. Biochem.* **39,** 929–976.
235. Schimke, R. T., Sweeney, E. W., and Berlin, C. M. (1965). Studies of the stability *in vivo* and *in vitro* of rat liver tryptophan pyrrolase. *J. Biol. Chem.* **240,** 4609–4620.
236. Schloemer, R. H., and Garrett, R. H. (1974). Nitrate transport system in *Neurospora crassa*. *J. Bacteriol.* **118,** 259–269.
237. Schrader, L. E., Ritenour, G. L., Eilrich, G. L., and Hageman, R. H. (1968). Some characteristics of nitrate reductase from higher plants. *Plant Physiol.* **43,** 930–940.
238. Schroder, J., Betz, B., and Halbrock, K. (1977). Messenger RNA controlled increase of phenylalanine ammonia lyase activity in parsley. Light-independent induction by dilution of cell suspension cultures into water. *Plant Physiol.* **60,** 440–445.
239. Shaner, D. L., and Boyer, J. S. (1976). Nitrate reductase activity in maize (*Zea mays* L.) leaves. II. Regulation by nitrate flux. *Plant Physiol.* **58,** 499–504.
240. Shaner, D. L., and Boyer, J. S. (1976). Nitrate reductase activity in maize (*Zea mays* L.)

leaves. II. Regulation by nitrate flux at low leaf water potential. *Plant Physiol.* **58**, 505–509.

241. Shain, Y., and Mayer, A. M. (1965). Proteolytic enzymes and endogenous trypsin inhibitor in germinating lettuce seeds. *Physiol. Plant.* **18**, 853–859.

242. Shain, Y., and Mayer, A. M. (1968). Activation of enzymes during germination—trypsin-like enzyme in lettuce. *Phytochemistry* **7**, 1491–1498.

243. Sinclair, T. R., and deWit, C. T. (1975). Comparative analysis of photosynthate and nitrogen requirements in the production of seeds by various crops. *Science (Washington, D.C.)* **189**, 565–567.

244. Sinclair, T. R., and deWit, C. T. (1976). Analysis of the carbon and nitrogen limitations to soybean yield. *Agron. J.* **68**, 319–324.

245. Smith, M. A., Criddle, R. S., Peterson, L. W., and Huffaker, R. C. (1974). Synthesis and assembly of ribulose bisophosphate carboxylase enzyme during greening of barley plants. *Arch. Biochem. Biophys.* **165**, 494–504.

246. Soong, T.-S. T., Feller, U. K., and Hageman, R. H. (1977). Changes in activities of proteolytic enzymes during senescence of detached corn (*Zea mays* L.) leaves as function of physiological age. *Plant Physiol.* (Suppl.) **59**, 112.

247. Sorger, G. J. (1966). Nitrate reductase electron transport systems in mutant and wild type strains of *Neurospora*. *Biochim. Biophys. Acta* **118**, 484–494.

248. St. Angelo, A. J., and Ory, R. L. (1970). Properties of a purified proteinase from hempseed. *Phytochemistry* **9**, 1933–1938.

249. Steward, F. C., and Bidwell, R. G. S. (1966). Storage pools and turnover systems in growing and non-growing cells: experiments with ^{14}C-sucrose, ^{14}C-glutamine, and ^{14}C-asparagine. *J. Exp. Bot.* **17**, 726–741.

250. Steward, F. C., Bidwell, R. G. S., and Yemm, E. W. (1958). Nitrogen metabolism, respiration, and growth of cultured plant tissues. *J. Exp. Bot.* **9**, 11–49.

251. Sun, S. M., Mutschler, M. A., Bliss, F. A., and Hall, T. C. (1978). Protein synthesis and accumulation in bean cotyledons during growth. *Plant Physiol.* **61**, 918–923.

252. Sundblom, N., and Mikola, J. (1972). On the nature of the proteinases secreted by the aleurone layer of barley grain. *Physiol. Plant.* **27**, 281–284.

253. Sussman, M., and Sussman, R. R. (1965). The regulatory program for UDP galactose polysaccharide transferase activity during slime mold cytodifferentiation: requirement for specific synthesis of ribonucleic acid. *Biochim. Biophys. Acta* **108**, 463–473.

254. Tanaka, Y., and Uritani, I. (1976). Immunochemical studies on fluctuation of phenylalanine ammonia-lyase activity in sweet potato in response to cut injury. *J. Biochem.* **79**, 217–219.

255. Tavares, J., and Kende, H. (1970). The effect of 6-benzylamino-purine on protein metabolism in senescing corn leaves. *Phytochemistry* **9**, 1763–1770.

256. Thomas, H. (1975). Regulation of alanine aminotransferase in leaves of *Lolium temulentum* during senescence. *Z. Pflanzenphysiol.* **74**, 208–218.

257. Thomas, H. (1975). Leaf growth and senescence in grasses. *Rep. Welsh Plant Breed. Stn.*, pp. 133–138.

258. Thomas, H. (1977). Ultrastructure, polypeptide composition and photochemical activity of chloroplasts during foliar senescence of a non-yellowing mutant genotype of *Festuca pratensis* Huds. *Planta* **137**, 53–60.

259. Thomas, H., and Stoddart, J. L. (1975). Separation of chlorophyll degradation from other senescence processes in leaves of a mutant genotype of meadow fescue (*Festuca pratensis* L.). *Plant Physiol.* **56**, 438–441.

260. Timasheff, S. N., and Garbunoff, M. J. (1967). Conformation of proteins. *Annu. Rev. Biochem.* **36**, 13–54.

261. Tischner, R., and Huttermann, A. (1978). Light-mediated activation of nitrate reductase in synchronous *Chlorella*. *Plant Physiol.* **62,** 284–286.

262. Travis, R. L., Huffaker, R. C., and Key, J. L. (1970). Light-induced development of polyribosomes and the induction of nitrate reductase in corn leaves. *Plant Physiol.* **46,** 800–805.

263. Travis, R. L., Jordan, W. R., and Huffaker, R. C. (1969). Evidence for an inactivating system of nitrate reductase in *Hordeum vulgare* L. during darkness that requires protein synthesis. *Plant Physiol.* **44,** 1150–1156.

264. Travis, R. L., Jordan, W. R., and Huffaker, R. C. (1970). Light and nitrate requirements for induction of nitrate reductase activity in *Hordeum vulgare*. *Physiol. Plant.* **23,** 678–685.

265. Trayser, K. A., and Colowick, S. P. (1961). Properties of crystalline hexokinase from yeast. III. Studies on glucose–enzyme interaction. *Arch. Biochem. Biophys.* **94,** 169–176.

266. Trewavas, A. (1972). Determination of the rates of protein synthesis and degradation in *Lemna minor*. *Plant Physiol.* **49,** 40–46.

267. Trewavas, A. (1972). Control of the protein turnover rates in *Lemna minor*. *Plant Physiol.* **49,** 47–51.

268. vanLoon, L. C., and Haverkort, A. J. (1977). No increase in protease activity during senescence of oat leaves. *Plant Physiol.* (Suppl.) **59,** 113.

269. Varner, J. E. (1964). Gibberellic acid controlled synthesis of α-amylase in barley endosperm. *Plant Physiol.* **39,** 413–415.

270. Varner, J. E., and Chandra, G. R. (1964). Hormonal control of enzyme synthesis in barley endosperm. *Proc. Natl. Acad. Sci. USA* **52,** 100–106.

271. Varner, J. E., and Meuse, R. (1972). Characteristics of the process of enzyme release from secretory plant cells. *Plant Physiol.* **49,** 187–189.

272. Varner, J. E., and Schidlovsky, G. (1963). Intracellular distribution of proteins in pea cotyledons. *Plant Physiol.* **38,** 139–144.

273. Vickery, H. B., Pucker, G. W., Schoenheimer, R., and Rittenberg, D. (1940). The assimilation of ammonia nitrogen by the tobacco plant: a preliminary study with isotopic nitrogen. *J. Biol. Chem.* **135,** 531–539.

274. Vigil, R. L., and Ruddat, M. (1973). Effect of gibberellic acid and actinomycin D on the formation and distribution of rouch endoplasmic reticulum in barley aleurone cells. *Plant Physiol.* **51,** 549–558.

275. Waechter, C. J., and Lennarz, W. N. (1976). The role of polyphenol-linked sugars in glycoprotein synthesis. *Annu. Rev. Biochem.* **45,** 95–112.

276. Walker-Simmons, M., and Ryan, C. A. (1977). Wound-induced accumulation of trypsin inhibitor activities in plant leaves. Survey of several plant genera. *Plant Physiol.* **59,** 437–440.

277. Wallace, W. (1973). A nitrate reductase inactivating enzyme from the maize root. *Plant Physiol.* **52,** 197–201.

278. Wallace, W. (1975). A re-evaluation of the nitrate reductase content of the maize root. *Plant Physiol.* **55,** 774–777.

279. Wallace, W. (1975). Effects of a nitrate reductase inactivating enzyme and NAD(P)H on the nitrate reductase from higher plants and *Neurospora*. *Biochim. Biophys. Acta* **377,** 239–250.

280. Wellman, E., and Schopfer, P. (1975). Phytochrome-mediated *de novo* synthesis of phenylalanine ammonia-lyase in cell suspension cultures of parsley. *Plant Physiol.* **55,** 822–827.

281. Wells, J. R. E. (1965). Purification and properties of a proteolytic enzyme from French beans. *Biochem. J.* **97**, 228–235.

282. Wells, J. R. E. (1968). Characterization of three proteolytic enzymes from French beans. *Biochim. Biophys. Acta* **167**, 388–398.

283. Willemot, C., and Stumpf, P. K. (1967). Fat metabolism in higher plants. XXXIV. Development of fatty acid synthetase as a function of protein synthesis in aging potato tube slices. *Plant Physiol.* **42**, 391–397.

284. Wilson, W. D., and Foster, J. F. (1971). Conformation-dependent limited proteolysis of bovine plasma albumin by an enzyme present in commercial albumin preparations. *Biochemistry* **10**, 1772–1780.

285. Wittenbach, V. A. (1977). Induced senescence of intact wheat seedlings and its reversibility. *Plant Physiol.* **59**, 1039–1042.

286. Wray, J. L., and Filner, P. (1970). Structural and functional relationships of enzyme activities induced by nitrate in barley. *Biochem. J.* **119**, 715–725.

287. Youle, R. J., and Huang, A. H. C. (1978). Albumin storage proteins in the protein bodies of castor bean. *Plant Physiol.* **61**, 13–16.

288. Youle, R. J., and Huang, A. H. C. (1978). Evidence that the castor bean alergens are the albumin storage proteins in the protein bodies of castor bean. *Plant Physiol.* **61**, 1040–1042.

289. Zielke, H. R., and Filner, P. (1971). Synthesis and turnover of nitrate reductase induced by nitrate in cultured tobacco cells. *J. Biol. Chem.* **246**, 1772–1779.

290. Zucker, M. (1971). Induction of phenylalamine ammonia-lyase in *Xanthium* leaf discs. Increased inactivation in darkness. *Plant Physiol.* **47**, 442–444.

291. Zucker, M. (1972). Light and enzymes. *Annu. Rev. Plant Physiol.* **23**, 133–156.

292. Zwar, J. A., and Jacobson, J. V. (1972). A correlation between a ribonucleic acid fraction selectively labeled in the presence of gibberellic acid and amylase synthesis in barley aleurone layers. *Plant Physiol.* **49**, 1000–1006.

293. Dalling, M. J., Tang, A. B., and Huffaker, R. C. (1983). Evidence for the existence of peptide hydrolase activity associated with chloroplasts isolated from barley mesophyll protoplasts. *Z. Pflanzenphysiol.*, in press.

294. Helsel, D. B., and Frey, K. J. (1978). Grain yield variations in oats associated with differences in leaf area duration among oat lines. *Crop Sci.* **18**, 765–769.

295. Larkins, B. A. (1981). Seed storage proteins. Characterization and biosynthesis. *In* "Biochemistry of Plants. A Comprehensive Treatise" (P. K. Stumpf and E. Conn, eds.-in-chief), Vol. 6 (A. Markus, ed.), pp. 450–485. Academic Press, New York.

296. Larkins, B. A. (1983). Genetic engineering of seed storage proteins. *In* "Genetic engineering of plants: an agricultural perspective" (T. Kosuge and C. P. Meredith, eds.), pp. 93–118. Plenum, New York.

297. Miller, B. L., and Huffaker, R. C. (1981). Partial purification and characterization of endoproteinases from senescing barley leaves. *Plant Physiol.* **68**, 930–936.

298. Miller, B. L., and Huffaker, R. C. (1982). Hydrolysis of ribulose-1,5-bisphosphate carboxylase by endoproteinases from senescing barley leaves. *Plant Physiol.* **69**, 58–62.

299. Thayer, S. S., and Huffaker, R. C. (1983). Vacuolar localization of endoproteinases EP1 and EP2 in barley mesophyll cells. *Plant Physiol.* (Suppl.) **72**, 118.

300. Wittenbach, V. A., Lin, W., and Hebert, R. R. (1982). Vacuolar localization of proteases and degradation of chloroplasts in mesophyll protoplasts from senescing primary wheat leaves. *Plant Physiol.* **69**, 98–102.

CHAPTER FOUR

Distribution of Metabolites

J. S. PATE

I. Introduction

To survive and grow a multicellular plant must exchange a range of dissolved substances between its constituent parts. In higher plants, in which path lengths of transport are great and in which specialized vascular systems exist for transport to occur, solute exchange and distribution comprise a veritable complex of events and processes. This chapter pieces together the basic structural and functional elements of transport activity and discusses current understanding of how metabolite partitioning is related to the growth and development of the whole plant.

The chapter commences with a general description of the major avenues of transport in the higher plant, dealing especially with the structural features and anatomical constraints associated with transport ac-

Plant Physiology
A Treatise
Vol. VIII: Nitrogen Metabolism

ISBN 0-12-668608-4

tivity. On the basis of this a generalized scheme is drawn up describing how metabolites exchange and circulate within the plant body. A second section then deals with the identity of solutes moving in xylem and phloem, relying heavily on published analyses of the composition of the liquids recovered from transport pathways. Separate elements of the plant's transport system are then described, dealing in sequence with the upward transport of solutes from roots in xylem, the lateral exchange of solutes between the ascending xylem stream and tissues of the stem, and the export in phloem from leaves of solutes recovered from xylem or generated in the photosynthetic activity of the leaf. The use of isotope-feeding experiments to study the likely origin of metabolites in transport fluids is discussed. The following section presents case histories of the significance of xylem and phloem transport in the functioning and nutrition of two contrasting plant organs. One of these, the root nodule of the legume, exemplifies the situation in a metabolically active organ relying for its functioning on intensive exchange of metabolites with the rest of the plant. The other example, the developing fruit, describes a situation in which import through xylem and phloem is geared primarily to the laying down of storage reserves in seeds. Following these case histories, a concluding section deals with the partitioning of metabolites within the whole plant. It demonstrates how empirically based modeling techniques can be used to describe quantitatively the flow profiles of carbon- and nitrogen-containing metabolites in organ exchanges by xylem and phloem.

Annual grain legumes[1] feature prominently in the material dealt with in the chapter. These have been favored objects of study in the author's laboratory for many years, and by concentrating on a group of familiar species it is hoped to provide an element of continuity and authenticity in the text. One grain legume, white lupine (*Lupinus albus*) receives special mention. It has proved valuable in studies of transport because it bleeds freely from cut phloem and xylem, enabling one to examine the solutes transported during growth at a range of locations in the plant body. Also, being a legume, it provides an opportunity for studying the interrelationships of carbon and nitrogen assimilation and partitioning in symbiotically effective plants fully autotrophic for these two elements. This feature is exploited particularly in the sections dealing with the modeling of plant transport systems.

[1]Legumes grown for their dried seeds. (Eds.)

II. Transport Systems

A. PRINCIPAL PATHWAYS OF TRANSPORT

Two distinct pathways of transport exist for solute transfer in vascular plants. The first of these utilizes the extraprotoplasmic compartment (apoplast) of the plant and consists essentially of the mass flow of solutes and water from the root to regions of transpirational loss in the shoot system. Over short distances, this pathway encompasses cell walls and water-filled intercellular spaces of plant tissues; over longer distances, tracheids and vessels of xylem are utilized—cell types specially designed for low-resistance, bulk flow of liquids. The second form of transport comprises solute flow within the cytoplasmic phase (symplast) of the plant, including transfer between cells over short distances by means of plasmodesmata and, over greater distances, using the highly specialized sieve elements of phloem. Like vessels and tracheids of xylem, phloem conducting elements are adapted structurally for bulk flow of liquids, and this flow occurs mainly from photosynthetically active "source" organs to "sink" organs in which photosynthate is consumed in growth and metabolism or storage. Concepts of phloem transport based on mass flow through sieve tubes picture the sugar loaded into phloem at sites of photosynthesis as the "driving force" in translocation. This concentrated phloem fluid emanating from photosynthesizing leaves not only contains products of photoassimilation but is also likely to carry out of the leaf solutes that have entered the leaf in the xylem or that are mobilized from temporary or long-term reserves in the leaf.

In addition to the operation of these two pathways, a general requirement exists for continuous exchange of solutes between symplast and apoplast. In most cases the net fluxes involved will be from apoplast to symplast, because the latter compartment contains the bulk of the plant's solutes, especially those utilized in growth and maintenance of living cells. Important locations for apoplast-to-symplast transfer are the stem and mature leaves, in which solutes may be abstracted from the ascending xylem stream and transferred to the upward- and downward-moving translocation streams of phloem. Another example, this time involving a net flux of solute from symplast to apoplast, operates in roots, enabling solutes arriving in excess from the shoot via phloem to pass to the ascending xylem stream and thus "cycle" through the root and back to the shoot system. With this system and the complementary cycling of solutes from xylem to phloem in stem and leaves, the potential exists for a solute to "circulate" freely between root and mature parts of the shoot and to continue to do so until sequestered by a growing organ or site of storage in root or shoot. The plant's capacity for circulating inorganic solutes has been well known since the earliest studies using isotopically labeled solutes such as $^{32}PO_4^{3-}$ and $^{35}SO_4^{2-}$ (16). It is now known that the same principle applies to a wide range of organic metabolites.

A most important example of transport involving exchange between apoplast and symplast compartments concerns the transfer of metabolites to developing seeds. Cytoplasmic discontinuities exist between the diploid tissue of the parent plant and the triploid endospermic tissue of the seed, and between endosperm or perisperm of the seed and the diploid tissue of the new sporophyte (embryo), so that transfer to each distinct symplastic compartment must be routed through an intervening apoplast. Major fractions of the vegetative plant's mineral reserves and photosynthate are mobilized via these symplast–apoplast and apoplast–symplast junctions, and because the volume and surface areas of seed tissues involved are small relative to the bulk of parent tissue from which the solutes are being mobilized, solute fluxes across the junctions must be particularly high.

B. Structural Considerations Relating to Solute Transport

It is a general rule that virtually every part of the shoot and root of the higher plant differentiates xylem and phloem and that these two systems accompany one another throughout every ramification of the plant's vascular network (52, 55). Exceptions to this rule are few but noteworthy. They include nectaries (69), seeds (79), the veinlets of fruit walls (161), and the phloem anastomoses between primary vascular strands of stem internodes (1), in which phloem is not accompanied by xylem, and specialized organs such as hydathodes (70) and certain root haustoria (64), in which xylem is the only vascular tissue present. In other situations, in which xylem and phloem are both present, the relative proportions of the two classes of conducting elements may vary with location in the plant. For example, in the stem base of our experimental species, *Lupinus albus*, the cross-sectional area of xylem vessels and tracheids exceeds that of sieve elements (J. S. Pate, unpublished), as might be expected of a site carrying all of the water transpired by the plant, whereas in the slowly transpiring fruit of the same species, fed predominantly by phloem transport (166), conducting elements of xylem occupy

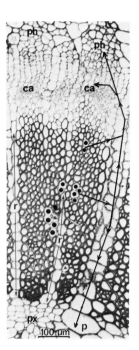

FIG. 1. Vascular tissue (transection) of portion of stalk of a 4-week-old fruit of *Lupinus albus*. ph, Phloem; ca, cambium; r, ray; p, pith parenchyma; px, protoxylem. Xylem conducting elements are marked with solid circles; other cells of xylem are fibers and ray parenchyma. Arrows suggest possible routings for metabolites abstracted from xylem. Modified from Pate *et al.* (163).

only 0.27–0.50% of the fruit stalk's cross-sectional area, compared with values of 0.66–1.13% for sieve elements (163). In this last instance, the bulk of the secondary xylem consists of fibers, which give the fruit stalk strength but do not increase its conducting capacity (Fig. 1).

Because phloem and xylem systems accompany one another throughout most of the plant's vascular system, a quite vast potential must exist for lateral exchange of solutes between the two long-distance transport channels. In the primary vasculature of stem and root and in the minor veins of the vasculature of leaves and fruits, xylem elements lie adjacent to sieve elements, often with one or only a few layers of parenchyma separating xylem conducting elements from the nearest sieve elements. This is most noticeable in minor veins of leaves (Figs. 2 and 3), in which close proximity of xylem and phloem is combined with a large total length of veins to provide an enormous capacity for xylem-to-phloem transfer (148). In secondarily thickened stems of dicotyledons and gymnosperms the functional conducting elements of xylem become progressively separated from phloem by cambium and associated undifferentiated cells. However, radial rows of ray parenchyma traverse the gap between mature parts of xylem and phloem, and this tissue

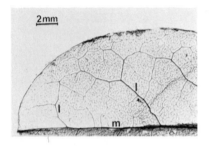

FIG. 2. Low-power view of cleared leaflet of *Lupinus albus* showing vasculature of leaf. m, Midrib vein; l, lateral veins. J. Kuo and J. S. Pate (unpublished).

might well provide an important avenue for solutes absorbed from xylem to be transferred to cambium, pith, or phloem. These suggestions are annotated in Fig. 1.

In many species of vascular plants, certain parenchymatous cells adjacent to conducting elements differentiate into a unique type of cell bearing irregular, fingerlike ingrowths of secondary wall material. Penetrating deep into the parent protoplast, these ingrowths have the effect of increasing the surface area of plasma membrane of the cells, and, in most instances, the walls bearing ingrowths abut on or lie adjacent to conducting elements (Figs. 4 and 5).

It has been proposed that these specialized *transfer cells* represent an adaptation facilitating apoplast–symplast exchanges of solutes within specific regions of the plant (66, 67, 156). Circumstantial evidence in favor of this concept comes from the ubiquity of such cells among plants of all of the major taxa, their ultrastructural features (which suggest involvement in transport), and their invariable association with locations in the plant at which particularly intensive fluxes of solutes are likely to take place or have been demonstrated to take place.

Unfortunately, there have been relatively few instances in which it has

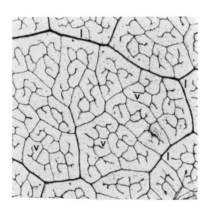

FIG. 3. High-power view of cleared leaflet of *Lupinus albus*. v, Free-ending veinlets; l, lateral veins. J. Kuo and J. S. Pate (unpublished).

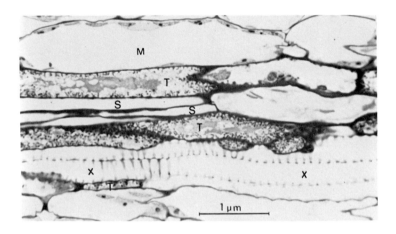

FIG. 4. Transfer cells associated with minor vein of leaf of *Pulicaria* sp. T, Transfer cells; S, sieve elements; X, xylem elements; M, leaf mesophyll cells. Material prepared and photographed by B. E. J. Gunning and J. S. Pate.

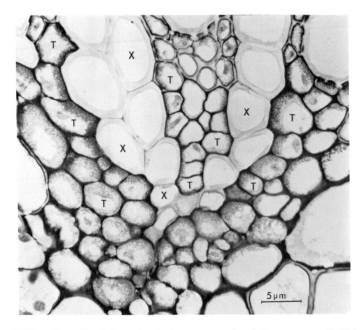

FIG. 5. Transfer cells of departing leaf trace at node of *Pisum arvense*. X, Xylem elements; T, transfer cells. Note that the most dense displays of wall ingrowths on transfer cells are associated with walls in closest proximity to xylem elements Material prepared and photographed by B. E. S. Gunning and J. S. Pate.

proven possible to measure fluxes across the enlarged plasma membranes of the transfer cells, but, where this has been done (69), transfer rates per unit area of plasma membrane of a transfer cell have proved to be of the same order of magnitude as those recorded for nonspecialized cells. So the surface-area amplification of the cell resulting from its wall ingrowths appears to give the transfer cell a considerable advantage over a smooth-walled cell in terms of uptake per unit volume of cell or tissue. This adaptation would convey particular benefit in relatively confined situations within the plant's transport systems, and this, indeed, is where transfer cells tend to be located.

A relatively common location for transfer cells is the minor veins of leaves (Fig. 4) (155), where they are envisaged as loading the symplast with photosynthate and solutes absorbed from the dilute apoplastic fluid supplied by the xylem (66, 156).

In root nodules (157) and certain glands (156), transfer cells occur in a position suggesting activity in the secretion of solutes to the apoplast or external surface of the plant. The enlarged plasma membranes might be as important in the retrieval of certain essential substances from the exudate as in the actual secretion process however. In nectaries, for example, a role has been suggested for the cells in reabsorption of amino acids from nectar (67). Transfer cells also abound at symplast–apoplast junctions within reproductive structures of a wide range of plants (67) and must act there in ferrying solutes across cytoplasmic discontinuities between nuclear generations (see previous discussion).

A very common site for transfer cells is in the vascular tissues of nodes, particularly adjoining xylem conducting elements of departing leaf traces (Fig. 5) or flanking the margins of the "gap" in the vasculature caused by the departure of such traces (26, 66, 68, 215). Transfer cells in these locations might function in the uptake of solutes from conducting tissues and in the subsequent transfer of these solutes to axillary structures borne at the node. This would be of value, say, in the early development of axillary buds, especially before vascular connections with the parent node have become established (158). A second interpretation of the role of nodal transfer cells implies a more general role in solute abstraction. McNeil *et al.* (124) have demonstrated especially active uptake of certain xylem-fed amino acids by the transfer cells of the departing leaf traces of *Lupinus albus,* and they have suggested that the strategy of restricting abstraction of solutes largely to strands already committed to leaves would allow a specific internode of the stem to absorb significant amounts of xylem solutes without greatly lowering the concentrations of solutes in the body of xylem fluid passing to higher regions of the shoot. With absorption associated principally with departing leaf traces, the potential would also exist for the cells to play a role in regulating the supply of solutes to individual leaves.

A final, most interesting situation for transfer cells is where they are present as modified parenchyma lining the xylem of the root. This has been well studied in the legume genera *Glycine* and *Phaseolus* (110, 115), and a function in reabsorption of Na^+ from xylem fluids has been suggested (see 66).

In summary, then, the transfer cell is envisaged as a highly versatile module with the potential for intensive short-distance transfer of a wide range of solutes. The types of solutes that transfer cells transport, and whether they act in a secretory or absorptive mode will depend largely on the cells' location within the plant, but the invariable occurrence of the cells in key bottleneck regions of the plant's transport system implies a generalized role in solute transport, complementary to that mediated by the long-distance transport systems.

It now becomes possible to incorporate the structural information assembled so far into a model outlining the basic transport systems of the whole plant (Fig. 6). In this scheme, epitomizing the situation in an annual herbaceous species, the basic functional units of the plant are depicted as root, stem, mature leaf, and an apical "sink" region of the shoot representing meristems, young vegetative organs, and developing reproductive structures. Sites of primary assimilation of metabolites are denoted by the symbol A (Fig. 6), including the leaf as a site of photoassimilation of CO_2 and the root and leaf as possible sites for the assimilation of nitrogen and other elements absorbed by the root system. Principal loading points for phloem and xylem are given the letter L, unloading sites the letter U, and regions where interchange of materials between xylem and phloem are likely to be important are given the symbol I. The model indicates where transfer cells might be located and marks the positioning of their wall ingrowths. Likely directions of flux of metabolites are shown by arrows, the flow paths referring essentially to the movement of major organic metabolites within the plant.

The scheme depicted in Fig. 6 embraces a circulatory pathway for solutes between the root and mature parts of the shoot (labeled 1–6), a unidirectional flow path for solutes in xylem and phloem to apical sink regions of the shoot (pathways 7 and 8), and a lateral uptake system in mature organs of root and stem (9–12) that provides these parts with assimilates for storage and secondary thickening or for re-export to other regions of the plant. Root-derived solutes abstracted by lower regions of the stem and transferred to phloem are likely to be swept back to the root with the descending stream of translocate from the lower leaves, whereas those transferred from xylem to phloem in stem tissue

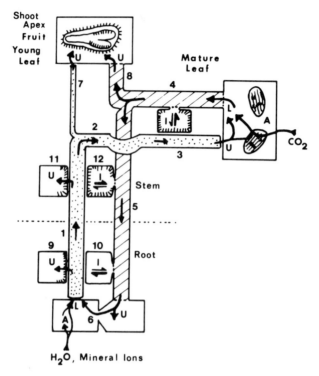

Shoot
Apex
Fruit
Young
Leaf

Mature
Leaf

Stem

Root

H_2O, Mineral Ions

Fig. 6. Principal elements of transport system of herbaceous higher plant. Locations for loading and unloading of xylem (▣) and phloem (▨) are labeled L and U, respectively; regions of synthesis of carbon- and nitrogen-containing assimilates are labeled A; and regions of metabolite exchange between xylem and phloem are labeled I. Possible locations for transfer cells are indicated, and the positioning of wall ingrowths on the transfer cells is marked. Major routings for metabolites: 1–6, circulatory system between mature parts of root and shoot; 7–8, undirectional flow from mature organs to apical sinks of shoot; 9–12, lateral abstraction of metabolites by stem. Redrawn and modified from Pate (148).

nearer the apex are pictured as joining the phloem stream from upper leaves feeding apical parts of the shoot. The capacity thus exists for xylem-borne solutes to exchange with phloem in a manner that is partly independent of the current loading of assimilates into phloem in leaves. The significance of this and other arrangements within the plant's transport system will be considered later in the chapter.

C. Vascular Anatomy and the Translocation of Assimilates

The vascular architecture of a plant species, especially the distribution of vascular traces to leaves and the pattern of vascular connections at

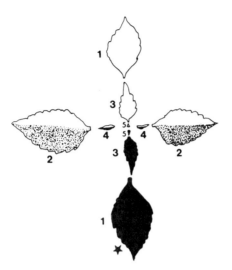

FIG. 7. Effects of vascular architecture on photosynthate flow from leaves of *Coleus*, a genus with opposite decussate leaves. (See legend of Fig. 8 for more information.)

nodes and within internodes of the stem, has quite obvious bearing on how assimilates are translocated. When such information is combined with study of the distribution of isotopically labeled assimilates within the plant body after feeding of specific photosynthetic surfaces with $^{14}CO_2$, it becomes possible to gain an accurate picture of which vascular connections are in use and which of these are of greatest importance in nourishing specific regions of the plant.

The literature within this area is vast, so attention will be restricted to the more important generalizations, using investigations of only a few species to illustrate the points in question.

Possibly the simplest pattern of assimilate distribution is that displayed by plants with opposite decussate pairs of leaves (e.g., *Coleus* and *Perilla*). Photosynthate from a mature leaf is likely to flow as shown in Fig. 7, with little or no traffic from one side of the stem to the other because of the absence of effective vascular connections at the nodes. The same principle applies to upward movement of nutrients in xylem, as shown for the growth studies on rooted cuttings of *Coleus* performed by Caldwell (31). A somewhat similar situation is found in species with alternately opposite leaves, for example, the pea (*Pisum sativum*) (148). Here again the strongest transport connections are between organs inserted on the same side of the stem, but some transfer to the opposite side of the stem can take place via vascular connections involving the lateral leaf traces to the leaves (27).

The great majority of herbaceous dicotyledons display spirally ar-

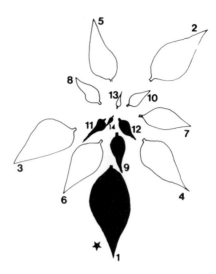

FIG. 8. Effects of vascular architecture or photosynthetic flow from leaves of tobacco or sugar beet, species with a spiral arrangement of leaves. Figures 7 and 8 illustruate the likely distribution of ^{14}C assimilates within a shoot after feeding a specific leaf (see black star) with $^{14}CO_2$. Dark and dotted areas would receive ^{14}C assimilates from the fed leaf; unshaded areas would receive little or no label.

ranged leaves, with a fairly constant angle of divergence (e.g., 72°, 137°, etc.) in the insertion of successive leaves on the stem. In these instances leaves of the same orthostichy (i.e., positioned on the same vertical sector of the stem) are interconnected by vascular traces and therefore tend to exchange assimilates with one another rather than with leaves of other orthostichies. If a terminal inflorescence is present, the flowers on it are usually inserted in a spiral pattern following on from that of the leaves, so that foliar surfaces tend to nurse sets of fruits belonging to the same orthostichy. Examples of species showing a geometry of assimilate flow of this nature are tobacco (*Nicotiana* spp.) (100), sugar beet (*Beta vulgaris*) (102), lupine (*Lupinus albus*) (141, 167), and sunflower (*Helianthus annuus*) (114) (see also Fig. 8).

The situation in herbaceous and arborescent monocotyledons and in trees is especially difficult to study because of the large number of leaf traces, the bewildering pattern of vascular anastomoses at the nodes, and the finding that traces from a leaf may travel to a node below the node bearing the leaf before entering the main vascular network [see Hartt and Kortschak (82) for sugar cane (*Saccharum officinale*)]. In woody species of gymnosperms and dicotyledons the situation is likely to be complicated by the capacity of cambium and rays to abstract assimilates moving in xylem and phloem (96), by the marked seasonality in the long-distance movement of assimilates (77, 111, 112, 121, 178, 202) and by the demonstration that not all conducting elements may be functional at any one time and, if functional, may not necessarily work with the same efficiency (223). Furthermore, in deciduous trees, source–sink relationships can be reversed during a growing season, storage parenchyma of stem, for instance, acting as a sink for assimilates when leaves are present, but as a source of assimilates at bud break in spring.

It is a fairly general rule that assimilates from leaves of seedlings and young vegetative plants are distributed widely and fairly evenly by means of an essentially integrated transport system (50, 114, 126, 179, 184). Then, as further growth takes place and distances between specific source and sink regions increase, distribution of assimilates may become markedly stratified, with upper leaves supplying the shoot apex and adjacent young leaves, middle leaves translocating to both shoot apex and root, and the lower stratum of leaves supplying only the lower stem and root system (29, 114, 145, 178, 192). Similar principles apply to the zoning of nutrition within lateral shoots, as originally shown for soybean (*Glycine max*) (15). On both main and lateral shoots, however, anatomical constraints may still dictate that the nearest and strongest sink within a particular nutritional zone will not necessarily obtain the largest share of assimilates from a particular leaf.

The onset of reproduction imposes new, overriding constraints on patterns of transport, due essentially to the high competing power of developing fruits for phloem-borne assimilates. In species of indeterminate growth in which a sequential pattern of flowering involving axillary flowers or inflorescences is displayed [e.g., cotton (*Gossypium*), squash (*Cucurbita*), pea (*Pisum*), and many other woody species], the photosynthetic surfaces at a node become deeply committed to nourishing the fruit(s) borne at that node (3, 58, 76, 111, 114). In terminal-flowering species (e.g., sunflower and cereals) photosynthetic organs embodied in the fruit or inflorescence and the leaves just below that inflorescence supply the bulk of the photosynthate required by the developing fruits and seeds. This is well demonstrated in cereals, in which flag leaves, glumes, and awns may collectively provide the filling grain with up to 70% of its requirement for dry matter (53, 116, 122, 208).

Several species exhibit only a slight tendency for nutrition to become stratified during the vegetative phase of growth. Instead, photosynthetic surfaces feed assimilates into a common "pool," from which the various sink organs draw more in proportion to their overall "sink strength" than in relation to their physical or anatomical "nearness" to specific source organs. This situation applies to potato (*Solanum tuberosum*) (135), white clover (*Trifolium repens*) (75), tomato (*Lycopersicon esculentum*) (105), and probably quite widely among monocotyledons, as the studies on sugar cane (82) and ryegrass (*Lolium* spp.) (126) suggest.

An advantage of a fully integrated source–sink system is that it has the potential to adapt readily to change in supply and demand of sources and sinks. Thus shaded or defoliated stolons of white clover (75, 83) and shaded organs of ryegrass (126) and wheat (*Triticum*) (176) are able to abstract photosynthate from other parts of the parent plant though not initially fed by these parts. This may be of some significance in pasture species in which grazing and canopy shading may foster drastic changes in the source–sink relationships of a plant's parts during a growing season. Responses of tomato plants to partial defoliation during glass-house culture also bear evidence of a capacity to reorder assimilate parti-tioning during stress. Trusses of fruits whose nurse leaves have been removed can then be fed assimilates from leaves much higher up the canopy, although under normal circumstances in a fully leafed plant blossom leaves have prime responsibility for supplying their subtended fruits (105). Like other members of the Solanaceae, the tomato possesses external and internal phloem systems, and study of distribution of ^{14}C assimilates within these systems suggests inherent versatility in partition-ing of assimilates (24). The same probably applies to members of other plant families possessing external and internal phloem [e.g., the aquatic plant *Trapa natans* (40)].

It is also possible that patterns of assimilate movement may be violated in plants showing a high degree of vertical or horizontal zonation of nutrition. Thus in sugar beet the marked restriction in assimilate flow to leaves within the same orthostichy can be changed to a more general pattern of distribution if all fully expanded leaves are removed from one side of the plant (102). Similarly, removal of certain source leaves below an inflorescence of sunflower causes the remaining leaves to commence to supply assimilates to fruits with which they have no direct vascular connection (114). It is not clear in these instances whether the repro-gramming of transport involves new vascular connections being forged or whether already existing phloem anastomoses are called into operation.

The processes of secondary thickening, with the development of inter-fascicular cambium and a complete ring of associated conducting ele-ments, must inevitably give the shoot a much greater capacity for lateral transfer of assimilates than was ever possible using its primary vascula-ture. In early flower development of the lupine (*Lupinus angustifolius*), for example, when the inflorescence stalk is not secondarily thickened, assimilates flow from main stem leaves to specific developing flowers along precise routes related to plant phyllotaxis. Later, however, during fruiting, the commitments between specific leaves and specific fruits become progressively less marked as secondary thickening permits the assimilates from a leaf to be shared more evenly within the inflorescence (P. Farrington and J. S. Pate, unpublished).

To summarize, there exist in higher plants highly ordered transport links for assimilates involving anatomical linkages related to the pattern of insertion of organs on a shoot. In addition to this, physiological attributes affect assimilate distribution, in particular the physical "nearness" of a sink to its respective sources and the competitiveness of that sink with other sinks for current assimilates. During plant growth, flow profiles for assimilates change continuously as organs age and their activity as sources or sinks waxes or wanes. These effects, compounded with ontogenetic changes in the assimilatory performance of the whole plant, make detailed study of the relationships between sources and sinks of the plant a particularly difficult task.

III. Identity of Transported Solutes

The direct and logical method for describing the solutes moving in xylem and phloem of plants is to collect the liquids present in conducting elements and identify the various solutes present. Unfortunately, not all species of plants lend themselves to this type of examination, and in those that do the information obtained must be interpreted with caution because techniques for collection of transported fluids may give rise to several types of artifact. Nevertheless, the approach is a useful one and when combined with isotope labeling studies provides valuable information on how specific metabolites arise in transport channels and how they are subsequently distributed in the plant body.

A. Solutes of Xylem

The principal techniques available for collecting liquids from xylem involve deliberate extraction of "tracheal" sap from stem segments by vacuum (20), liquid displacement (65), or centrifugation (63) or by the collection of sap bleeding from xylem under the influence of root pressure (21). Vacuum extraction is the most commonly used technique for woody species, the bleeding-sap technique the one generally used for herbaceous species. However, vacuum extraction of solid stems of certain herbaceous plants has proved successful (168), and xylem sap bleeding under root pressure can be obtained in quantity from deciduous trees in autumn or spring simply by lopping off branches or boring holes in stems (49).

The major contaminants of tracheal sap or bleeding sap are likely to arise from cut cells, but in certain situations phloem or laticifers might liberate solutes to the supposed xylem samples. Concentrations of solutes in bleeding sap are usually much higher than in tracheal sap, as is to be expected from the lower water flux through the xylem of a decapitated plant bleeding under root pressure than in the xylem of a whole transpiring plant. Differences in relative composition between bleeding sap and tracheal sap have been recorded (81, 136), and gradients in relative concentrations of solutes in tracheal sap occur from base to top of stems (39). The latter is likely to be due to progressive release or absorption of solutes as the xylem stream ascends the stem. These complications and the demonstrations that the solute composition of the xylem fluid of a plant may vary markedly with plant age and season, with time of day, and with nutritional status of the plant (149) suggest that it would be unwise to generalize on xylem transport in a species without analyzing a number of samples of sap collected under a wide range of physiological and nutritional conditions.

When collecting root bleeding sap it is usual to discard the first drops that form so as to avoid contamination from cut cells. Also, by restricting collection periods to as short a time as possible, starvation reactions in the decapitated plant are likely to be minimized (21, 149). The vacuum extraction technique should involve only mild vacuum, or, if feasible, rely solely on gravity displacement of xylem contents (60). These precautions will tend to avoid displacing liquids from compartments outside the xylem elements. The most commonly used technique for collecting tracheal sap involves cutting short segments off the free end of the shoot while applying vacuum to the other end (20). Sap should not be collected after the segment has been reduced in length to less than 5–10 cm, as solutes released on cutting of the stem might easily be sucked through the segment to the collecting tracheal sap.

Sampling xylem sap by root pressure is not necessarily confined to plants decapitated at ground level. In pea, tomato, and balsam (*Impatiens*), for instance, xylem bleeding sap can be obtained from the petioles of defoliated plants (170), and, in legumes, detached root nodules bleed a xylem exudate that has proved useful in identifying the products of nitrogen fixation exported to the host plant in the xylem (157, 162).

Xylem exudates sampled by these methods are acidic (pH 5.2–6.5) and, usually, have a total dry matter content of 0.5–40 mg ml^{-1}; up to a third or so of this is likely to be in inorganic form (21). The major cations are likely to be K^+, Ca^{2+}, Mg^{2+}, and Na^+, the major anions PO_4^{3-}, Cl^-, SO_4^{2-}, and NO_3^-. (101). Trace elements are present at levels suggesting full mobility in xylem (149), although there is evidence that certain elements such as iron and copper are transported in chelated form as citrate or are attached to other organic compounds (37, 88, 199). Sulfur may be in organic form as methionine, cysteine, and glutathione (144), and some phosphorus of xylem may also be organic (e.g., as phosphorylcholine) (18). Nevertheless, SO_4^{2-} and PO_4^{3-} are usually the dominant forms of sulfur and phosphorus present.

Nitrogenous compounds constitute a major fraction of the solutes of xylem fluid. A range of compounds has been encountered, including amides (glutamine, asparagine, and substituted amides), amino acids, ureides (allantoin and allantoic acid), and alkaloids, but usually only one or two of these compounds predominate and at least one major solute is a compound with a low carbon–nitrogen ratio (146, 149). The significance of this in the carbon economy of a root assimilating nitrogen will be discussed later. Nitrate may constitute a significant fraction of the xylem nitrogen of certain species but not of others. This is also discussed later.

Sugars and other forms of carbohydrate (e.g., sugar alcohols) are usually absent or present in only low amount in xylem fluid of herbaceous species, but these compounds achieve high concentrations (up to 20 mg ml^{-1}) at certain seasons in trees, shrubs, and woody climbers (11, 81). Cases in point are the high levels of sugars recorded in xylem sap of sugar maple (*Acer saccharum*) (186) and the high concentration of sorbitol observed in xylem of apple (*Malus sylvestris*) (78, 213). In both instances the compounds are at highest level in winter, or after bud break in spring.

Organic acids, especially citrate and malate (113, 188, 199), have been recorded as xylem constituents of several species, although levels are usually low in comparison with nitrogenous solutes and major cations such as K^+.

Several substances exhibiting growth-regulating activity have been recovered from xylem, although few identifications have been made of the actual auxins, gibberellins, cytokinins, and abscisins involved (30, 104, 188). Exceptions to this are the recovery of the cytokinin N^6-(4^2-isopentenyl)adenosine from xylem of *Acer saccharum* (71), of indole-3-acetic acid (IAA) from xylem of *Ricinus* (74), and of a cytokinin nucleotide in the exudate of *Yucca* (203).

B. SOLUTES OF PHLOEM

Collection of the contents of sieve elements of phloem is by no means as easy as from the conducting elements of xylem. Phloem tissue is delicate, difficult to locate precisely by cutting, and, unlike xylem, becomes readily blocked with slime or caliose when damaged. All existing techniques rely on turgor release of contents when sieve elements are cut or punctured.

One technique utilizes feeding aphids. After encouraging the aphids to feed on selected regions of the plant, they are anesthetized in CO_2 and their bodies severed from their mouthparts. Phloem exudate is then collected from the stylet tips left embedded in the tissue after removal of the insect body (133, 221). Volumes of phloem exudate obtained are low, but exudation occurs over a long period, and sectioning of tissues shows that the stylet tips actually penetrate sieve elements (25, 54). Hence, the exudate is likely to be a reliable sample of the contents of conducting elements of phloem. Unfortunately, the technique is limited by the small number of suitably large species of aphids and by the narrow range of plant species on which they will feed. Moreover, even if an aphid is on its normal host plant it is likely to be highly selective in its feeding habits, tending to favor senescent leaves or young apical regions of shoots. It would therefore be difficult to obtain samples of phloem exudate from old stem tissue, roots, and fruits. The aphid technique has been applied with particular success to species of *Salix, Tilia, Juniperus, Picea,* and *Heracleum* (32).

The second principal technique involves bleeding of phloem contents from incisions in the outer bark of trees or from superficial cuts in nonwoody tissues such as fruit pods, inflorescence stalks, petioles, or roots. The technique is essentially restricted to species (Table I) that are sluggish in sealing their sieve pores with callose, and in these exudation may cease after only a few minutes, as in *Lupinus* (167), or continue for days, as in the case of *Yucca* and certain palms (46). In certain herbaceous species, for example, *Ricinus* (127), phloem exudates can be obtained at any point along the length of a stem. In others, such as *Lupinus* (152, 162), petioles, fruits, and fruit stalks bleed as well as main and lateral stems, enabling phloem streams to be sampled at a range of points between the various sources and sinks of the plant.

Artifacts associated with collection of phloem sap from incisions in plant tissue are numerous and can be sufficiently serious to cast doubt on the value of the technique. First, nonmobile, structural constituents may be released from sieve elements during cutting, the most important of these being p-protein—the structural protein of the sieve elements

TABLE I

EXAMPLES OF PLANTS KNOWN TO BLEED SPONTANEOUSLY
FROM CUT PHLOEM AND LOCATIONS WHERE IT IS POSSIBLE
TO OBTAIN THE EXUDATE

Woody species	
Incisions in bark	Over 400 species known to bleed, mainly deciduous trees
Cut distal tips of fruits	*Spartium, Jacksonia, Genista*
Arborescent monocotyledons	
Cut inflorescence stalks or floral apices	*Yucca, Cocos, Arenga, Corypha*
Herbaceous dicotyledons	
Incisions in stem	*Curcurbita, Ricinus, Lupinus*
Cut distal tips of fruit	*Lupinus*
Cut root phloem	*Beta*
Cut inflorescence stalks	*Pisum, Brassica*
Cut petioles	*Phaseolus, Cucurbita*

(41). Protein contamination of this nature reaches particularly high levels in *Cucurbita* phloem exudates. In addition to p-protein, a wide variety of enzymes and certain lectins have been identified in phloem exudates (103, 119, 217), and these too are likely to be static components rather than translocated materials. Second, osmotic attraction of water to cut sieve elements is likely to dilute the exuding sap and thus lead to solute concentrations in the exudate being significantly lower than those obtaining in the intact sieve element. Symptomatic of this is the progressive dilution of solutes in sap with time after cutting (220). However, a falloff in solute concentration is not always observed after cutting; for example, phloem of *Ricinus* (127) and *Lupinus* (167) shows an exponential decline in bleeding rate but not in sucrose level in the exudate. A third major source of artifact concerns contamination of phloem exudate with solutes released from adjacent tissues, or the possible release of ions to the exudate following irrigation of cut tissues with the K^+-rich phloem contents. This class of artifact would produce special problems when using phloem-bleeding techniques to monitor movement of trace elements. Finally, more subtle forms of artifact relate to loss of phloem functioning following cutting or to wounding-induced changes in the source–sink relationships of the phloem tissue being sampled (149). Effects of this nature are likely to be minimized if collection periods are short and if the amount of cut tissue is small relative to the translocatory capacity of the section of stem or other plant part where the sap is being sampled.

There is evidence from analyses of phloem exudates that the composition of translocation streams can differ significantly with time of day, season, location on the plant, plant age, and nutrition (118, 191, 209, 219, 221). Hence, as with xylem-sap analyses, many samples should be analyzed before generalizing on the behavior of a species.

Unlike xylem sap, phloem exudates are invariably alkaline (pH 7.6–8.4) and possess a high level of carbohydrate (5–250 mg ml^{-1}). Nitrogenous solutes and most ions are usually present at 10–20 times the concentration in xylem exudates of the same species (49, 72, 149), but the principal organic solute or solutes of nitrogen of phloem of a species are usually identical with those of its xylem. If nitrate is present at high level in xylem, it is usually absent or at much lower concentration in phloem (41, 125, 168). Calcium and certain other sparingly mobile elements (e.g., iron and manganese) may also be at low level in phloem relative to xylem (92). Calcium, zinc, and manganese may be present in phloem as organic complexes, according to Goor and Wiersma (61, 62), and the proportion of sulfur and phosphorus in organic form is usually higher in phloem than in xylem, probably reflecting reduction of inorganic forms of these elements in leaves prior to loading of phloem (18, 41). Sugar phosphates and nucleotides may comprise significant fractions of the phosphorus of phloem (217). The presence of growth substances in phloem exudates has been recorded for a variety of species, including substances that promote or inhibit flowering, B-class vitamins, and substances showing auxin-, gibberellin-, cytokinin-, and abscisin-like properties.

Finally, it is interesting to note that a technique has been developed for recovering exudates from plants that do not normally bleed from phloem. This exploits the observation that if the cut distal ends of attached petioles of *Perilla* leaves are placed in ethylenediamenetetraacetic acid (EDTA) solution after removal of the leaf lamina, bleeding of solutes is enhanced. ^{14}C-Labeling studies suggest that a major fraction of the leakage products is from phloem (106). The technique has been extended to pod leakage studies on soybean (56), and again ^{14}C feeding suggests that assimilates being translocated are among the leakage products. Extensive contamination from tissues other than phloem is likely with this technique, so results must be viewed with caution.

IV. Working Units of the Plant's Transport System

The generalized scheme for the transport system of the plant developed in Section II depicted a series of interlocking functional units involving long- and short-distance transport in apoplast and symplast. It is now time to deal in turn with some of these units and to examine the bearing each has on the distribution and circulation of metabolites within the whole plant.

A. ORIGIN AND EXPORT OF METABOLITES FROM ROOTS IN XYLEM

When xylem sap is collected in the manner described in Section III, it is by no means clear how the various ions and organic compounds it contains have originated and been released to xylem. In most situations, at least in herbaceous species, the major fraction of the inorganic component of xylem is likely to represent recent uptake from the rooting environment, with storage pools of the root or circulatory processes within the plant making relatively minor contributions under normal circumstances. However, it has been suggested that significant proportions of the SO_4^{2-}, PO_4^{3-}, and K^+ ions of xylem circulate between shoot and root via phloem and xylem, whereas the fact that shoot stumps of decapitated plants continue to bleed ions from their xylem for some time after transfer to distilled water (2, 209) bears witness to the extent to which vacuolar pools of roots contribute when the root no longer has access to an external source of ions.

These principles apply also to organic solutes, but in this case uptake from the rooting medium is unlikely to occur or make only a minor contribution to the xylem. There may also be complications resulting from metabolic transformations prior to loading of the xylem, as, for example, in the case of the organic phosphorus and sulfur-containing amino compounds that become labeled in studies involving feeding of $^{32}PO_4^{3-}$ (136) or $^{35}SO_4^{2-}$ (144) to roots. The same applies to the bulk of the nitrogen of xylem, for example, the appearance in xylem of ^{15}N-labeled amino compounds after $^{15}NO_3^-$ has been fed to roots (99, 140). Similarly, if ammonium is the nitrogen source to the root, organic nitrogen, not ammonia, is the principal form released to the xylem (206), whereas in symbiotic systems fixing atmospheric nitrogen, $^{15}N_2$ feeding experiments have demonstrated that organic compounds, usually amide or ureide (86), are the compounds that export the assimilated nitrogen. When nodulated legumes assimilate $^{14}CO_2$ in photosynthesis of their shoots, the carbon of the amino compounds of their root bleeding sap becomes labeled within an hour or so of applying the $^{14}CO_2$, indicating that recently translocated carbohydrate is utilized by the ammonia-assimilating systems of the nodule (86, 143).

Roots of different plant species vary widely in their ability to reduce nitrate (147). In those with relatively weak nitrate reductase activity in their roots [e.g., cocklebur (*Xanthium strumarium* var. [*X. pensylvanicum*])] (207) and cotton (*Gossypium*) (180)], upward of 95% of the nitrogen leaving the root in xylem is as unreduced nitrate. Any nitrate reduction that might occur in roots of these species is likely to be utilized in growth of the root (180) insufficient to generate an exportable surplus of organic solutes of nitrogen. Indeed, it has been argued that phloem transport from the shoot might provide roots of such species with a large fraction of their nitrogen requirement for growth (207).

Species with active nitrate reductase systems in their roots have almost all of their xylem nitrogen present in organic form if low levels of nitrate are presented to roots, but they bear evidence of an increasing spillover of free nitrate to the shoot if the level of nitrate in the medium is raised (139, 168, 206). With high levels of nitrate in the medium, the nitrate reductase system of the root is likely to be saturated, and roots are then likely to store or export increasing fractions of the absorbed nitrate in unreduced form. The passage of nitrate through the root by extracellular pathways that circumvent the cytoplasmically located reductase systems may be an important aveune of transport of nitrogen to the shoot in such circumstances.

It is not easy to identify which precise fractions of the xylem's nitrogen have arisen from processes other than assimilation of recently absorbed nitrogen. However, a nonassimilatory source of nitrogen for xylem loading is readily identifiable in roots of seedling legumes grown without combined nitrogen and without root nodules (169, 211). The compounds present in xylem sap of such plants (e.g., valine, leucine, and aspartic acid in *Pisum* seedlings) are different from those present (asparagine and glutamine in *Pisum* seedlings) in the xylem if a source of nitrogen is present (e.g., nitrate, ammonium, or urea) or if symbiotic nitrogen fixation is taking place (169). In older plants, proteolysis in roots, discharge from pools of nitrogen established earlier in the life of the root, and the cycling of nitrogen through the root might all act in contributing nitrogenous solutes to xylem. If these nonassimilatory sources were to make a large contribution to the exported nitrogen, the ratio of nitrate to amino nitrogen in xylem would provide misleading information on the proportion of the incoming nitrate being handled by the root reductase. It becomes evident from these complications that detailed analyses of time courses of labeling of root pools and xylem sap following application of $^{15}NO_3^-$ are essential if accurate information on a root's uptake, assimilation, and turnover of nitrogen is to be obtained (99).

The opportunity presents itself in the nodulated legume for distinguishing that fraction of the xylem's nitrogen resulting from current fixation in nodules from that cycled through the roots via phloem and xylem. First, the products of fixation can be directly identified by analyzing xylem-bleeding sap of detached nodules (69, 131), and second, the relative amounts of nitrogen from this and other nonassimilatory sources can be assessed by collecting root-bleeding sap from above and below the zone of nodules of plants growing in minus-nitrogen water culture. In an experiment of this kind on white lupine (J. S. Pate and D. B. Layzell, unpublished), it was found that the concentration of organic nitrogen in the exudate from the top of the root (i.e., including the exports of nodules and root tissues) was approximately 10 times that collected from the root zone below the nodules. Because similar levels of nonnitrogenous solutes were present in the two samples, it was concluded that 90% or so of the nitrogen collected at ground level had originated from the recently fixed nitrogen of root nodules, the remainder being from the release of nitrogen by the supporting root system. It would be interesting to know how other legumes and nonlegumes behave in this respect.

Organic acids in xylem have been suggested to be a major form in which photosynthetically fixed carbon circulates in plants, and a role of shoot-derived organic acids in root biosynthesis has been suggested (113). If $^{14}CO_2$ is fed to the root surroundings of certain legumes (e.g., Psophocarpus or Vigna), however, xylem sap becomes labeled in malate, citrate, and a range of amino compounds, some of these products attributable to CO_2 fixation in roots, others to CO_2 fixation by nodules (K. Hillman, R. Rainbird, C. A. Atkins, and J. S. Pate, unpublished). This suggests de novo synthesis in root and nodules of organic acids by carboxylating enzymes, such as phosphoenolpyruvate (PEP) carboxylase. In fact, this enzyme has been known to be active in root nodules of several legumes (36, 117, 210).

The presence of sugars in xylem of deciduous woody species at the beginning of a growing season may have special nutritional significance in providing bursting buds and expanding leaves with a ready source of available carbohydrate via the xylem. Sauter et al. (186), studying sugar maple (Acer saccharum), have shown that the late-winter maximum in sucrose concentration in xylem coincides with the mobilization of carbohydrate from ray tissues of root and stem, and it is possible that gibberellins released to xylem at this time might promote mobilization from the ray tissue, in a manner akin to the action of gibberellins in mobilizing starch from endosperm of cereals. Other metabolites are also present at peak concentration in spring sap of deciduous trees [e.g., allantoin and allantoic acid in Acer and other genera (183)]. These may also arise largely from mobilization of stored reserves of stem and root rather than from the assimilation of newly absorbed nitrogen by the rooting medium.

Turning finally to the growth substances of xylem, one cannot as yet make definite pronouncements regarding origin and function. Demonstrations of changes in composition and relative concentrations with plant age and environmental stress, however, do suggest that levels of growth substances in xylem may be controlled by factors within the root, even if not all of the compounds present have been actually synthesized in below-ground parts (34, 175).

B. LATERAL EXCHANGE OF XYLEM SOLUTES IN STEMS

There is much evidence to suggest that stem tissues of herbaceous and woody plants have the ability to abstract solutes selectively from the ascending transpiration stream. The first class of evidence comes from analysis of samples of bleeding sap collected at different levels of a shoot system. The samples are usually found to be progressively more dilute in total solutes with height up the stem (39, 170, 189), and selectivity in withdrawal from the xylem of leaf traces is indicated from the changes observed in relative composition of the sap collected from base to top of stem; for example, the abstraction of asparagine is more pronounced than that of homoserine and aspartic acid from stem xylem of *Pisum* (28). Comparisons of composition of root-bleeding sap and guttation fluid of barley (*Hordeum vulgare*), chickweed (*Stellaria media*), and of the associated hemiparasite *Odontites verna* have yielded further examples of species-specific abstraction patterns from xylem (64). Host and hemiparasite tap the same xylem source of solutes, and nitrate, a major xylem solute, is found to escape to guttation fluid of *Stellaria* but not to the corresponding guttation fluid of the attached hemiparasite, whereas shoots of the parasite are noticeably more effective in removing amino compounds from xylem than are shoots of either of the associated hosts.

A second source of evidence for lateral abstraction of xylem solutes relates to experiments involving feeding of specific labeled substrates to the base of cut shoots through the transpiration stream. A high degree of selectivity in uptake by stems is again evident, as illustrated by the great diversity of autoradiographic images obtained when shoots of white lupine are fed through the xylem with different ^{14}C-labeled amino compounds (124) (Fig. 9). Using uniformly labeled ^{14}C forms of normal xylem constituents as radiosubstrates and applying these singly to shoots

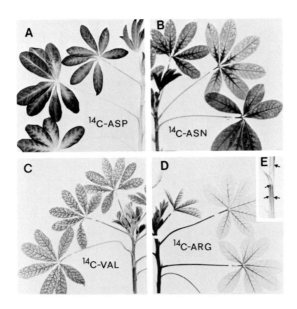

FIG. 9. (A–D) Autoradiographs of shoots of *Lupinus albus* fed via the transpiration stream for 20 min with various ¹⁴C-labeled amino acids dissolved in unlabeled xylem sap. A 20-min chase in unlabeled xylem sap was given before the shoots were harvested. (E) Detail of autoradiograph of part of stem of shoot fed [¹⁴C]valine. Note concentration of absorbed label in leaf traces (arrows). Data from McNeil *et al.* (124).

in diluted xylem sap, arginine is seen to be very effectively absorbed (up to 78% of the applied label) by stems, whereas asparagine, glutamine, valine, and serine are absorbed with only moderate effectiveness. In contrast, the dicarboxylic amino acids aspartate and glutamate largely escape the stem retrieval system, with 80–90% of the ¹⁴C applied as these compounds reaching the leaves and concentrating there in the mesophyll rather than in veins, as with the other amino acids (124). Histoautoradiography of absorbed label in stems of various species has indicated that xylem parenchyma is especially active in uptake from the xylem stream [e.g., in *Pinus* seedlings (214), *Pisum* shoots (165), and in seedlings of *Senecio* (158)]. Departing leaf traces may become intensely labeled when ¹⁴C-labeled amino acids are fed to cut shoots (Fig. 9), suggesting that the transfer cells located in the traces might have a special role in uptake from xylem.

The final class of evidence for lateral uptake of metabolites from xylem comes from experiments in which labeled solutes are percolated through detached segments of stem. The studies have involved tomato (*Lycopersicon*) (12, 13, 47) and apple (*Malus*) (88–91) and have examined the nature of the uptake process for amino compounds and the response of the system to pH and water flux through the xylem. The experiments have suggested that absorption by the stem involves three distinct processes: diffusion of the amino compound through the extracellular space surrounding xylem, exchange of the compound with the fixed ionic groups of adjacent cell walls, and a metabolically implemented accumulation of the compound by cells of the stem. The diffusion and absorption phenomena, essentially physical processes, are shown to be reversible to the extent that percolation, say with strongly acidic solutions, can displace label from the segment. The degree of adsorption is likely to be determined largely by the ionic charge of the introduced solute, cationic forms of nitrogen (e.g., arginine) adsorbing strongly on the predominantly negatively charged wall sites, and anionic forms of nitrogen (e.g., dicarboxylic amino acids) being prevented from doing so by co-ion exclusion (35, 88). Uptake into living cells lining the xylem appears to be determined by membrane-based selectivity, discrimination between D- and L-forms of an amino acid and between related protein and nonprotein amino acids having been demonstrated (12, 47). Entry of the amino acid into metabolically active compartments of the stem is proven from the appearance of new labeled products in the absorbing tissues (165) and from the finding that $^{14}CO_2$ may be recovered from labeled stem segments (47). Studies on the effect of the percolation rate of an amino acid through stem segments have concluded that the uptake mechanism of the stem is likely to remove only a small fraction of the ascending solute under flow conditions and solute concentrations simulating those of a rapidly transpiring shoot (13).

In studies of the uptake of glutamine, alanine, and glutamic acid from xylem of tomato stem segments, van Bel and Hermans (14) have shown that relative uptake rates of different amino acids are altered significantly by pH of the percolation fluid. Increasing the pH from 4.7 to 8.0, for example, increases the uptake ratio of glutamine:glutamic acid from 1.8 to 3.3, whereas an increase of pH from 3.7 to 6.6 leads to a decrease in the uptake ratio for alanine: glutamine from 3.7 to 1.6. Maximum uptake for glutamine, the principal nitrogenous solute of tomato (43, 95), takes place at pH 5.2, a value close to the isoelectric point of this amide and approximating the natural reaction of tomato xylem sap (14). On the other hand, uptake of glutamic acid by tomato stem segments shows a sharp maximum at pH 4.6 and is much reduced at the normal xylem pH 5.2 (14), suggesting that much of the compound might normally escape the stem retrieval system and supply leaves with nitrogen [cf. the situation demonstrated for aspartic acid in transpiring shoots of *Lupinus* (Fig. 9) (124)].

Although much evidence points to a net uptake of metabolites from xylem into stems, circumstances are recorded in which stem tissue appears to release specific metabolites to xylem. This is likely to be a common event during senescence or seasonal mobilization from rays, but it may also take place during the normal daily functioning of relatively young stem tissue. The $^{14}CO_2$ feeding studies on nonsenescent grapevine (*Vitis*) shoots by Hardy (80) support this contention in showing that [^{14}C]glutamine and [^{14}C]malate are released to stem xylem near the point of insertion of the fed leaf but are subsequently reabsorbed higher up the stem and metabolized to other substances. A parallel case has been reported for groundsel (*Senecio vulgaris*) seedlings by Pate *et al.* (158). Nitrate absorbed from the xylem by hypocotyls is apparently reduced in this region, and the reduction products, principally glutamine, are then released back to the xylem for transfer to the seedling shoot.

It is likely that two-way traffic of metabolites exists between the xylem stem tissues of a wide range of plants and that this activity is intimately involved in the processes leading to the progressive emptying and filling of stem segments during the growth of the shoot.

C. EXPORT OF SOLUTES FROM MATURE LEAVES IN PHLOEM

Measurements of net daily gains of CO_2 by leaves and of the amounts of carbon that they lose in respiration and growth have enabled estimates to be made of the rate at which leaves consume and export photosynthate during their lives. Generally, export of photosynthate is not in evidence until the leaf is more than one-third to one-half expanded, and during the subsequent transition to full autotrophy for carbon a period occurs when export and import of carbohydrate may proceed simultaneously (129, 197, 198, 200, 201). Export rate rises rapidly until the leaf is fully expanded, continues at a near maximum rate until the leaf loses an appreciable amount of its chlorophyll and protein, and then declines rapidly as leaf senescence is approached (58, 97). Simple sugars (sucrose and, less commonly, raffinose and stachyose) or sugar alcohols (e.g., sorbitol, mannitol, and dulcitol) are the principal compounds carrying photosynthetically fixed carbon from the leaves of virtually all species studied (48, 73, 149, 222), and, because of the high concentration and osmotic value of these compounds, the rate at which they are synthesized and released to phloem is likely to set the tempo for translocation of other solutes (33, 128).

Not all carbon fixed in photosynthesis is transported as sugars or sugar alcohols. In soybean (*Glycine max*) leaves, for example, serine is translocated in large amount (138), whereas this amino acid, aspartic acid, glycine, and alanine carry up to 10% of the ^{14}C recovered in phloem sap of plants of *Spartium* (167), pea, and lupine (120, 189) after the feeding of $^{14}CO_2$ to photosynthetic surfaces of the shoot. Organic acids may also be shown to be labeled and exported in $^{14}CO_2$ feeding experiments (114, 196), though the proportion of ^{14}C carried in these compounds is usually low compared with that in carbohydrate or amino compounds. Experiments feeding leaves with $^{32}PO_4{}^{3-}$ have demonstrated a variety of labeled products in the resulting phloem sap (17, 108), indicating that incorporation of phosphate into phosphorylated compounds may be a major function of the photosynthesizing leaf. In a similar fashion, feeding of $^{15}NO_3{}^-$ to leafy shoots of *Lupinus* via xylem (168) leads to a range of ^{15}N-labeled amino compounds being exported in phloem, a result in keeping with the presence of a nitrate reductase in leaves of the species (7).

Because uptake of carbon dioxide proceeds side by side with transpirational loss of water through stomata, a positive correlation is to be expected between the daily photosynthetic activity of the leaf and its ability to attract ions and metabolites from the rest of the plant via the xylem.

By comparing the gross composition of incoming xylem fluid with that of the outflowing stream of phloem translocate, important conclusions can be reached concerning the interaction of photosynthesis and xylem intake in leaf metabolism and transport activity. Elements such as phosphorus and sulfur, which are present predominantly in organic form in phloem but largely in inorganic form in xylem, are obviously likely to have been subject to reduction or metabolic incorporation following arrival in the leaf. On the other hand, mineral ions such as K^+, Mg^{2+}, and certain organic solutes of nitrogen that are present in quantity in both xylem and phloem are likely to have passed in unchanged form to phloem, either through the xylem-to-phloem transfer system described earlier (Section II) for the minor veins of leaves or by a longer route via mesophyll. Further elements (e.g., Ca, Mn, Zn, Fe, and B) characteristically present at low level in phloem relative to that in xylem tend to accumulate in leaves during aging as supply in xylem continues to exceed mobilization through phloem (94, 98, 123). For obvious reasons these elements are classified in the literature as "immobile" or "sparingly mobile" in plants (51). Finally, there are those metabolites that are present at high level in phloem, but absent or present at only trace level in xylem. The obvious inference is that the substances in question have been synthesized within the leaf. Falling into this category are nucleotides (109) and B vitamins of phloem sap (218) and the variety of growth-regulating substances recorded for phloem (149). Compounds implicated in the promotion or inhibition of flowering are generally stated to be phloem mobile (38), but it is debatable whether their movement out of leaves is always restricted to functional phloem elements. Photoperiodically induced leaves of *Xanthium,* for instance, synthesize and transmit flower-promoting stimuli when as little as one-eighth fully expanded (185), a stage of development when the leaf is unlikely to be exporting through its phloem.

The relationships in the leaf between the export of sugar and the export in phloem of solutes imported by xylem have received scanty attention. Some indication can be obtained of the synchronization of these processes during the diurnal functioning of a leaf from studies on phloem-bleeding species such as the annual lupines *Lupinus albus* (191) and *L. angustifolius* (93), however. In these plants, as in many other species, pools of sugar and starch build up in the leaves during the day and then provide sucrose for phloem loading during the subsequent night. Because the levels of most mineral elements in the phloem sap rise and fall in phase with changes in sucrose level in the sap, the daily cycle of mineral export from the leaf in phloem is closely tied to carbohydrate export. During the day, when transpirational delivery exceeds phloem export, a buildup of minerals is observed. Then, during the following night when transpiration is at a minimum, these pools of minerals in the leaf are drawn upon for phloem loading (93).

Carbohydrate and amino compounds are quantiatively the most important nutrients exported from leaves, so study of the extent of coupling in export of these carbon- and nitrogen-containing substances during leaf aging is of great importance in understanding the mechanisms whereby leaves release assimilates to roots and heterotrophic parts of the shoot. In plants in which nitrogen is delivered to leaves mainly as nitrate, the nitrate reductase activity of the leaf, the storage of free nitrate in leaf tissues, and the incorporation in the leaf of reduced nitrogen into protein and pools of soluble nitrogen are all likely to be important in determining rates of export of nitrogen via phloem. Unfortunately, however, there are no detailed reports of phloem sap composition for species exhibiting a shoot-dominated pattern of nitrate reduction, so it is not clear whether NO_3^- ever cycles through leaves of such plants or whether only reduced forms of nitrogen are involved in phloem translocation.

In those plants in which a variety of organic solutes carry reduced nitrogen from roots and stems to leaves and in which some free nitrate may reach the leaves through xylem, the interrelationships of photosynthesis and nitrate reduction and of export of photosynthate and nitrogenous solutes become especially complex and difficult to resolve [see studies on *Capsicum* (195)].

Symbiotically fed legumes, relying solely on their nodules for nitrogen, present a somewhat simpler system than in corresponding nitrate-fed plants of the same species, for without complications due to reduction of nitrate, routes and sources of transfer to phloem can be more readily determined (162).

The translocatory system for amino compounds in nodulated lupine (*Lupinus albus*) has been examined in detail in this respect by feeding [14]C- or ^{15}N-labeled amino compounds singly to shoots in dilute xylem sap and subsequently analyzing phloem exudates for labeled metabolites (8, 190). Combining data from such studies with information on the labeling of phloem sap following feeding of $^{14}CO_2$ to leaves, a quantitative picture has been built up of the source agencies and metabolic conversions involved in loading the translocation stream (149). In the scheme that has been proposed (Fig. 10), the amino compounds involved in the transfer reactions are grouped according to the phloem compounds to which they donate carbon, and relative magnitudes are indicated for the donation of carbon each amino compound makes to phloem through photosynthesis and xylem-to-phloem transfer. The scheme for lupine highlights the importance of direct transfer from xylem in providing asparagine, glutamine, threonine, and valine for phloem loading, the ineffective transfer of certain amino compounds (e.g., aspartic acid) from xylem to phloem, and the importance of photosynthesis in generating exportable quantities of carbon in the form of amino compounds such as serine, alanine, aspartic acid, glutamic acid, and glycine. Judging from the overall complexity of the flow patterns shown in Fig. 10, one would suspect that it will prove difficult to pinpoint the regulatory agencies controlling nitrogen flow within the lupine system.

As well as the sources of amino compounds just discussed, turnover of protein throughout the mature life of the leaf and net loss of soluble nitrogen and protein during leaf senescence may donate nitrogen to the translocatory process of the leaf. Isotopic studies of these processes are particularly difficult, so one must rely for the present on evidence from changes in phloem sap composition with leaf age to assess the possible significance of net mobilization of nitrogen from leaves in phloem translocation of nitrogen. In both woody (172, 217, 221) and herbaceous species (162), the onset of leaf senescence signals a massive mobilization of nitrogen from leaves, as evidenced by a rise at this time in absolute levels of amino compounds in phloem exudates and a concommitant

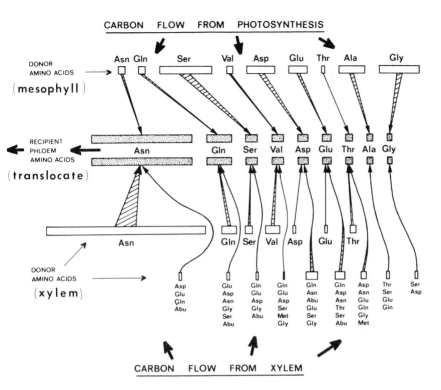

FIG. 10. Carbon flow from photosynthesis and from xylem amino acids to the amino compounds of the phloem stream passing from leaves to fruits of *Lupinus albus*. The extent of traffic of carbon via each amino acid is indicated by the width of the rectangles representing that amino acid in the upper and lower horizontal rows of the figure. Rectangles representing amino acids of phloem (central block of figure) are drawn in proportion to their abundance on a carbon basis. Redrawn and modified from Pate (149).

lowering of the C/N ratio of the export stream of the leaf. Balance sheets for nitrogen in annual grain legumes suggest substantial benefit to developing fruits and filling seeds from the nitrogen released from leaves (164), so the overall efficiency of the mobilization process is likely to have considerable impact on plant productivity. The timing of mobilization would also be important in view of the deleterious effect that nitrogen loss from leaves is likely to have on photosynthetic performance. Indeed, authors such as Sinclair and de Wit (193) have termed high-protein seed crops, such as grain legumes, as potentially self-destructive in view of the high demand for nitrogen from vegetative parts of the plant to satisfy the seed-filling process.

V. Case Histories of Transport Activity in Organ Functioning

Each part of the plant has special arrangements to facilitate the exchange of metabolites with the rest of the plant, and knowledge of the quantities and types of metabolites transported and of their relevance to an organ's function provides information essential to a fuller understanding of the allocation systems for assimilates within a whole plant. In this section two contrasting plant structures will be selected—the root nodule of a legume and a developing fruit—and these used to illustrate the relevance of transport exchanges in growth, nutrition, and functioning of plant organs.

A. Transport and the Functioning of Nitrogen-Fixing Nodules of Legumes

Nodules function by importing sugar from the plant shoot via phloem and exporting nitrogen-rich metabolic products of nitrogen fixation back to the shoot via xylem.

Considerable information is available on the nodule's economy of carbon and nitrogen (4, 164). Data for a variety of species of annual grain legumes suggest that the major share (40–55%) of the carbohydrate imported by the nodule acts as source of carbon skeletons for formation of fixation products, a much smaller fraction (9–20%) is built into dry matter during nodule growth, and the remainder (30–40%) can be accounted for as CO_2 released in nodule respiration (164). Consumption by these three classes of metabolism totals the equivalent of 4–10 mg of carbon (10–25 mg carbohydrate) per milligram of nitrogen fixed by the nodule (117, 151), and, depending on the host species and the stage of growth at which it has been studied, the population of nodules on a plant consumes the equivalent of 10–30% of the current net photosynthate of the plant. Bearing in mind that nodules usually constitute only a few percent of the fresh weight of an annual legume (87, 142, 159), it becomes clear that these highly specialized organs engage in particularly intensive exchange of metabolites with the host plant.

The vascular system of the nodule consists of a network of vascular strands surrounding a central core of bacterial tissue. Each strand consists of a central cylinder of xylem and phloem conducting elements, a concentric ring of pericycle cells [which is some species are modified into transfer cells (157)], and an outer endodermal layer with well-defined Casparian thickenings. The endodermis effectively separates the apoplast (extracellular compartment) of the vascular tissue from the apoplast of the surrounding cortex and bacterial tissue.

A model of functioning of the nodule's export system (156) envisages a twofold role for the pericycle cells (transfer cells) of the nodule, first in ferrying sugars from phloem to the bacterial tissue of the nodule via the nodule symplast and, second, in the secretion of amino compounds to the apoplast of the vascular strands of the nodule. By lowering the water potential of the bundle apoplast with nitrogenous solutes and thereby attracting water osmotically across the endodermal boundary to the vascular strands, the secretory process of pericycle tissue motivates export from the nodule. As a result, the amino compounds resulting from nitrogen fixation are flushed out of the nodule through the xylem connection with the host plant. Evidence in favor of an osmotically operated transport system comes from the observation that if freshly detached nodules are placed on damp filter paper they will bleed a fluid from the cut proximal ends of the xylem region of their vascular tissue. The bleeding fluid carries nitrogenous solutes at concentrations several times higher than in the donor bacterial tissue (157), implying that an "active" secretory process is involved in the release of solutes from the pericycle cells.

A special feature of the model of functioning just described is the economy of use of water in export from the nodule. Assuming that fixation products leave the nodule in the xylem at the same concentration observed in bleeding sap of detached nodules (i.e., 1.5% w/v), and assuming that mass flow of sugars occurs into the nodule as a 14% w/v solution, it has been calculated that nodules of pea fixing nitrogen with an efficiency of 10 mg of carbohydrate per milligram of nitrogen fixed would acquire approximately one-fifth of their water requirement for export by intake through phloem (131). The remainder would presumably come from surface absorption or, as indicated by the experiments of Sprent (194) by lateral abstraction of water from the parent root. In any event the amount of water utilized by nodules is small compared with that moving through the root. In nodulated pea roots, for example, the flux of water through the xylem of nodules is estimated to be only 7% of that moving up through the xylem of the whole root system.

Many authors have concluded that legume root nodules possess high rates of respiration, although most of the experimental estimates of respiration have been made indirectly or by means of comparisons of the specific activity of respiration of freshly detached roots and nodules (5). A system has been devised in which cuvettes are enclosed around clusters of attached nodules of intact plants raised in water culture, and a continuous recording is made of the amounts of CO_2 and H_2 that they

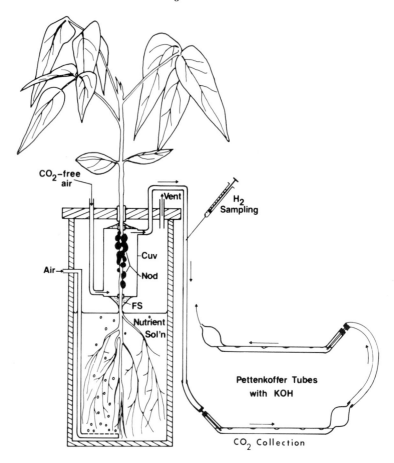

FIG. 11. Cuvette system for studying CO_2 and H_2 evolution by the attached nodules of a legume. Cuv, Cuvette; Nod, nodule cluster; FS, sealing compound attaching cuvette to root system. Modified from Layzell *et al.* (117).

evolve as they fix nitrogen (117) (Fig. 11). It has proved possible to compare the carbon economy of nodules of various host–*Rhizobium* associations, either on the basis of such experimentally obtained data and carbon utilization in transport and growth or on the basis of theoretical costings for specific elements of nodule functioning, especially hydrogen evolution and nitrogen fixation by nitrogenase. The theoretical treatments carry the basic assumptions that nodules utilize sugar by oxidative phosphorylation, yielding 6 ATP per CO_2 (mole/mole), that the stoichiometry of ATP utilization during reduction of N_2 to NH_3 or $2 H^+$ to H_2 involves an activation value of 4 ATP per $2 e^-$, and that during recycling of H_2 by hydrogenase a yield of 3 ATP is obtained per H_2 converted to 2 H^+. The costs of synthesis of organic fixation products from NH_3 are computed on the basis that rates of synthesis of amino acids, amides, and ureides are in the proportions suggested from analysis of nodule-bleeding sap and that the metabolic routes involved in synthesis of these compounds are as suggested by Atkins *et al.* (5). Finally, the ATP equivalents of the growth and maintenance components of respiration of the nodule are estimated using the values for plant tissues suggested by Penning de Vries (173).

The analyses require knowledge of the C/N ratios of imported phloem translocate and exported xylem fluid, and it is assumed that nitrogen carried into the nodule as phloem translocate is incorporated into nodule dry matter, whereas the remainder of the nodule's nitrogen requirement for growth comes from direct incorporation of fixed products.

A scheme (Fig. 12) incorporating experimentally obtained data is shown for nodules of cowpea (*Vigna unguiculata*) and lupine (*Lupinus albus*) associations (117). The comparison between the associations is expressed in terms of the utilization by the nodule of 100 units by weight of carbon as carbohydrate from the host plant. The data are expressed in terms of units by weight of N_2 fixed, or H_2 and CO_2 evolved, and of nitrogen and carbon exported from the nodule or incorporated into nodule dry matter.

The data indicate that nodules of the cowpea association are more efficient in terms of milligrams of carbon used per milligram of N_2 fixed than are those of the lupine association, and that the use of ureides (C/N = 1) by cowpea rather than the amides (C/N = 2 or 2.5) exported by lupine is a principal factor in the better economy of carbon in the cowpea nodule. However, a major difference between the two symbioses is also evident in amounts of CO_2 respired per unit of nitrogen fixed, and the scheme of Fig. 12 carries the suggestion that this difference might be explained in terms of ability of the nodules to fix respired CO_2. Figure 12 suggests processes of nodule functioning that might consume ATP and thus affect output of respired CO_2, including items for transport, ammonia assimilation, maintenance respiration, and reactions associated with the nitrogenase and hydrogenase systems of the nodule. Differences in any one or more of these items might also be responsible for the observed differences in carbon economy of the two associations.

The suggestion that nodules might conserve carbon by carboxylation reactions has support from the demonstration of active PEP carboxylase activity in nodule extracts and of the ability by detached nodules of several legumes to fix $^{14}CO_2$ applied to their gaseous environment (210). In the case of the cowpea and white lupine symbioses used in the

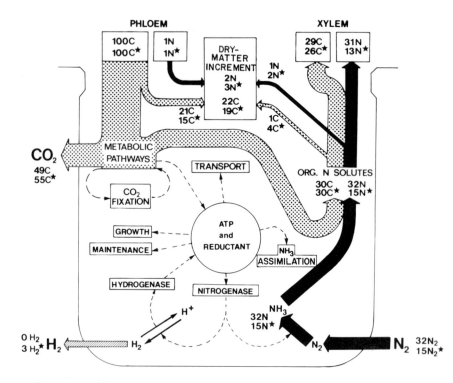

Fig. 12. Partitioning and utilization of translocated carbon by nitrogen by fixing nodules of two legume associations: Cowpea–*Rhizobium* CB756 and White lupine–*Rhizobium* WU425. All budget items are expressed on a weight basis and in terms of a net intake by the nodule of 100 mg of carbon as translocated carbohydrate. Experimental data on carbon and nitrogen economy are given as weights (in milligrams) of carbon and nitrogen transported or consumed by the nodule. Values for the cowpea association are given above those for the lupine association (values for carbon and nitrogen in lupine marked by a star). The scheme includes items of the nodule's budget (dotted lines), which, on theoretical grounds, would have consumed or produced ATP and reductant during nodule functioning and would therefore have had bearing on the observed CO_2 output of the nodule. The possibility of fixation of respired CO_2 by the carboxylase systems of the nodules is suggested. Redrawn from data in Layzell *et al.* (117).

study summarized in Fig. 12, PEP carboxylase and $^{14}CO_2$ fixation activities were considerably higher in the cowpea nodules than in lupine nodules (117), a finding in accord with the difference in observed CO_2 output of the two associations.

Despite these strategies for economy in use of carbohydrate in nitrogen fixation, the high rates of metabolism of nodules incurs a relatively high rate of consumption of photosynthate compared with that of most other plant organs, and because nodules of most legumes do not engage in extensive storage of carbohydrate, their fixation of nitrogen is unavoidably susceptible to restriction in availability of carbohydrate to the nodules. However, in certain species (e.g., pea) pools of sugar and starch build up in nodules during the day, and these are utilized to supplement a dwindling supply of translocate to the nodule at night (132). In other species (e.g., lupine) the starch reserves of the nodule tend to be of greater size and capable of sustaining several days of N_2 fixation in the absence of photosynthate from the shoot (J. S. Pate, unpublished). Even with such reserves the link between fixation and current availability of translocated sugar must still remain a dominant pace-setting relationship in nodule functioning. So, when seeking the influences that control the rate at which translocate is made available to or is attracted to the nodule, it may become possible to discover how symbiotic functioning is regulated within the general framework of resource allocation within the plant.

B. TRANSPORT OF METABOLITES TO FRUITS

Regardless of their morphology and whether they form above or below ground, all fruits are heavily dependent on their parent plant for metabolites and are relatively ineffective in transpiration and in the associated attraction of xylem-borne solutes. These features and the fact that seeds accumulate high concentrations of ergastic substances relative to their contents of dry matter and water lead to the conclusion that phloem, not xylem, is the principal avenue for import of inorganic and organic solutes by the fruit (21, 32, 129, 137). Experimental data supporting this generalization are not hard to find.

Possibly the most convincing evidence of a phloem-based nutrition of fruits comes from the resemblances to be found between composition of phloem exudates and of the recipient fruits. The studies of van Die (44), for example, on fruits of the palms *Arenga saccharifera, Cocos nucifera,* and *Phoenix dactylifera* show that the nutrient balance of phloem exudates from severed inflorescences matched closely the hypothetical nutrient fluid required to implement the gains by the fruit of dry matter, total ash, K, Mg, P, and N. A notable exception, however, is calcium, which builds up in fruits to a level 10 times that expected from delivery through phloem. This calcium is mainly located in the vascular bundles of the husk and is therefore likely to have entered through xylem (44).

The close fit between dry-matter levels in the tissues of a swelling fruit and the dry-matter content of phloem sap has been generally interpreted as evidence that mass flow in phloem might supply a fruit with all the water incorporated into tissues during growth and that any additional water carried into the fruit through xylem and phloem would be lost in transpiration or, in the case of undergound fruits, return to the soil via the cuticle or periderm of the fruit (32, 41). Alternatively, as suggested by Münch (137) and Ziegler (216, 217) excess water entering through phloem might leave the fruit through xylem, possibly carrying with it certain solutes back to the parent plant.

It is possible, of course, that transport exchanges between fruit and parent plant are bidirectional, because certain xylem and phloem strands might be engaged in export from the fruit, whereas others import metabolites in the generally accepted fashion. A system of this nature has been suggested for xylem transport in fruits of *Phaseolus vulgaris* (134) and, under certain circumstances, in respect of phloem transport in tomato fruits, according to the ^{14}C-labeling studies of Walker *et al.* (205).

Further evidence of nutrition predominantly via phloem comes from the low ratios for Ca/K and Na/K in fruits, especially in underground fruits (217). Values for these ratios in fruits are characteristically closer to those of phloem sap than of xylem. Nevertheless, xylem transport may still play a highly significant role in the delivery of certain minerals, as, for example, the elements Ca, Mn, Fe, Cu, and Na in fruits of *Lupinus albus* (160). In this and other species there is evidence of intake of Ca, Mg, and Mn by routes other than conventional transport through xylem and phloem (22, 44), the classic instance of a nonvascular component in mineral supply being for the underground fruits of peanut (*Arachis hypogaea*). Development of fruits of this species requires absorption of calcium from the rooting medium by the surface of the fruit as a supplement to the meager levels of calcium supplied through phloem and xylem of the fruit stalk (19). In aboveground fruits, reduction of transpiration can induce deficiency of elements supplied mainly through xylem (21, 212), although increasing the level of the deficient element in the rooting medium can offset any reduction in transpiration by elevating levels in the xylem, thereby increasing the proportional intake by the fruit through this route [e.g., see study on manganese nutrition of lupine fruits (94)].

A principal aim of studies of nutrition of fruits is to identify the sources of carbon for fruit development, especially the photosynthetic surfaces involved and the relative contributions they make to the fruits during growth. A commonly employed technique is to use $^{14}CO_2$ feeding studies to identify the source–sink routings for photosynthate and to assess the proportion of exported photosynthate donated from each sink to specific fruits, matching this information with the carbon economy of sink and source organs as determined by measurements of their dry-matter gains or losses during growth and their net exchanges of carbon in photosynthesis and respiration.

The most fully documented carbon budgets undoubtedly relate to the fruiting of cereals, especially wheat (*Triticum*). As mentioned in an earlier section, the flag leaf and photosynthetic surfaces of the inflorescence of cereals comprise major sources of current photosynthate to developing grains. A second related source is the soluble carbohydrate reserves of the culms, particularly the internodes of inflorescence stalks, these being a sink for assimilates early in fruiting and then a source of assimilates for the fruit heads once the grain-filling process gathers momentum (10, 59, 171, 182). According to various authors, from 2 to as much as 70% of the carbon utilized by the grain may come from carbohydrate reserves of stems, the process assuming particular importance under conditions of drought, when the flag leaves may fail to photosynthesize effectively. A final source of assimilates for grain filling comes from mobilization of metabolites from lower leaves and stems and non-reproductive tillers. These appear to be responsible particularly for mobilizing nitrogen to the grain, especially under low-fertility conditions (53).

Similar studies on the carbon economy of developing green fruits of legumes have also demonstrated the importance of postanthesis assimilates for fruit and seed development. The main sources of photosynthate for these organs have been outlined in an earlier section. Green legume fruits have the potential to make effective contributions to their carbon economy through photosynthesis of their carpel walls [see studies of *Phaseolus vulgaris* (42), *Glycine max* (177), and *Pisum sativum* (57)]. Our experimental species, *Lupinus albus*, is somewhat exceptional in this regard, photosynthesizing by means of its seeds as well as its pod wall (4). Much of the photosynthetic activity of the legume fruit is apparently directed toward fixation of CO_2 escaping to the fruit gas space from the respiring pod wall and seeds. Certain structural features facilitate conservation of respired CO_2, notably the relatively high translucence of the pod wall, the poor ventilation of the pod and an attendant buildup of high levels of CO_2 (0.1–2% v/v) in the fruit gas cavity (57, 84), and the presence of chloroplasts in the mesocarp and inner epidermis of the fruit (161) and in the developing seeds (4). The ribulosebisphosphate (RuBP) carboxylase and PEP carboxylase activity of pods and seeds has been associated with photoassimilation of CO_2 by the fruit (6, 85).

The significance of respiratory and photosynthetic exchanges between fruit tissues and the external and internal atmosphere of the fruit has been assessed during fruit development of *Pisum sativum* and related to the overall economy of carbon in the fruit (Fig. 13) (57). During the first half of development (0–18 days after anthesis) net photosynthetic gain of carbon from the external atmosphere is as great as is the assimilation of CO_2 from the pod interior, but in the second half of development (18–36 days after anthesis) recycling of endogenously produced CO_2 proves much more important than net gain of carbon from the outside atmosphere. Assuming that pod respiration in the day equals that at night, the estimated photosynthetic performance of the pod over the period 0–18 days is 10.8 units by weight of carbon (as CO_2) from the pod's respiration—3.7 units of carbon from the respiring seeds and 4.6 units of carbon from the outside atmosphere (Fig. 13A)—a total conservation of 19.1 units of carbon per 100 units of carbon imported from the parent plant. Over the period 18–36 days these values changed to 5.5, 8.6, and 1, respectively (Fig. 13B), a photosynthetic contribution of 15.1 units of carbon for every 100 units of carbon imported from the parent-plant. Scoring these values in proportion to the net carbon intake for the two periods of growth, photosynthetic activity is estimated to benefit a fully illuminated fruit to the extent that it requires 16% less translocate than if laying down an equal amount of dry matter in continuous darkness. Adequate illumination is clearly essential for legume fruits to function at full efficiency, and the same is likely to apply to green fruits of a wide range of other plants.

A more comprehensive assessment of the nutrition of a developing legume fruit has proved possible from studies on our experimental species, *Lupinus albus*, taking advantage of the ability to monitor the levels of translocated solutes moving into the fruit in xylem and phloem. Combining this information with data on CO_2 exchange, transpiration, and growth increments of carbon and nitrogen in the fruit, a model has been constructed (Fig. 14) to account quantitatively for transport exchanges in xylem and phloem and utilization of carbon, nitrogen, and water during fruit development (166). The model suggests that for every 100 units of carbon imported from the parent plant by a lupine fruit, 52 end up in seed dry matter, 37 in nonmobilizable material (largely cellulose and lignin of the pod walls), and the remaining 11 units are lost as CO_2 to the atmosphere. Phloem supplies virtually all (98%) of the carbon entering the fruit from the parent plant and 89% of the nitrogen. Xylem supplies the remaining carbon and nitrogen and 60% of the fruit's total water intake. Xylem transport is of greater prominence in fruit nutrition in early growth than later, when the seeds start to lay down storage reserves. It would be interesting to know how the fruits of other species compare in these respects.

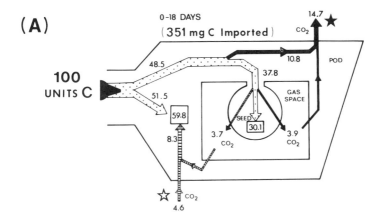

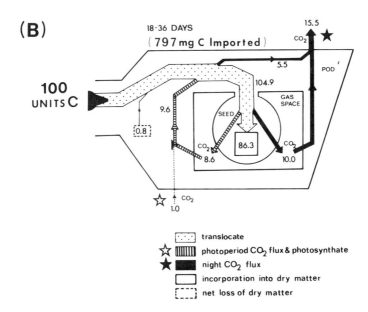

FIG. 13. Economy of carbon in a developing fruit of *Pisum sativum* (cv. Greenfeast). Items of the fruit's carbon budget are expressed relative to an intake of 100 units by weight of carbon as translocate through the fruit stalk. Two stages in fruit development are considered, 0–18 (A) and 18–36 (B) days after anthesis, the fruit being fully ripe by day 36. Amounts of carbon imported during each of these stages are indicated. Redrawn with modifications from Flinn *et al.* (57).

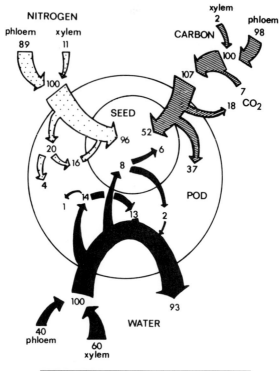

FIG. 14. Economy of carbon, nitrogen, and water in a developing fruit of *Lupinus albus*.
Proportional intake (by weight: C:N:H₂O = 12:1:600) of these commodities through
xylem and phloem is indicated, and the overall relationships between weight of imports is
shown. Transpiration ratio (22.5 ml/g) refers to the amount (in milliliters) of water transpired per gram of fruit dry matter produced. Import ratios (w/w) for sucrose and amino
compounds in xylem and phloem are detailed in the tabulation. All data refer to the
complete (12-week) growth cycle of the fruit. Redrawn from data in Pate *et al.* (166).

Sucrose carries 90% of the carbon utilized by fruits of white lupine,
the amide asparagine 50–70% of the nitrogen (152). Sucrose is therefore likely to act as primary source of synthesis of the carbohydrate, oil,
and fibrous components of the seeds, asparagine the nitrogen source for
protein synthesis. A study of the relationship between intake and utilization of asparagine during fruit growth by Atkins *et al.* (8) has shown that

rates of supply of the compound greatly exceed direct incorporation of the amide into protein and any temporary storage of the amide that may occur in pod and seeds during fruit development. Depending on the age of the fruit, from 44 to 88% of the entering asparagine is metabolized, and an asparaginase demonstrated in crude extracts of seeds is implicated in utilization of the amide prior to its amide nitrogen being incorporated into other nitrogenous compounds. Feeding of fruiting shoots with [U-^{14}C, amide-^{15}N]asparagine through the transpiration stream shows that virtually all of the applied label enters the fruit as unmetabolized asparagine. In young seeds the ^{14}C is traced largely to soluble, nonamino compounds, the ^{15}N to the ammonia, glutamine, and alanine of the endospermic fluid, whereas in older seeds the carbon and nitrogen of the asparagine is contributed to a variety of amino acid residues of seed protein and to other compounds of the seed.

Several investigators of fruit nutrition have attempted to identify the mode of origin of the major metabolites donated to fruits in phloem and xylem. As might be expected, the most definitive results have come from species whose phloem yields an exudate, especially where phloem sap has been analyzed for isotopically labeled solutes after feeding an appropriately labeled source to donor organs of the plant. The general picture emerging for photosynthate contributions is that some 90% of the fixed carbon travels to fruits as carbohydrate, the remainder being attached to amino acids and organic acids [e.g., see data on *Yucca flaccida* (45), *Ricinus communis* (72), *Pisum sativum* (120), and *Lupinus albus* (167)]. A similar proportioning of photosynthetically fixed ^{14}C has been demonstrated in other species [e.g., *Vicia faba* (107) and tomato (204)], from which it is not possible to collect phloem sap but in which whole fruit stalks have been analyzed for labeled assimilates. In these instances metabolic products of nonvascular tissues might have been extracted, so identification of solutes in transit to fruits becomes difficult.

Where it has been possible to measure the specific activity of solutes of phloem sap after ^{14}CO$_2$ application to a source leaf, it has been shown [e.g., for *Pisum sativum* (120) and *Yucca flaccida* (48)] that the specific activity of the sucrose of the exudate is consistently higher than that of other metabolites. The time courses of labeling of inflorescence-stalk phloem sap of *Yucca* reported by van Die and Tammes (46) are particularly instructive in showing that although the specific activities of the sucrose and amino fractions of the phloem exudate reach maxima at the same time after application of ^{14}CO$_2$ to the source leaves, the intensity of labeling of amino acid carbon reaches a value of only 5–12% of that of sucrose. Later in the time course, specific activity of the carbon in nitrogenous solutes falls less rapidly than that of sucrose, suggesting a rela-

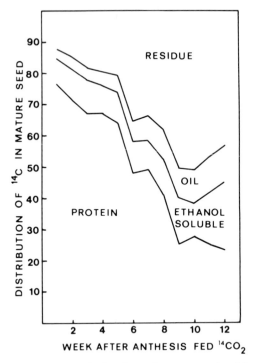

FIG. 15. Labeling of major organic fractions of mature seeds of *Lupinus angustifolius* after feeding a pulse of $^{14}CO_2$ to vegetative parts of the plant at different times after anthesis of the plant's primary inflorescence. Data of J. S. Pate, C. A. Atkins, and M. W. Perry (unpublished).

tively slow release of labeled amino compounds from the source leaf, possibly as a result of protein turnover. In our $^{14}CO_2$ labeling studies of *Lupinus albus,* a similar picture has emerged, but one that is complicated by marked differences in specific activity of different amino acids. The compounds produced photosynthetically in the leaf (e.g., valine, glycine, serine, and alanine) tend to achieve much higher specific activities than those (e.g., amides) cycling through leaves via the root system (see Section IV,C and Fig. 10).

Few attempts have been made to exploit species that bleed from phloem to identify which assimilates donate the carbon of preanthesis photosynthate to the fruit. However, a study of distribution of ^{14}C in phloem sap and fruits of *Lupinus angustifolius* after feeding pulses of $^{14}CO_2$ has shown that the earlier the $^{14}CO_2$ pulse is fed to the plant during its life cycle, the higher the proportion of ^{14}C in phloem sap carried by metabolites other than sucrose. The final labeling patterns of seed components reflect this effect by showing higher proportions of ^{14}C recovered attached to protein in plants fed $^{14}CO_2$ early in plant growth and much lower proportions in the case of ^{14}C fixed during fruiting (Fig. 15). Nitrogenous solutes are therefore likely to be responsible for mobilizing significant amounts of preanthesis carbon to fruits, a feature reflecting the ability of the plant to mobilize the carbon and nitrogen from the protein of its vegetative parts to a higher degree than the carbon of carbohydrate-based materials, most of which will be embodied in the vegetative organs. A similar conclusion is evident from the $^{14}CO_2$ and $^{15}NO_3^-$ labeling studies of peas (153). The ^{14}C and ^{15}N assimilated during fruiting is translocated to fruits from vegetative parts with equally high (74–76%) efficiency, but for assimilates formed before flowering, efficiencies of transfer are 51% for ^{15}N and only 2% for ^{14}C. Accordingly, the nitrogen assimilated before flowering of pea provides approximately one-fifth of the seeds' requirements for nitrogen, whereas the carbon fixed before flowering provides less than 3% of the seeds' intake of carbon (154).

VI. Modeling the Transport and Utilization of Carbon- and Nitrogen-Containing Metabolites in a Whole Plant

Before attempting to define the factors that might regulate metabolite distribution in a plant, it is desirable that information be made available on the amounts of various solutes moving between different regions of the plant and for assessments to be made of the extent to which xylem and phloem participate in such traffic. Earlier sections of the chapter, using nodulated white lupine as an experimental system, have indicated how transport exchanges can be modeled for individual plant organs such as root nodules, fruits, and leaves. In this section, essentially the same procedure will be adopted for describing the flow patterns within whole plants.

A. THE MODELING TECHNIQUE

The basic data required for construction of empirically based models of plant transport are measurements of carbon and nitrogen increments in dry matter of plant parts, C/N weight ratios of the xylem and phloem fluids serving specific plant organs, and information on the net exchanges of carbon as CO_2 between plant parts and the surrounding atmosphere. The form of model currently used for white lupine separates the plant into belowground parts (root and nodules), mature shoot axis (stem and petioles), leaflets, reproductive parts (if present), and apical vegetative parts consisting of leaves and shoot segments not self-supporting in terms of photosynthesis. The set of transport fluids examined is used to monitor exchange between these regions of the plant: root-bleeding (xylem) sap to sample solutes released from roots; petiole phloem sap to analyze the stream of translocate leaving leaflets; and fruit- or inflorescence-stalk phloem sap, stem-top phloem sap, and stem-base phloem sap to sample, respectively, the assimilate streams entering reproductive parts, vegetative apices, and roots. The roles played in modeling by these kinds of data are displayed schematically in Fig. 16. By restricting the modeling exercise to relatively short (10-day) intervals in the life of the plant, the results obtained are not unduly complicated by ontogenetic changes in assimilatory activity and partitioning of assimilates, and by making comparisons between models constructed for plants of differing age, nutritional status, or environmental treatment, information can be gained on how the resource management of the plant can become modified by internal or external agency.

Studies of CO_2 exchange in white lupine (159, 166) have indicated that fruits more than 1 week old and leaflets more than 60% expanded are the only parts of the plant deriving net gains of CO_2 from the atmosphere during daytime. Stem and petioles and shoot apices are found to maintain their tissues at or close to CO_2 compensation point during daytime and are therefore judged as relying on leaflets to support night respiration and any gains of carbon made as dry matter. The CO_2 output of belowground parts is monitored by attaching Pettenkoffer assemblies to a gas stream flowing through the rooting medium, the flow rate of this stream being adjusted to achieve an ambient CO_2 level around the enclosed root equal to that of plants raised normally in sand culture (5).

The respiratory output of nodules can be measured using the cuvettes described earlier, so that the carbon consumption of the root's complement of nodules can be costed separately from that of the root. The measured value for night respiration of the whole shoot is allocated to shoot parts on the assumption that CO_2 loss per unit weight of dry matter (maintenance component of respiration) of a plant part is 1.8 times its loss of carbon in the synthesis of a gram of dry matter (growth component of respiration) (82, 83).

Using these assumptions, the net photosynthesis of the leaflets is measured indirectly by summation of the following three items:

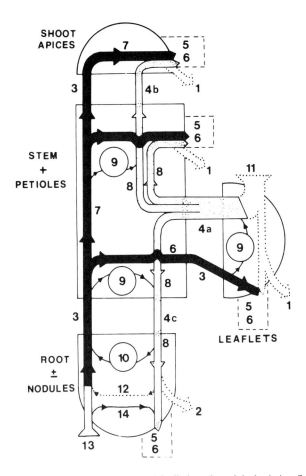

FIG. 16. Data used in construction of empirically based models depicting flow of carbon and nitrogen in white lupine (*Lupinus albus*). J. S. Pate, D. B. Layzell, and C. A. Atkins [unpublished; see also Pate *et al.* (162) for further details of modeling technique]. Primary data: 1, night respiration of shoot; 2, day plus night respiration of root with or without nodules; 3, C/N ratio of xylem sap; 4, C/N ratio of phloem sap (4a, petioles; 4b, stem top; 4c, stem base); 5, carbon increments of plant parts; 6, nitrogen increments of plant parts. Derived data: 7, carbon and nitrogen flow in xylem (■); 8, carbon and nitrogen flow in phloem (▨); 9, xylem-to-phloem transfer; 10, phloem-to-xylem transfer; 11, carbon net photosynthate; 12, carbon for carbon skeletons in nitrogen assimilation; 13, nitrogen uptake–fixation; 14, direct incorporation of assimilated nitrogen by root.

1. Net gain of carbon by the plant in dry matter
2. Respiratory loss of carbon from the shoot at night
3. Respiratory loss from the nodulated root day and night

A typical set of data for C/N weight ratios of transport fluids is shown for the vegetative to early fruiting phases of growth of white lupine in Fig. 17. Xylem delivers metabolites with an average C/N ratio of 1.9, a value close to that of its principal organic solutes asparagine and glutamine. The C/N ratios for phloem vary from 20 to 27 for fruits, 33 to 38 for phloem sap entering vegetative apices of the shoot, 51 to 78 for stem base, and 65 to 127 (consistently the highest values) for phloem exudate from the petioles of leaves. Analysis of these phloem samples has indicated that variations in C/N ratio with sampling location and plant age are almost entirely due to variations in the ratio of sucrose to amides in the phloem sap (162). Samples of phloem sap supplying roots, shoot apices and fruits at all sampling times show lower C/N ratios than that of the translocate currently leaving leaves (petiole phloem sap), and this is shown to be principally the result of an enrichment by stem tissue of the upward- and downward-moving streams of translocate with amino compounds, specially the amides glutamine and asparagine and the amino acids valine, serine, and lysine (162).

Construction of the models involves allocating the net intake of carbon and nitrogen by the plant to various organs and transport activities. It is assumed that transport exchanges take place as mass flow in xylem or phloem, that the solutes moving in phloem and xylem dispense carbon and nitrogen at the C/N ratio of the nearest relevant xylem sap or phloem sap sampling point, and that solutes arriving in a sink organ in xylem or phloem are equally available within that organ for support of growth and respiration. Calculations are then made to determine the mixtures of phloem and xylem streams that meet precisely the recorded amounts of carbon and nitrogen utilized by specific parts of the system (162). Equations (1)–(4) are employed when making calculations relating to the exchange of carbon or nitrogen in xylem or phloem between an organ (A) and the rest of the plant (B):

Carbon flow in phloem from B to A ($C_{\vec{p}}$):

$$C_{\vec{p}} = \left[\frac{(C/N_p)}{(C/N)_p - (C/N)_x} \right] [(\Delta C_A + C_{Res\ A} - C_{Fix\ A})$$
$$+ (C/N)_x(N_{Fix\ A} - \Delta N_A)] \tag{1}$$

Nitrogen flow in phloem from B to A ($N_{\vec{p}}$):

$$N_{\vec{p}} = \frac{C_{\vec{p}}}{(C/N)_p} \tag{2}$$

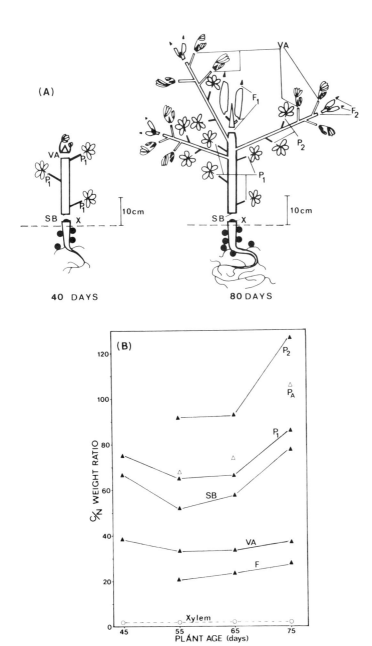

Fig. 17. (A) Sampling points for phloem and xylem sap used in studying the flow of carbon- and nitrogen-containing metabolites in white lupine (*Lupinus albus*). Phloem sap: VA, P, F, SB. Xylam sap: X. (B) C/N weight ratios of solutes of the phloem and xylem sap samples collected at the sites shown in (A). Data of this kind are used in construction of the models shown in Figs. 16, 18, and 19. P_A is the average C/N value for petiole phloem sap samples collected from all leaves of a shoot. Redrawn with modification from Pate *et al.* (162).

Nitrogen flow in xylem from B to A ($N_{\vec{x}}$):

$$N_{\vec{x}} = (\Delta N_A - N_{Fix\ A}) - N_{\vec{p}} \tag{3}$$

Carbon flow in xylem from B to A ($C_{\vec{x}}$):

$$C_{\vec{x}} = N_{\vec{x}}(C/N)_p \tag{4}$$

$(C/N)_x$ and $(C/N)_p$ denote the C/N weight ratios of xylem and phloem transport fluids, respectively, ΔC and ΔN the carbon and nitrogen increments in dry matter, $C_{Res\ A}$ the net respiratory loss of carbon from A, $C_{Fix\ A}$ the net photoperiod gain of carbon by A (applied only to mature leaflets and fruits), and $N_{Fix\ A}$ the direct nitrogen gain through N_2 fixation (relevant only to root nodules).

Values of positive sign denote net import by A (e.g., xylem import of leaflets, phloem intake by nodules, and xylem and phloem intake by vegetative shoot apices and fruits), and negative values denote net export from A (e.g., xylem export of nodules and phloem export of leaflets) (162).

B. Models for Carbon and Nitrogen Flow in
   White Lupine

A detailed model illustrating the flow profiles for carbon and nitrogen in xylem and phloem of a 60- to 70-day-old nodulated white lupine is shown in Fig. 18. Amounts (in milligrams) of carbon and nitrogen transported, produced, or consumed by plant parts are indicated by the pairs of numerical values in the figure. Net intake of nitrogen through dinitrogen fixation in root nodules, carbon losses of plant parts in respiration, and net intake of carbon as net photosynthate by leaves are also included, enabling information to be obtained on the efficiency of utilization of net photosynthate for different purposes during growth.

The form of presentation illustrated in Fig. 18 uses lines of differing thickness to denote the fractions of the plant's assimilated carbon or nitrogen committed for specific purposes during the study interval. Black lines denote flow of nitrogen, stippled lines flow of carbon, and special symbols are used for distinguishing xylem transport, phloem transport, carbon dioxide exchanges with the atmosphere, increments of carbon and nitrogen in dry matter, and xylem-to-phloem transfer of carbon and nitrogen.

To give an indication of the types of information available from the technique, some of the principal predictions arising from the model of

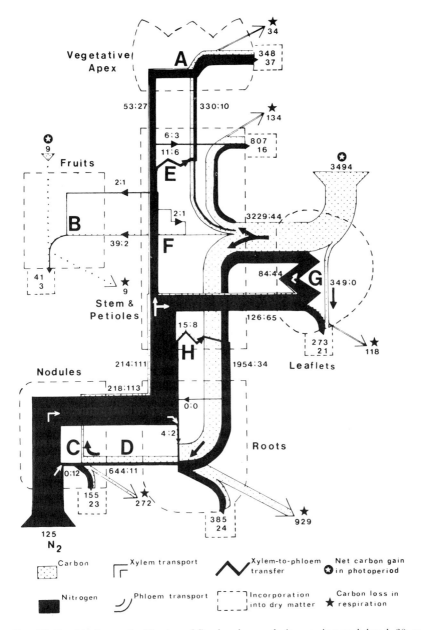

FIG. 18. Partitioning and utilization of fixed carbon and nitrogen in a nodulated, 60- to 70-day-old white lupine (*Lupinus albus*) plant relying solely on root nodules for its supply of nitrogen. The numbers refer to carbon and nitrogen (in milligrams per plant) transported and utilized by plant parts during the 10-day period. Values of carbon appear above or to the left of those for nitrogen. Flow of carbon and nitrogen in the xylem and phloem is indicated by lines of thickness proportional to the fractions of the plant's net photosynthate and fixed nitrogen moving along specific transport pathways. See text for significance of the annotations A to H. Redrawn with modifications from Pate *et al.* (162).

the 60- to 70-day-old nodulated white lupine plant (see items A–H, Fig. 18) are

A. Vegetative apices depend on xylem for 73% of their nitrogen and 14% of their carbon.
B. Fruits acquire 38% of their nitrogen and only 6% of their carbon through the xylem, that is, they are less dependent on xylem than are vegetative apices.
C. Approximately one-third (34%) of the carbon supplied as net photosynthate to the nodule returns to the shoot attached to the amino compounds generated during ammonia assimilation in the nodule.
D. Over one-half (52%) of the nitrogen incorporated into growing nodules results from direct incorporation of fixed nitrogen into nodule tissue. The remainder is supplied with phloem translocate from the shoot.
E. The transfer of nitrogenous solutes from xylem to phloem in upper regions of the stem provides vegetative apices with 60% of the nitrogen they receive through phloem, or with 16% of their total intake of nitrogen (see also Fig. 12).
F. Xylem-to-phloem transfer in the stem provides fruits with half of the nitrogen they import by the phloem, or with the equivalent of one-third of their total intake of nitrogen.
G. The equivalent of 68% of the nitrogen received by leaflets in the xylem is transferred to phloem and exported from the leaflets with photosynthate.
H. Xylem-to-phloem transfer in the lower stem provides the nodulated root with 17% of its net intake of nitrogen.

Viewing the findings of the model in terms of efficiency of operation of the whole plant, a 60- to 70-day-old nodulated lupine plant is seen to convert 57% of the carbon of its net photosynthate into dry matter, losing the remainder as respired carbon dioxide, most (62%) of this loss from the nodulated root. It should be noted that photorespiration of the shoot is not considered in the present modeling exercise, so it is not possible to appraise fully plant efficiency in harnessing photosynthesis to growth and metabolism.

As the plant being considered is a legume, costs relating to the fixation of nitrogen are noteworthy. The equivalent of 18% of the plant's net photosynthate is delivered as translocate to nodules of the 60- to 70-day-old plant, and, on average, 28 mg of carbon as net photosynthate are generated by the leaflets per milligram of nitrogen fixed by the nodules. These values, though relating to only one stage of development, allow one to view in a more general context the detailed costings for nodule processes outlined in Section IV,B.

A logical extension of the modeling exercise has been to compare the patterns of assimilate allocation of the nodulated plant relying on symbiotically fixed nitrogen with the performance of comparably aged, nonnodulated plants relying on nitrate for their nitrogen supply (162). Relevant information is provided in Fig. 19, which also includes a model constructed for a set of similarly aged, nonnodulated plants whose nitrate supply had been withdrawn immediately prior to the 10-day interval during which information for the model was gathered. These plants and the plants continuing to receive nitrate had been raised on a level of nitrate (5 mM NO_3^-) that permitted the plants to take up nitrogen at a rate closely matching that of nodulated plants of the same age. At this level of nitrate in the medium, roots of white lupine reduce 90% of the nitrate they absorb from the rooting medium (7), so the costs of assimilating nitrate by the plants are almost entirely borne by a reductase system linked to the respiration of the roots. At this level of nitrate, also, the plants store less than 1% of their nitrogen as nitrate, so, in the models depicted (Fig. 19), the amount of nitrate assimilated by the plant is equated with the amount of nitrogen absorbed by the plant during the growth interval (7, 162).

The models (Fig. 19) indicate that the transport exchanges of nitrate-fed plants match closely those of nodulated plants assimilating nitrogen at a closely similar rate, the only major difference between the two forms of nutrition being the lower respiratory loss from belowground parts of nitrate-fed than from the nodulated roots of symbiotically fed plants. Stems of nitrate-fed plants also show a somewhat greater capacity to abstract nitrogen from the ascending xylem stream, so that a lesser proportion of assimilated nitrogen cycles through leaves of the nitrate-fed plants than of the symbiotic plants.

The model illustrating plant behaviour after withdrawal of nitrate (see 5 → 0 mM NO_3^- plants, Fig. 19C) shows that the resulting reduction in nitrogen assimilation is not accompanied by a matching reduction in net photosynthesis, so that the C/N ratio of the dry matter laid down in these starved plants is much higher than that of plants continuing to receive an adequate supply of nitrogen. More important, the pattern of distribution of nitrogen is altered radically by incipient nitrogen deficiency, roots receiving the equivalent of 62% of the nitrogen increment effected by the starved plants, compared with only 36% by plants continuing to receive nitrate, or 28% in the case of nodulated plants (see Fig. 19). Most of the nitrogen donated to roots of the nitrate-starved plants is likely to come from mobilization from lower leaves, because these are the main

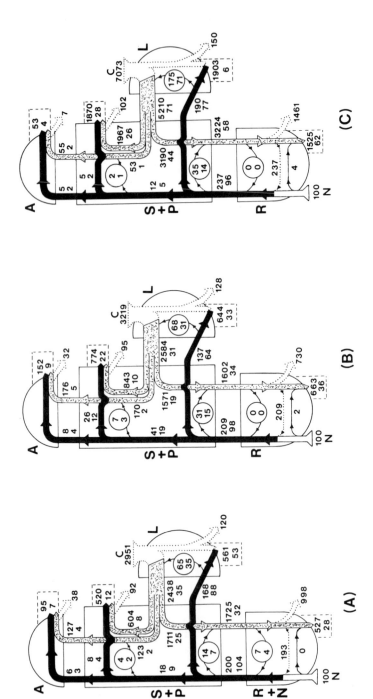

Fig. 19. Models for flow and utilization of carbon and nitrogen in plants of white lupine (*Lupinus albus*). (A) Nodulated plants relying solely on nodules for nitrogen supply; 52–62 days; nitrogen fixed = 83.5 mg of nitrogen; net photosynthate = 2463 mg of carbon. (B) Nonnodulated plants with 5 m*M* NO$_3^-$ as source of nitrogen. 55–65 days. Nitrate assimilated = 85.5 mg of nitrogen; net photosynthate = 2750 mg of carbon. (C) Nonnodulated plants grown for 55 days on 5 m*M* NO$_3^-$ and then (55–65 days) fed with culture solution containing no nitrate. 55–65 days. Nitrate assimilated = 30.2 g of nitrogen; net photosynthate = 2136 mg of carbon. Data from J. S. Pate, D. B. Layzell, and C. A. Atkins (1979).

leaves supplying the roots and are the first to yellow during nitrogen deficinecy. Upper leaves, in contrast, tend to retain their nitrogen but become less active in translocating assimilates (J. S. Pate, unpublished), so that vegetative apices, principal sinks for assimilates from upper leaves, are likely to be particularly affected during nitrogen deficiency. This is borne out by the values shown in the model, less than 1% of the plant's net photosynthate and 4% of its nitrogen increment reaching apical parts of the shoot during nitrate starvation, compared with values of 6 and 9%, respectively, for plants still receiving nitrate.

C. General Predictions from the Models and Future Prospects for the Technique

An instructive feature of the modeling exercise is the light it sheds on the relationships between mature parts of the shoot and the growing organs to which they donate carbon- and nitrogen-containing assimilates. Nitrogen delivered from roots is acquired initially mainly by stem and fully expanded leaflets, stem tissue presumably gaining significant amounts of this nitrogen by lateral uptake from xylem (see Section III), leaves largely through their transpirational activity. Transfer to phloem is pictured as making much of this nitrogen available for translocation to other growing regions that are weakly active in transpiration. Nitrogen cycled through leaflets is allocated essentially in proportion to the amounts of photosynthate exported to various sink organs from the leaves, whereas the nitrogen abstracted by stems is envisaged as being loaded onto the passing phloem streams in a manner that is to an extent independent of the rates of photosynthate flow. By enriching significantly the various streams of translocate with nitrogen, the stem is shown to have the potential to alter differentially the C/N ratio of phloem translocate to different organs and thus provide certain sink organs, such as fruits and vegetative apices, with a supply of nitrogen relative to carbon far in excess of that which is possible from their direct xylem intake or consumption of photosynthate.

What is not known at this stage is how the systems for cycling of nitrogen from leaves interact during growth with the stem exchange systems for nitrogen. Mutual shading of leaves in a developing foliar canopy, for example, is likely to induce changes in rates of transpiration and in the source and sink activities of different strata and types of shoot organs, and it would be particularly instructive to determine how the rationing systems of nitrogen in stem and leaf respond to such changes. Judging from the high degree of constancy of protein (nitrogen) level in seeds of lupine grown under a wide range of contrasting nutritional and environmental influences (J. S. Pate, unpublished), these translocatory systems of the vegetative plant collectively provide the developing fruit with highly standardized mixtures of metabolites at each stage of its development (163, 166). It is impossible to evaluate how this is accomplished until a more sophisticated series of models is developed incorporating values for carbon and nitrogen flow from different age groups of leaves and stem segments.

The heavy dependence of roots on nitrogen cycled through the shoot system is an important aspect of the models for white lupine, and, judging from isotope studies on other legumes (140, 153), this may prove to be a widespread feature of legume transport systems. Both nodules and roots benefit from phloem-borne nitrogen, nodules being pictured as receiving half, roots upward of 90% of their nitrogen as translocate from the shoot. Indeed, in certain situations (e.g., Fig. 19A) the nitrogen supplied to the root as phloem translocate exceeds the nitrogen increment of the root, and this extra nitrogen is then pictured as being spilled back over to the shoot in the xylem (Fig. 19A) or being excreted, as suggested from studies on legumes demonstrating release of fixed nitrogen to the rooting medium (150).

It remains to be seen whether nonlegumes subsisting on combined nitrogen exhibit patterns of carbon and nitrogen partitioning similar to those of white lupine. In a series of studies on the nitrogen nutrition of cotton (*Gossypium*) plants, Radin (180) suggested that this may not be the case for this plant. Cotton reduces nitrate mainly in its shoots (181) and transports nitrogen to its shoots in xylem mainly as nitrate, yet, judging from *in vivo* nitrate reductase assays, the capacity of the root to assimilate nitrate appears to be sufficient to maintain a large fraction of the organic nitrogen requirements of the root's growth (180). It is unfortunate that phloem exudates have not been collected for cotton, because information on the assimilates donated to roots from shoots would enable one to examine the complementary role of phloem in nourishment of roots.

Disappointingly, the study of resource allocation and usage by plants is clearly still in its infancy. In particular, research has yet to combine the type of approach outlined here with more detailed examinations of the costings of biosynthesis and growth at tissue and organ level, as developed in the pioneer studies of Penning de Vries (173). Such a dual approach could be used to study how transport phenomena interact with the growth and physiological activity of the sources and sinks they serve and how they become tuned and modified in response to external and internal influences within the plant. The role of growth-regulating substances in coordination of these responses may well prove to be central, judging from the many experimental findings which have shown a role for these compounds in metabolite transport and utilization. Much broader integrative hypotheses may therefore be expected to result from further study, embracing the respective roles of metabolism, transport, and hormone-based phenomena in a wide spectrum of physiological activity related to plant growth and development.

References

1. Aloni, R., and Sachs, T. (1973). The three-dimensional structure of primary phloem systems. *Planta* **113**, 345–353.
2. van Andel, O. A. M. (1953). The influence of salts on the exudation of tomato plants. *Acta Bot. Neerl.* **2**, 445–521.
3. Ashley, D. A. (1972). ^{14}C-Labelled photosynthate translocations and utilization in cotton plants. *Crop Sci.* **12**, 69–74.
4. Atkins, C. A., and Flinn, A. M. (1978). Carbon dioxide fixation in the carbon economy of developing seeds of *Lupinus albus* L. *Plant Physiol.* **62**, 486–490.
5. Atkins, C. A., Herridge, D. F., and Pate, J. S. (1978). The economy of carbon and nitrogen in nitrogen-fixing annual legumes—experimental observations and theoretical considerations. *In* "Isotopes in Biological Dinitrogen Fixation," pp. 211–242. Int. Atomic Energy Agency, Vienna (IAEA–AG–92/13).
6. Atkins, C. A., Kuo, J., Pate, J. S., Flinn, A. M., and Steele, T. W. (1977). Photosynthetic pod wall of pea (*Pisum sativum* L.). Distribution of carbon dioxide–fixing enzymes in relation to pod structure. *Plant Physiol.* **60**, 779–786.
7. Atkins, C. A., Pate, J. S., and Layzell, D. B. (1979). Economy of C and N in a nodulated and non-nodulated (NO₃-grown) legume. *Plant Physiol.* **63**, 1083–1088.
8. Atkins, C. A., Pate, J. S., and Sharkey, P. J. (1975). Asparagine metabolism—key to the nitrogen nutrition of developing legume seeds. *Plant Physiol.* **56**, 807–812.
9. Atkins, C. A., Pate, J. S., and Layzell, D. B. (1979). The assimilation and transport of nitrogen in non-nodulated (NO₃-grown) *Lupinus albus* L. *Plant Physiol.* **64**, 1078–1082.
10. Austin, R. B., Edrich, J. A., Ford, M. A., and Blackwell, R. D. (1977). The fate of the dry matter, carbohydrates and ^{14}C lost from the leaves and stems of wheat during grain filling. *Ann. Bot. (London)* **41**, 1309–1321.
11. Beever, D. J. (1970). The relationship between nutrients in extracted xylem sap and the susceptibility of fruit trees to silverleaf disease caused by *Stereum purpureum* (Pers.) Fr. *Ann. Appl. Biol.* **65**, 85–92.
12. van Bel, A. J. E. (1974). Different translocation rates of ^{14}C-L-α-Alanine (U) and tritiated water in the xylem vessels of tomato plants. *Acta Bot. Neerl.* **23**, 715–722.
13. van Bel, A. J. E. (1974). The absorption of L-α-alanine and α-amino-iso-butyric acid during their movement through the xylem vessels of tomato stem segments. *Acta Bot. Neerl.* **23**, 305–313.
14. van Bel, A. J. E., and Hermans, H. P. (1977). pH dependency of the uptake of glutamine, alanine and glutamic acid in tomato internodes. *Z. Pflanzenphysiol.* **84**, 413–418.
15. Belikov, I. F. (1955). The local utilization of photosynthetic products in soybeans. *Dokl. Akad. Nauk SSSR Ser. Biol.* **102**, 379–81.
16. Biddulph, O. (1959). Translocation of inorganic solutes. *In* "Plant Physiology" (F. C. Steward, ed.), Vol. 2, pp. 553–603. Academic Press, New York.

17. Bieleski, R. L. (1969). Phosphorus compounds in translocating phloem. *Plant Physiol.* **44,** 497–502.
18. Bieleski, R. (1973). Phosphate pools, phosphate transport and phosphate availability. *Annu. Rev. Plant Physiol.* **24,** 225–252.
19. Bledsoe, R. W., Comar, C. L., and Harris, H. C. (1949). Absorption of radioactive calcium by the peanut fruit. *Science (Washington, D.C.)* **109,** 329–330.
20. Bollard, E. G. (1953). The use of trachael sap in the study of apple-tree nutrition. *J. Exp. Bot.* **4,** 363–368.
21. Bollard, E. (1960). Transport in the xylem. *Annu. Rev. Plant Physiol.* **11,** 141–166.
22. Bollard, E. G. (1970). The physiology and nutrition of developing fruits. *In* "The Biochemistry of Fruits and Their Products" (A. C. Hulme, ed.), Vol. 1, pp. 387–425. Academic Press, New York and London.
23. Bollard, E. G., and Butler, G. W. (1966). Mineral nutrition of plants. *Annu. Rev. Plant Physiol.* **17,** 77–112.
24. Bonnemain, J. L. (1965). Sur le transport diurne des produits d'assimilation lors des la floraison chez la tomate. *C. R. Hebd. Seances Acad. Sci. Ser. D* **260,** 2054–2057.
25. Bornman, C. H., and Botha, C. E. G. (1973). The role of aphids in phloem research. *Endeavour* **32,** 129–133.
26. Bowes, B. G. (1973). Observations on xylem transfer cells in the leaf traces of *Linum usitatissimum* L. (Flax). *Z. Pflanzenphysiol.* **68,** 319–322.
27. Brennan, H. (1966). "Patterns of Translocation in Ageing Plants of Field Pea, *Pisum arvense* L." M.Sc. Thesis, Queen's Univ., Belfast, Northern Ireland.
28. Brennan, H., Pate, J. S., and Wallace, W. (1964). Nitrogen containing compounds in the shoot system of *Pisum arvense* L. I. Amino compounds in the bleeding sap from petioles. *Ann. Bot. (London)* **28,** 527–540.
29. Brown, K. J. (1968). Translocation of carbohydrate in cotton:movement to the fruiting bodies. *Ann. Bot. (London)* **32,** 703–13.
30. Burrows, W. J., and Carr, D. J. (1969). Effects of flooding the root system of sunflower plants on the cytokinin content in the xylem sap. *Physiol. Plant.* **22,** 1105–1112.
31. Caldwell, J. (1961). Further evidence of polar movements of nutrients in plants. *Nature (London)* **190,** 1028.
32. Canny, M. J. (1973). "Phloem Translocation." Cambridge Univ. Press, Cambridge.
33. Canny, M. J. (1975). Mass Transfer. *Encycl. Plant Physiol. New Ser.* **1,** 139–153.
34. Carr, D. J., and Burrows, W. J. (1966). Evidence of the presence of substances with kinetin-like activity. *Life Sci.* **5,** 2061–2077.
35. Charles, A. (1953). Uptake of dyes into cut leaves. *Nature (London)* **171,** 435–436.
36. Christeller, J. T., Laing, W. A., and Sutton, D. W. (1977). Carbon dioxide fixation by lupin root nodules. I. Characterization, association with phosphoenolpyruvate carboxylase, and correlation with nitrogen fixation during nodule development. *Plant Physiol.* **60,** 47–50.
37. Clarke, R. B., Tiffin, L. O., and Brown, J. C. (1973). Organic acids and iron translocation in maize genotypes. *Plant Physiol.* **52,** 147–150.
38. Cleland, C. F., and Ajami, A. (1974). Identification of the flower-inducing factor isolated from aphid honeydew as being salicyclic acid. *Plant Physiol.* **54,** 904–906.
39. Cooper, D. R., Hill-Cottingham, D. G., and Shorthill, M. J. (1972). Gradients in the nitrogenous constituents of the sap extracted from apple shoots of different ages. *J. Exp. Bot.* **23,** 247–254.
40. Couillault, J., and Bonnemain, J. L. (1973). Transport et distribution des assimilates chez *Trapa natans* L. *Physiol. Veg.* **11,** 45–53.
41. Crafts, A. S., and Crisp, C. L. (1971). "Phloem Transport in Plants." Freeman, San Francisco, California.

42. Crookston, R. K., O'Toole, J., and Ozbun, J. L. (1974). Characterisation of the bean pod as a photosynthetic organ. *Crop Sci.* **14**, 708–712.
43. van Die, J. (1961). Synthesis and translocation of glutamine and related substances in the tomato plant. *Proc. K. Ned. Akad. Wet. Ser. C* **64**, 375–381.
44. van Die, J. (1974). The developing fruits of *Cocos nucifera* and *Phoenix dactylifera* as physiological sinks importing and assimilating the mobile aqueous phase of the sieve tube system. *Acta Bot. Neerl.* **23**, 521–540.
45. van Die, J., and Tammes, P. M. L. (1964). Studies on phloem exudation from *Yucca flaccida* Haw. II. The translocation of assimilates. *Acta Bot. Neerl.* **13**, 84–90.
46. van Die, J., and Tammes, P. M. L. (1975). Phloem exudation from monocotyledonous axes. *Encycl. Plant Physiol. New Ser.* **1**, 196–222.
47. van Die, J., and Vonk, C. R. (1967). Selective and stereospecific absorption of various amino acids during xylem translocation in tomato stems. *Acta Bot. Neerl.* **16**, 147–152.
48. van Die, J., Vonk, C. R., and Tammes, P. M. L. (1973). Studies on phloem exudation from *Yucca flaccida* Haw. XII. Rate of flow of ^{14}C-sucrose from a leaf to the wounded inflorescence top. Evidence for a primary origin of the major part of the exudate sucrose. *Acta Bot. Neerl.* **22**, 446–451.
49. van Die, J., and Willemse, P. C. M. (1975). Mineral and organic nutrients in sieve tube exudate and xylem vessel sap of *Quercus rubra* L. *Acta Bot. Neerl.* **24**, 237–239.
50. Doodson, J. K., Manners, J. G., and Myers, A. (1964). The distribution pattern of ^{14}C assimilated by the third leaf of wheat. *J. Exp. Bot.* **15**, 96–103.
51. Epstein, E. (1972). "Mineral Nutrition of Plants. Principles and Perspectives." Wiley, New York.
52. Esau, K. (1965). "Plant Anatomy," 2nd ed. Wiley, New York.
53. Evans, L. T., Wardlaw, I. F., and Fischer, R. A. (1975). Wheat. *In* "Crop Physiology, Some Case Histories" (L. T. Evans, ed.), pp. 101–149. Cambridge Univ. Press, London.
54. Evert, R. F., Eschrich, W., Eichhorn, S. E., and Limbach, S. T. (1973). Observations on penetration of barley leaves by the aphid *Rhopalosiphum maidis* (Fitch). *Protoplasma* **77**, 95–110.
55. Fahn, A. (1974). "Plant Anatomy," 2nd ed. Pergamon, New York.
56. Fellows, R. J., Egli, D. B., and Leggett, H. E. (1978). A pod leakage technique for phloem translocation studies in soybean [*Glycine max* (L.) Merr.]. *Plant Physiol.* **62**, 812–814.
57. Flinn, A. M., Atkins, C. A., and Pate, J. S. (1977). The significance of photosynthetic and respiratory exchanges in the carbon economy of the developing pea fruit. *Plant Physiol.* **60**, 412–418.
58. Flinn, A. M., and Pate, J. S. (1970). A quantitative study of carbon transfer from pod and subtending leaf to the ripening seeds of the field pea (*Pisum arvense* L.) *J. Exp. Bot.* **21**, 71–82.
59. Gallagher, J. N., Biscoe, P. V., and Scott, R. K. (1975). Barley and its environment. V. Stability of grain weight. *J. Appl. Ecol.* **12**, 319–336.
60. Gessner, F. (1965). Untersuchungen über den Gefäss-Saft tropischer Lianen. *Planta* **64**, 186–190.
61. Goor, B. J., and Wiersma, D. (1974). Redistribution of potassium, calcium, magnesium and manganese in the plant. *Physiol. Plant.* **31**, 163–168.
62. Goor, B. J., and Wiersma, D. (1976). Chemical forms of manganese and zinc in phloem exudates. *Physiol. Plant.* **36**, 213–216.
63. Gottleib, D. (1943). The presence of a toxin in tomato wilt. *Phytopathology* **36**, 126–135.
64. Govier, R. N., Brown, J. G. S., and Pate, J. S. (1968). Hemiparasitic nutrition in

angiosperms. II. Root haustoria and leaf glands of *Odontites verna* (Bell.) Dum. and their relevance to the abstraction of solutes from the host. *New Phytol.* **67**, 963–972.

65. Gregory, C. F. (1966). An apparatus for obtaining fluid from xylem vessels. *Phytopathology* **56**, 463.

66. Gunning, B. E. S. (1977). Transfer cells and their roles in transport of solutes in plants. *Sci. Prog. (Oxford)* **64**, 539–568.

67. Gunning, B. E. S., and Pate, J. S. (1974). Transfer cells. *In* "Dynamic Aspects of Plant Ultrastructure" (A. W. Robards, ed.), pp. 441–480. McGraw-Hill, London.

68. Gunning, B. E. S., Pate, J. S., and Green, L. W. (1970). Transfer cells in the vascular system of stems. Taxonomy association with nodes structure. *Protoplasma* **71**, 147–171.

69. Gunning, B. E. S., Pate, J. S., Minchin, F. R., and Marks, I. (1974). Quantitative aspects of transfer-cell structure in relation to vein loading in leaves and solute transport in legume nodules. *Symp. Soc. Exp. Biol.* **28**, 87–126.

70. Haberlandt, G. (1914). "Physiological Plant Anatomy," 4th ed., transl. Macmillan, London.

71. Hall, R. H. (1973). Cytokinins as a probe of developmental processes. *Annu. Rev. Plant Physiol.* **24**, 415–444.

72. Hall, S. M., and Baker, D. A. (1972). The chemical composition of *Ricinus* phloem exudate. *Planta* **106**, 131–140.

73. Hall, S. M., Baker, D. A., and Milburn, J. A. (1971). Phloem transport of ^{14}C labelled assimilates in *Ricinus*. *Planta* **100**, 200–207.

74. Hall, S. M., and Medlow, G. C. (1974). Identification of IAA in phloem and root pressure saps of *Ricinus communis* L. by mass spectrometry. *Planta* **119**, 257–261.

75. Halliday, J., and Pate, J. S. (1976). The acetylene reduction assay as a means of studying nitrogen fixation in white clover under sward and laboratory conditions. *J. Br. Grassl. Soc.* **31**, 29–35.

76. Hansen, P. (1970). ^{14}C studies on apple trees. V. Translocation of labelled compounds from leaves to fruit and their conversion within the fruit. *Physiol. Plant.* **23**, 564–573.

77. Hansen, P., and Grausland, J. (1973). ^{14}C studies on apple trees. VIII. The seasonal variation and nature of reserves. *Physiol. Plant.* **28**, 24–32.

78. Hansen, P., and Grausland, J. (1978). Levels of sorbitol in bleeding sap and in xylem sap in relation to leaf mass and assimilate demand in apple trees. *Physiol. Plant.* **42**, 129–133.

79. Hardham, A. R. (1976). Structural aspects of the pathways of nutrient flow to the developing embryo and cotyledons of *Pisum sativum* L. *Aust. J. Bot.* **24**, 711–721.

80. Hardy, P. J. (1969). Selective diffusion of basic and acidic products of CO_2 fixation into the transpiration stream in grapevine. *J. Exp. Bot.* **20**, 856–862.

81. Hardy, P. J., and Possingham, J. V. (1969). Studies on translocation of metabolites in the xylem of grapevine shoots. *J. Exp. Bot.* **20**, 325–335.

82. Hartt, C. E., and Kortschak, H. P. (1963). Tracing sugar in the cane plant. Proc. 11th Congress of the I.S.S.C.T. Mauritius, 1962. Published as Paper No. 97 in Journal Series of the Experimental Station, Hawaiian Sugar Planters' Ass. Honolulu, Hawaii.

83. Harvey, H. J. (1970). Patterns of assimilate translocation in *Trifolium repens*. *In* "Proc. Symposium on White Clover Research, 1969. Queen's University, Belfast, Northern Ireland (J. Lowe, ed.) pp. 181–186. (Occasional Symposium No. 6. Published by The British Grassland Society.

84. Harvey, D. M., Hedley, C. L., and Keely, R. (1976). Photosynthetic and respiratory studies during pod and seed development in *Pisum sativum* L. *Ann. Bot. (London)* **40**, 993–1001.

85. Hedley, C. L., Harvey, D. M., and Keely, R. J. (1975). Role of PEP carboxylase during seed development in *Pisum sativum*. *Nature (London)* **258**, 352–354.

86. Herridge, D. F., Atkins, C. A., Pate, J. S., and Rainbird, R. (1978). Allantoin and allantoic acid in the nitrogen economy of the cowpea [*Vigna unguiculata* (L.) Walp.]. *Plant Physiol.* **62**, 495–498.

87. Herridge, D. F., and Pate, J. S. (1977). Utilization of net photosynthate for nitrogen fixation and protein production in an annual legume. *Plant Physiol.* **60**, 759–764.

88. Hill-Cottingham, D. G., and Lloyd-Jones, C. P. (1968). Relative mobility of some organic nitrogenous compounds in the xylem of apple shoots. *Nature (London)* **220**, 389–390.

89. Hill-Cottingham, D. G., and Lloyd-Jones, C. P. (1973). Seasonal variations in absorption and metabolism of carbon-14-labelled arginine in intact apple stem tissue. *Physiol. Plant.* **29**, 35–44.

90. Hill-Cottingham, D. G., and Lloyd-Jones, C. P. (1973). Metabolism of carbon-14-labelled arginine, citrulline and ornithine in intact apple stems. *Physiol. Plant.* **29**, 125–128.

91. Hill-Cottingham, D. G., and Lloyd-Jones, C. P. (1973). A technique for studying the adsorption, absorption and metabolism of amino acids in intact apple stem tissue. *Physiol. Plant.* **28**, 443–446.

92. Hocking, P. J., and Pate, J. S. (1977). Mobilization of minerals to developing seeds of legumes. *Ann. Bot. (London)* **41**, 1259–1278.

93. Hocking, P. J., Pate, J. S., Atkins, C. A., and Sharkey, P. J. (1978). Diurnal patterns of transport and accumulation of minerals in fruiting plants of *Lupinus angustifolius* L. *Ann. Bot. (London)* **42**, 1277–1290.

94. Hocking, P. J., Pate, J. S., Wee, S. C., and McComb, A. J. (1977). Manganese nutrition of *Lupinus* spp., especially in relation to developing seeds. *Ann. Bot. (London)* **41**, 677–688.

95. Hofstra, J. J. (1964). Amino-acids in the bleeding sap of fruiting tomato plants. *Acta Bot. Neerl.* **13**, 148–158.

96. Holl, F. B. (1975). Host plant control of the inheritance of dinitrogen fixation in the *Pisum–Rhizobium* symbiosis. *Euphytica* **24**, 767–770.

97. Hopkinson, J. M. (1964). Studies on the expansion of the leaf surface. IV. The carbon and phosphorus economy of a leaf. *J. Exp. Bot.* **15**, 125–137.

98. Humphries, E. C. (1958). Entry of nutrients into the plant and their movement within it. *Proc. Fert. Soc.* **48**, 1–31.

99. Ivanko, S. (1971). Metabolic pathways of nitrogen assimilation in plant tissues when ^{15}N is used as a tracer. *In* "Nitrogen-15 in Soil Plant Studies," pp. 141–149. Int. Atomic Energy Agency, Vienna.

100. Jones, H., Martin, R. V., and Porter, H. K. (1959). Translocation of ^{14}C in tobacco following assimilation of ^{14}C dioxide by a single leaf. *Ann. Bot. (London)* **23**, 493–508.

101. Jones, O. P., and Rowe, R. W. (1968). Sampling the transpiration stream in woody plants. *Nature (London)* **219**, 403.

102. Joy, K. W. (1964). Translocation in sugar beet. I. Assimilation of $^{14}CO_2$ and distribution of materials from leaves. *J. Exp. Bot.* **15**, 485–494.

103. Kauss, H., and Ziegler, H. (1974). Carbohydrate-binding proteins from the sieve tube sap of *Robinia pseudoacacia* L. *Planta* **121**, 197–200.

104. Kende, H. (1965). Kinetin-like factors in the root exudate of sunflower. *Proc. Natl. Acad. Sci. USA* **53**, 1302–1307.

105. Khan, A. A., and Sagar, G. R. (1966). Distribution of ^{14}C-labelled products of photosynthesis during the commercial life of the tomato crop. *Ann. Bot. (London)* **30**, 727–743.

106. King, R. W., and Zeevaart, J. A. D. (1974). Enhancement of phloem exudation from cut petioles by chelating agents. *Plant Physiol.* **53,** 96–103.

107. Kipps, A., and Boulter, D. (1973). Light control of periodicity of bleeding from roots of *Vicia faba. Phytochemistry* **12,** 767–768.

108. Kluge, M., Becker, D., and Ziegler, H. (1970). Untersuchungen über ATP und andere organische Phosphorverbindungen im Siebröhrensaft von *Yucca flaccida* and *Salix triandra. Planta* **91,** 68–79.

109. Kluge, M., and Ziegler, H. (1964). Der ATP gehalt der Siebröhrensaft von Laubbaumen. *Planta* **61,** 167–177.

110. Kramer, D., Lauchli, A., Yeo, A. R., and Gullasch, J. (1977). Transfer cells in roots of *Phaseolus coccineus:* ultrastructure and possible function in exclusion of sodium from the shoot. *Ann. Bot. (London)* **41,** 1031–1040.

111. Kriedemann, P. E. (1968). ^{14}C translocation patterns in peach and apricot shoots. *Aust. J. Agric. Res.* **19,** 775–780.

112. Kriedemann, P. E. (1969). ^{14}C translocation in orange plants. *Aust. J. Agric. Res.* **20,** 291–300.

113. Kursanov, A. L. (1961). The transport of organic substances in plants. *Endeavour* **20,** 19–25.

114. Kursanov, A. L. (1963). Metabolism and the transport of organic substances in the phloem. *Adv. Bot. Res.* **1,** 209–278.

115. Lauchli, A., Kramer, D., and Stelzer, R. (1974). Ultrastructure and ion localization in xylem parenchyma cells of roots. *In* "Membrane Transport in Plants" (U. Zimmermann and J. Dainty, eds.), pp. 363–371. Springer, New York.

116. Lawes, D. A., and Trehorne, K. J. (1971). Variation in photosynthetic activity in cereals and its implications in a plant breeding programme. I. Variation in seedling leaves and flat leaves. *Euphytica* **20,** 86–92.

117. Layzell, D. B., Rainbird, R. M., Atkins, C. A., and Pate, J. S. (1979). Economy of photosynthate use in N-fixing legume nodules; experimental and theoretical observations on two contrasting symbioses. *Plant Physiol.* **64,** 888–891.

118. Leckstein, P. M., and Llewellyn, M. (1975). Quantitative analysis of seasonal variation in the amino acids in phloem sap of *Salix alba* L. *Planta* **124,** 89–91.

119. Lehman, J. (1973). Untersuchungen am Phloemexsudat von *Cucurbita pepo* L. I. Enzymakitivitäten von Glykolyse, Garung and Citrät-Cyclus. *Planta* **114,** 41–50.

120. Lewis, O. A. M., and Pate, J. S. (1973). The significance of transpirationally derived nitrogen in protein synthesis in fruiting plants of pea (*Pisum sativum* L.). *J. Exp. Bot.* **24,** 596–606.

121. Loach, K., and Little, C. H. A. (1973). Production, storage, and use of photosynthate during shoot elongation in balsam fir (*Abies balsamea*). *Can. J. Bot.* **51,** 1161–1168.

122. Lupton, F. G. H. (1966). Translocation of photosynthetic assimilates in wheat. *Ann. Appl. Biol.* **57,** 355–364.

123. McIlrath, W. J. (1965). Mobility of boron in several dicotyledonous species. *Bot. Gaz. (Chicago)* **126,** 27–30.

124. McNeil, D. L., Atkins, C. A., and Pate, J. S. (1979). Uptake and utilization of xylem-borne amino compounds by shoot organs of a legume. *Plant Physiol.* **63,** 1076–1081.

125. MacRobbie, E. (1971). Phloem translocation. Facts and mechanisms. A comparative survey. *Biol. Rev. Cambridge Philos. Soc.* **46,** 429–481.

126. Marshall, C., and Sagar, G. R. (1965). The influence of defoliation on the distribution of assimilates in *Lolium multiflorum* Lam. *Ann. Bot. (London)* **29,** 365–370.

127. Milburn, J. A. (1974). Phloem transport in *Ricinus:* concentration gradients between source and sink. *Planta* **117,** 303–319.

128. Milburn, J. A. (1975). Pressure Flow. *Encycl. Plant Physiol. New Ser.* **1**, 328–353.
129. Milthorpe, F. L., and Moorby, J. (1969). Vascular transport and its significance in plant growth. *Annu. Rev. Plant Physiol.* **20**, 117–138.
130. Minchin, F. R. (1973). "Physiological Functioning of the Plant Nodule Symbiotic System of Garden Pea (*Pisum sativum* L. cv Meteor)." Ph.D. Thesis, Queen's Univ., Belfast, Northern Ireland.
131. Minchin, F. R., and Pate, J. S. (1973). The carbon balance of a legume and the functional economy of its root nodules. *J. Exp. Bot.* **24**, 259–271.
132. Minchin, F. R., and Pate, J. S. (1974). Diurnal functioning of the legume root nodule. *J. Exp. Bot.* **25**, 295–308.
133. Mittler, T. E. (1958). Sieve-tube sap via aphid stylets. In "The Physiology of Forest Trees" (K. V. Thimann, ed.), pp. 312–320. Ronald Press, New York.
134. von Mix, V. P., and Marschner, H. (1976). Effect of external and internal factors on the calcium content of paprika and bean fruits. *Z. Pflanzeneraehr. Bodenkd.* **139**, 551–563.
135. Moorby, J., and Milthorpe, F. L. (1975). Potato. In "Crop Physiology, Some Case Histories" (L. T. Evans, ed.), pp. 225–258. Cambridge Univ. Press, London.
136. Morrison, T. M. (1965). Xylem sap composition in woody plants. *Nature (London)* **205**, 1027.
137. Münch, E. (1930). "Die Stoffbewegungen in der Pflanze." Fischer, Jena.
138. Nelson, C. D. (1962). The translocation of organic compounds in plants. *Can. J. Bot.* **40**, 757–770.
139. Oghoghorie, C. G. O., and Pate, J. S. (1971). The nitrate stress syndrome of the nodulated field pea (*Pisum arvense* L.). Techniques for measurement and evaluation in physiological terms. *Plant and Soil*, Special Vol.
140. Oghoghorie, C. G. O., and Pate, J. S. (1972). Exploration of the nitrogen transport system of a nodulated legume using ¹⁵N. *Planta* **104**, 35–49.
141. O'Neill, T. B. (1961). Primary vascular organization of *Lupinus* shoots. *Bot. Gaz. (Chicago)* **123**, 1–9.
142. Pate, J. S. (1958). Nodulation studies in legumes. I. The synchronisation of host and symbiotic development in the field pea (*Pisum arvense* L.). *Aus. J. Biol. Sci.* **11**, 366–381.
143. Pate, J. S. (1962). Root exudation studies on the exchange of ¹⁴C-labeled organic substances between the roots and shoot of the nodulated legume. *Plant Soil* **17**, 333–356.
144. Pate, J. S. (1965). Roots as organs of assimilation of sulfate. *Science (Washington, D.C.)* **149**, 547–548.
145. Pate, J. S. (1966). Photosynthesizing leaves and nodulated roots as donors of carbon to protein of the shoot of the field pea (*Pisum arvense* L.). *Ann. Bot. (London)* **30**, 94–109.
146. Pate, J. S. (1971). Movement of nitrogenous solutes in plants. In "Nitrogen-15 in soil–plant studies," pp. 165–187. Int. Atomic Energy Agency, Vienna (IAEA–P1–341–13).
147. Pate, J. S. (1973). Uptake, assimilation and transport of nitrogen compounds by plants. *Soil Biol. Biochem.* **5**, 109–119.
148. Pate, J. S. (1975). Exchange of solutes between phloem and xylem and circulation in the whole plant. *Encycl. Plant Physiol. New Ser.* **1**, 451–473.
149. Pate, J. S. (1976). Transport in symbiotic systems fixing nitrogen. *Encycl. Plant Physiol. New Ser.* **2**, 278–303.
150. Pate, J. S. (1977). Functional biology of dinitrogen fixation by legumes. In "A Trea-

398   J. S. PATE

tise on Dinitrogen Fixation" (R. W. F. Hardy and W. S. Silver, eds.), Sect. 3, pp. 473–517. Wiley, New York.

151. Pate, J. S. (1979). Physiologist amongst grain legumes. *J. R. Soc. West. Aust.*

152. Pate, J. S., Atkins, C. A., Hamel, K., McNeil, D. L., and Layzell, D. B. (1979). Transport of organic solutes in phloem and xylem of a nodulated legume. *Plant Physiol.*

153. Pate, J. S., and Flinn, A. M. (1973). Carbon and nitrogen transfer from vegetative organs to ripening seeds of field pea (*Pisum arvense* L.). *J. Exp. Bot.* **24**, 1090–1099.

154. Pate, J. S., and Flinn, A. M. (1977). Fruit and seed development. *In* "The Physiology of the Garden Pea" (J. F. Sutcliffe and J. S. Pate, eds.), pp. 431–468. Academic Press, London.

155. Pate, J. S., and Gunning, B. E. S. (1969). Vascular transfer cells in angiosperm leaves. A taxonomic and morphological survey. *Protoplasma* **68**, 135–156.

156. Pate, J. S., and Gunning, B. E. S. (1972). Transfer cells. *Annu. Rev. Plant Physiol.* **23**, 173–196.

157. Pate, J. S., Gunning, B. E. S., and Briarty, L. (1969). Ultrastructure and functioning of the transport system of the leguminous root nodule. *Planta* **85**, 11–34.

158. Pate, J. S., Gunning, B. E. S., and Milliken, F. F. (1970). Function of transfer cells in the nodal regions of stems, particuarly in relation to the nutrition of young seedlings. *Protoplasma* **71**, 313–324.

159. Pate, J. S., and Herridge, D. F. (1978). Par6itioning and utilization of net photosynthate in a nodulated annual legume. *J. Exp. Bot.* **29**, 401–412.

160. Pate, J. S., and Hocking, P. J. (1978). Phloem and xylem transport in the supply of minerals to a developing legume fruit. *Ann. Bot. (London)* **42**, 911–921.

161. Pate, J. S., and Kuo, J. (1981). Anatomical studies of legume pods—a possible toc ʼn taxonomic research. *In* "Advances in Legume Systematics" (R. M. Polhill and P. H. Ragen, eds.), pp. 903–912.

162. Pate, J. S., Layzell, D. B., and McNeil, D. L. (1979). Modelling the transport and utilization of carbon and nitrogen in a nodulated legume. *Plant Physiol.*

163. Pate, J. S., Kuo, J., and Hocking, P. J. (1978). Functioning of conducting elements of phloem and xylem in the stalk of the developing fruit of *Lupinus albus* L. *Aust. J. Plant Physiol.* **5**, 321–336.

164. Pate, J. S., and Minchin, F. R. (1980). Comparative studies of carbon and nitrogen nutrition of selected grain legumes. *In* "Advances in Legume Systematics" (R. J. Summerfield and A. H. Bunting, eds.), pp. 105–114.

165. Pate, J. S., and O'Brien, T. P. (1968). Micro-autoradiographic study of the incorporation of labelled amino acids into insoluble compounds of the shoot of a higher plant. *Planta* **78**, 60–71.

166. Pate, J. S., Sharkey, P. J., and Atkins, C. A. (1977). Nutrition of a developing legume fruit. Functional economy in terms of carbon, nitrogen, water. *Plant Physiol.* **59**, 506–510.

167. Pate, J. S., Sharkey, P. J., and Lewis, O. A. M. (1974). Phloem bleeding from legume fruits. A technique for study of fruit nutrition. *Planta* **120**, 229–243.

168. Pate, J. S., Sharkey, P. J., and Lewis, O. A. M. (1975). Xylem to phloem transfer of solutes in fruiting shoots of legumes studied by a phloem bleeding technique. *Planta* **122**, 11–26.

169. Pate, J. S., and Wallace, W. (1964). Movement of assimilated nitrogen from the root system of the field pea (*Pisum arvense* L.) *Ann. Bot. (London)* **28**, 80–99.

170. Pate, J. S., Wallace, W., and van Die, J. (1964). Petiole bleeding sap in the examination of the circulation of nitrogenous substances in plants. *Nature (London)* **204**, 1073–1074.

171. Patrick, J. W. (1972). Distribution of assimilate during stem elongation in wheat. *Aust. J. Biol. Sci.* **25**, 455–467.
172. Peel, A. J., and Weatherley, P. E. (1959). Composition of sieve-tube sap. *Nature (London)* **184**, 1955–1956.
173. Penning de Vries, F. W. T. (1975). The cost of maintenance processes in plant cells. *Ann. Bot. (London)* **39**, 77–92.
174. Penning de Vries, F. W. T., Brunsting, A. H. M., and Van Laar, H. H. (1974). Products, requirements and efficiency of biosynthesis: a quantitative approach. *J. Theor. Biol.* **45**, 339–377.
175. Phillips, I. D. J., and Jones, R. L. (1964). Gibberellin-like activity in bleeding sap of root systems of *Helianthus annuus* detected by a new dwarf pea epicotyl assay and other methods. *Planta* **63**, 269–278.
176. Puckridge, D. W. (1968). Photosynthesis of wheat under field conditions. I. The interaction of photosynthetic organs. *Aust. J. Agric. Res.* **19**, 711–719.
177. Quebedeaux, B., and Chollet, R. (1975). Growth and development of soybean [*Glycine max* (L.) Merr.] pods. *Plant Physiol.* **55**, 745–748.
178. Quinlan, J. D. (1965). The pattern of distribution of ¹⁴C in a potted apple root stock following assimilation of ¹⁴C dioxide by a single leaf. *Annu. Rep. East Malling Res. St. England*, No. 41.
179. Quinlan, J. D., and Sagar, G. R. (1962). An autoradiographic study of the movement of ¹⁴C-labelled assimilates in the developing wheat plant. *Weed Res.* **2**, 264–273.
180. Radin, J. W. (1977). Contribution of the root system to nitrate assimilation in whole cotton plants. *Aust. J. Plant Physiol.* **4**, 811–819.
181. Radin, J. W., Sell, C. R., and Jordan, W. R. (1975). Physiological significance of the *in vivo* assay for nitrate reductase in cotton seedlings. *Crop Sci.* **15**, 710–713.
182. Rawson, H. M., and Evans, L. T. (1971). The contribution of stem reserves to grain development in a range of wheat cultivars of different height. *Aust. J. Agric. Res.* **22**, 851–863.
183. Reinbothe, H., and Mothes, K. (1962). Urea, ureides, and guanidines in plants. *Annu. Rev. Plant Physiol.* **13**, 129–150.
184. Rogan, P. G., and Smith, D. L. (1974). Patterns of translocation of ¹⁴C labelled assimilates during vegetative growth of *Agropyron repens* (L.) Beauv. *Z. Pflanzenphysiol.* **73**, 405–414.
185. Salisbury, F. B., and Ross, C. (1969). "Plant Physiology." Wadsworth, Belmont, California.
186. Sauter, J. J., Iten, W., and Zimmermann, M. H. (1973). Studies on the release of sugar into the vessels of sugar maple (*Acer saccharum*). *Can. J. Bot.* **51**, 1–8.
187. Schmid, W. E., and Gerloff, G. C. (1961). A naturally occurring chelate of iron in xylem exudate. *Plant Physiol.* **36**, 226–231.
188. Selvendran, R. R., and Sabaratnam, S. (1971). Composition of the xylem sap of tea plants (*Camellia sinensis* L.). *Ann. Bot. (London)* **35**, 679–682.
189. Sharkey, P. J. (1976). "Transport to and Nutrition of the Developing Lupin Fruit." Ph.D. Thesis, Univ. of Western Australia, Nedlands.
190. Sharkey, P. J., and Pate, J. S. (1975). Selectivity in xylem to phloem transfer of amino acids in fruiting shoots of white lupin (*Lupinus albus* L.). *Planta* **127**, 251–262.
191. Sharkey, P. J., and Pate, J. S. (1976). Translocation from leaves to fruits of a legume studied by a phloem bleeding technique. Diurnal changes and effects of continuous darkness. *Planta* **128**, 63–72.
192. Shibles, R. M., Anderson, I. C., and Gibson, A. H. (1975). Soybean. *In* "Crop Physiology, Some Case Histories" (L. T. Evans, ed.), pp. 151–189. Cambridge Univ. Press, London.

193. Sinclair, T. R., and de Wit, C. T. (1975). Photosynthate and nitrogen requirements for seed production by various crops. *Science (Washington, D.C.)* **189**, 565–567.

194. Sprent, J. I. (1972). The effects of water stress on nitrogen-fixing root nodules. IV. Effects on whole plants of *Vicia faba* and *Glycine max*. *New Phytol.* **71**, 603–611.

195. Steer, B. T. (1976). Rhythmic nitrate reductase activity in leaves of *Capsicum annuum* L. and the influence of kinetin. *Plant Physiol.* **57**, 928–932.

196. Taames, P. M. L., and van Die, J. (1964). Studies on phloem exudation from *Yucca flaccida* Haw. I. Some observations on the phenomenon of bleeding and the composition of the exudate. *Acta Bot. Neerl.* **13**, 76–83.

197. Thrower, S. L. (1962). Translocation of labelled assimilate in the soybean. II. The pattern of translocation in intact and defoliated plants. *Aust. J. Biol. Sci.* **15**, 629–649.

198. Thrower, S. L. (1967). The pattern of translocation during leaf ageing. *Symp. Soc. Exp. Biol.* **21**, 483–506.

199. Tiffin, L. O. (1970). Translocation of iron citrate and phosphorus in xylem exudate of soybean. *Plant Physiol.* **45**, 280–283.

200. Turgeon, R., and Webb, J. A. (1973). Leaf development and phloem transport in *Cucurbita pepo*: transition from import to export. *Planta* **113**, 179–191.

201. Turgeon, R., and Webb, J. A. (1976). Leaf development and phloem transport in *Cucurbita pepo*. Maturation of the minor veins. *Planta* **129**, 265–269.

202. Ursino, D. J., and Paul, J. (1973). The long term fate and distribution of ^{14}C photoassimilated by young white pines in late summer. *Can. J. Bot.* **51**, 683–687.

203. Vonk, C. R. (1974). Studies on phloem exudation from *Yucca flaccida* Haw. XIII. Evidence for the occurrence of a cytokinin nucleotide in the exudate. *Acta Bot. Neerl.* **23**, 541–548.

204. Walker, A. J., and Ho, L. C. (1977). Carbon translocation in the tomato: carbon import and fruit growth. *Ann. Bot. (London)* **41**, 813–823.

205. Walker, A. J., Ho, L. C., and Baker, D. A. (1978). Carbon translocation in the tomato; pathways of carbon metabolism in the fruit. *Ann. Bot. (London)* **42**, 901–909.

206. Wallace, W., and Pate, J. S. (1965). Nitrate reductase in the field pea (*Pisum arvense* L.). *Ann. Bot. (London)* **29**, 654–671.

207. Wallace, W., and Pate, J. S. (1967). Nitrate assimilation in higher plants with special reference to the cocklebur (*Xanthium pennsylvanicum* Wallr.). *Ann. Bot. (London)* **31**, 213–228.

208. Walpole, P. R., and Morgan, D. G. (1972). Physiology of grain filling in barley. *Nature (London)* **240**, 416–417.

209. Weatherley, P. E. (1969). Ion movement within the plant and its integration with other physiological processes. In "Ecological Aspects of the Mineral Nutrition of Plants" (I. H. Rorison, ed.), pp. 215–229. Blackwell, Oxford.

210. Wheeler, C. T. (1978). Carbon dioxide fixation in the legume root nodule. *Ann. Appl. Biol.* **88**, 481–484.

211. Wieringa, K. T., and Bakhuis, J. A. (1957). Chromatography as a means of selecting effective strains of *Rhizobia*. *Plant Soil* **8**, 254–260.

212. Wiersum, L. K. (1951). Water transport in the xylem as related to calcium uptake by groundnuts (*Archis hypogaea* L.). *Plant Soil* **3**, 160–169.

213. Williams, M. W., and Raese, J. T. (1974). Sorbitol in tracheal sap of apple as related to temperature. *Physiol. Plant.* **30**, 49–52.

214. Wooding, F. B. P., and Northcote, D. H. (1965). An anomalous wall thickening and its possible role in the uptake of stem-fed tritiated glucose by *Pinus pinea*. *J. Ultrastruct. Res.* **12**, 463–472.

215. Zee, S. Y., and O'Brien, T. P. (1971). Vascular transfer cells in the wheat spikelet. *Aust. J. Biol. Sci.* **24**, 35–49.

216. Ziegler, H. (1963). Der Fernstransport organischer Stoffe in den Pflanzen. *Natur-wissenschaften* **50**, 177–186.
217. Ziegler, H. (1975). Nature of transported substances. *Encycl. Plant Physiol. New Ser.* **1**, 59–100.
218. Ziegler, H., and Ziegler, I. (1962). The water soluble vitamins in the sieve tube sap of some trees. *Flora (Jena)* **152**, 257–278.
219. Zimmermann, M. H. (1960). Transport in the phloem. *Annu. Rev. Plant Physiol.* **11**, 167–189.
220. Zimmermann, M. H. (1961). Movement of organic substances in trees. *Science (Washington, D.C.)* **133**, 73–79.
221. Zimmermann, M. H. (1969). Translocation of nutrients. *In* "The Physiology of Plant Growth and Development" (M. B. Wilkins, ed.), pp. 113–147. McGraw-Hill, Maidenhead, England.
222. Zimmermann, M. H., and Ziegler, H. (1975). List of sugars and sugar alcohols in sieve tube exudates. Appendix III. *Encycl. Plant Physiol. New Ser.* **1**, 480–503.
223. Zimmermann, M. H. (1976). Structural requirements for optimal water conduction in tree stems. *In* "Tropical Trees as Living Systems" (P. B. Tomlinson and M. H. Zimmermann, eds.). Cambridge Univ, Press, London.

EPILOGUE:
INTEGRATION OF ENERGY, FORM, AND COMPOSITION

Inorganic nitrogen, water, and carbon dioxide (together with other essential elements) are evident raw materials out of which all autotrophic plants during growth fabricate their substance, their complexity, and their ability to function. And through the ability of plants that fix atmospheric nitrogen, whether these are free-living forms or live in symbiotic associations, the plant kingdom can utilize the elementary nitrogen of the air (see Chapter 1). But it is through the properties of proteins and nucleic acids (see Chapter 3), the most complex of nitrogen compounds, that life on earth is conceivable. The role of nitrogen compounds in the inheritance of information to chart the course of the chemical working of plants and to bring it into play via genes and proteins as enzymes needs no further comment at the end of this volume. But the integration of metabolism into the overall operation of plants as physical systems does merit some final comment.

As cells and organs grow they create the complexity within which they perform their various chemical functions according to the design their inheritance prescribes, but the resultant system must also function as a stable physical entity in its environment. At the outset free cells—whether zygotes or somatic cells—or cells in growing regions obligate part of their total energy and part of their available substance (originally in the form of nutrient salts, especially nitrate, phosphate, and sulfate, and sugar) to build improbable forms and structures that provide the milieux within which the chemical events of metabolism occur. But the energy so obligated to complexity appears as the reduced entropy of the organization, that is, the organelles, cells, and organs. Concurrently, part of the total energy available for physical use appears as the concentrations of solutes, organic and inorganic, which are quite different from anything in the environment or the ambient media, and as such they have a positive free-energy equivalent. These internal secretions represent, therefore, osmotic work. A complete physical accounting of all that happens during growth (see Volume VII, Chapter 4) would allow a moiety of energy to go into the structural complexity and improbability of the forms that develop and would allow a moiety of energy to be built into the complex improbable molecules that biochemistry or inheritance needs, especially those such as enzymes, proteins, and DNA that have special biological significance. And, as water is the universal ingredient in organisms, much of what happens to its solutes could be construed as changing the activity of the water in organisms compared to that in the environment. Viewed in this broad way the growth and development from zygotes or somatic cells or from shoot tips represent a progression of biophysical states in the transformation of inorganic matter, especiallly nitrogen, into the substance and form of plants, which converts molecules into morphology, and builds inanimate, random molecules into the composition of viable plants.

The descriptive biochemistry of plants is now well documented. The more attention can be narrowly confined to the detail of the reactions that occur, to the enzyme systems that catalyze them and to the individual substances so synthesized, the more satisfying the interpretations seem to be. If the detailed mechanisms of reactions and the ways in which energy from catabolism is used to drive synthetic metabolism are to be understood, they must be viewed, as in Chapter 2 of Volume VII, in the context of the ultrastructure or the membranes and organelles in which they occur, as described in Chapter 1 of that volume. But this is plant chemistry in terms of specific substances. In terms of metabolism (e.g., of carbon compounds), although the individual reactions need to be understood in detail, they only become comprehensible in terms of the overall metabolism of the whole plant. This approach was adopted in Chapters 3 and 4 of Volume VII, with the minimum of arbitrary distinction between classically independent functions such as respiration, photosynthesis, etc.

The treatment of nitrogen metabolism (Chapter 2 of this volume) has also comprehended the biochemical reactions collectively in relation to the organism, its organs, tissues, cells, and organelles that, in toto, provide the setting in which the biochemical events occur. If, however, descriptive biochemistry in plants is a well-documented book, its biophysical counterpart still needs to be written. But when written, and if all the appropriate terms are entered into the balance sheet, then it is an article of faith that no physical laws of the material universe will be breached. Hence, for nitrogen metabolism, as in all physiological events that critically involve the on-going organization (with its initial drive to fulfill through ontogeny the morphogenetic propensities of zygotes, totipotent cells, or growing regions), full understanding is limited to our ability to relate it all to a complete energy balance sheet. Such an analysis should cover all that happens as plants convert water, carbon dioxide, and light into molecular structures and create from sugar the negotiable phosphate bond energy which, through all the events of bioenergetics (Volume VII, Chapter 2) is spent in so many ways. In part, it is spent in the diversity of compounds (many containing nitrogen) that metabolism employs, or to build up the complexity of proteins, enzymes, and even genes, which not only carry the needed information but are also the means to implement its operation. In part, a moiety of the total energy is budgeted for the organelles, cells, and organs that are the operating parts of the organization at work. Here (Chapter 2), as in Chapter 4, a complete synthesis is achieved only as consideration is given to metabolism throughout the plant body with the demands that the diversity of organs makes upon translocation of metabolites and upon controls that integrate that diversity into a holistic pattern. It is here that plants and their environments are so intimately intertwined (see Chapter 2). Finally, a moiety of the total energy maintains the physical stability and the integrity and composition of each plant in its environment, notably by controlling the activity of its water—the universal solvent—through the free energy from its solutes.

However plants have evolved, their stability and continued existence as physical systems requires an integration of all aspects (chemical and morphogenetic) of their life and work. Heredity and genetics and the molecular biology of DNA, controlling proteins and enzymes, define tracks along which specific chemical events can move. But before these different tracks can become available for metabolic traffic, involving nitrogen as crucially as carbon, morphogenetic decisions, which are part of the organismal heritage, need to be implemented.

Inevitably, the concepts mentioned above will need to be elaborated as problems of cell physiology, growth, and development are reexamined in subsequent volumes.

F.C.S.

Author Index

Numbers in parentheses are reference numbers and indicate that an author's work is referred to, although his name is not cited in the text. Numbers in boldface show the page on which the complete reference is cited.

A

Abe, M., (1), 313(92), 314(92), **319, 324**
Abelson, J., 131(1), **186**
Abelson, P. H., **264,** 270(2, 3), 294(4), **319**
Aberhart, D. J., 69(2), **186**
Abu-Shakra, S. S., 291(4), 293(4), 294(4), **319**
Adams, D. O., 91(3), **186**
Aguilar-Santos, **212**
Aitchison, P. A., 188(292), **199**
Ajami, A., 363(38), **392**
Akasi, S., **250**
Akmakci, L. C., 30(55), **52**
Al-Baldawai, N. F., **253**
Albein, A. D., 308(90), **323**
Albrecht, S. L., 20(25), 21(25), 43(25), **50**
Aldag, D. R., **218**
Alfthan, M., **261, 262, 264, 265**
Allen, C. M., Jr., **236**
Allen, G. M., 77(4), **186**
Aloni, R., 338(1), **391**
Alscher, R., 282(5), 287(5), **319**
Altamura, M. R., **200**
Altschul, A. M., 63(5, 6), **186**
Amagase, S., 313(6), **319**
Ambler, R. P., **200**
Amhad, V. U., **218**
Amir, J., 279(183), **328**
Amzel, M. L., 123(7), **186**
Andersen, K., 11(69), 21(69), 24(69), **53**
Anderson, I. C., 347(192), **399**
Anderson, J. W., 303(7), 313(8), **319**
Anderson, M. B., 303(9), **320**
Anderson, N. G., 57(8, 9), **186**
Apgar, J., 131(111), **191**
Arai, S., 313(92), 314(92), **324**

Archer, M. C., **243, 258**
Arias, I. M., 274(10), **320**
Arjmand, M., **226**
Armbrust, K., **242**
Arnon, D. I., 10(1), **49**
Ashley, D. A., 347(3), **391**
Ashton, F. M., 307(11), 311(11), 313(11, 12), **320**
Aslam, M., 270(13), 297(16), 294(13), 295(16), 296(13, 14), 297(14, 16), **320,** 298(14, 15), 299(15), 300(193), **320, 328**
Atkin, R. K., 292(17), **320**
Atkins, C. A., 24(89), 25(89), **54,** 297(50), **322,** 338(166), 242(124), 343(9), 352(152), 355(86), 357(117), 358(124), 359(124), 360(124), **396,** 362(7), 363(93), 364(8), 366(4, 5, 117), 367(5), 368(117), 369(5, 117), 370(117), 373(4–6, 57), 374(166), 375(57), 376(8, 152, 166), 380(5, 166), 387(7), 390(166), **391, 393, 395, 396, 398**
Atkinson, D. E., 98(10), 105(10), 109(10), 148(10), **186**
Atkinson, Y. E., 138(26), **187**
Attridge, T. H., 302(18, 19), **320**
Audrain, L., 317(101), **324**
Audus, L. J., 171(11), **186**
Aue, W. A., 63(90), **190**
Austin, R. B., 343(10), 373(10), **391**
Ausubel, F. M., 13(76), 14(76), 22(76), **53**
Aviado, D. M., **259**
Avila, E. J., **257**
Axelrod, B., 101(258), **198, 238**
Azari, P. R., 317(20), **320**
Aziz, S. A., **257**

Index to Plant Names

Numbers in this index designate the pages on which reference is made in the text to the plant in question. No reference is made in the index to plant names included in the titles that appear in the reference lists. In general, when a plant has been referred to in the text sometimes by its common name, sometimes by its scientific name, all such references are listed in the index after the scientific name; cross reference is made, under the common name, to this scientific name. However, in a few instances in which a common name as used cannot be referred with certainty to a particular species, the page numbers follow the common name.

Subject Index

Names of chemicals in the Appendixes to Chapter 2 are not listed in this index, but all chemical names mentioned elsewhere are included.

A

Acetate phosphotransferase, nitrogen fixation, 25
α-Aceto-α-hydroxybutyrate, amino acid metabolism, 89
Acetolactate, amino acid metabolism, 92
Acetyl CoA, amino acid metabolism, 92
Acetylene reduction
  assay, 12
  overestimation of nitrogen fixation, 46
N-Acetylgalactosamine, inhibition of *Rhizobium* association, 33
N-Acetylglucosamine, glycoprotein, 308
γ-Acetylglutamate, from glutamate, 71
N-Acetylglutamate, from glutamate, 71
O-Acetylhomoserine, from aspartate, 89
N^ε-Acetyl- and methyllysines, from aspartate, 89
O-Acetylserine, cyanide assimilation, 75
Acid, protease, 313–315
Actinomycetes, nitrogen-fixing, 39
Activating factor, 13
Adenine
  breakdown, 102
  deaminase, 121
  formation, 97
Adenosine deaminase, 121
Adenosine diphosphate deaminase, 121
Adenosine monophosphate amidase, 121
Adenosine triphosphate
  carbamoyl phosphate, 77
  hydrogenase, 368
  nitrogen fixation, 11, 25, 368
  synthesis, 97
S-Adenosylmethionine
  from methionine, 91
  from ornithine, 87
Adenylic acid, synthesis, 97

5'-Adenylic acid deaminase, 121
Adenylosuccinate
  purine breakdown, 102
  purine synthesis, 97
Aerotaxis, 30
Aesculus growth factor, 177
Agmatine
  from arginine, 87
  breakdown, 120
Alanine
  amino acid metabolism, 78
  ammonia assimilation, 75
  dehydrogenase, 75, 121
  family, 92
  from aspartate, 73
  from glutamate, 71
  gluconeogenesis, 118
  metabolic products, 75
  protein, 83
β-Alanine
  from aspartate, 89
  from ornithine, 87
  pyrimidine breakdown, 100
  transaminase, 101
Albumin, 307, 309
  degradation, 316
Alkaloid
  from phenylalanine, 85
  xylem, 351
Allantoic acid
  nitrogen fixation, 24
  purine breakdown, 102
  xylem, 351
Allantoicase, 102
Allantoin
  nitrogen fixation, 24
  purine breakdown, 102
  xylem, 351
Allergen, 307

β-Cyanoalanine, 75
  amino acid metabolism, 78
  cyanide assimilation, 75
  hydratase, 75
  from phenylalanine, 85
  from serine, 94
  synthetase, 75
Cyanogenic glycosides, 75
Cyclic AMP synthesis, 97
Cycloalliin, from cysteine, 91
Cycloheximide
  nitrate-transport system, 301
  RuBPcase regulation, 291
  sulfate-transport system, 301
Cystathionine
  from aspartate, 89, 91
  from serine, 94
L-Cysteine
  amino acid metabolism, 78
  from aspartate, 91
  cyanide assimilation, 75
  protein, 83
  from serine, 94
  sulfinate dehydrogenase, 122
  synthesis, 94
Cytidine deaminase, 121
Cytidylic acid synthesis, 96
Cytochrome, succinate cycle, 95
Cytokinesin, purine breakdown, 102
Cytosine
  breakdown, 100
  deaminase, 121

D

Day length, nitrogen metabolism,
  155–157
Defoliation, and transport, 348
Degenerate code, 124
Dehydratase, amino acid synthesis, 108
Dehydrogenase, amino acid metabolism,
  107
5-Dehydroquinic acid, amino acid
  metabolism, 84
5-Dehydroshikimate, amino acid
  metabolism, 84
3-Deoxy-D-arabinoheptulosonic-7-phosphate,
  amino acid metabolism, 84
2-Deoxyglucose, inhibition of *Rhizobium*
  association, 33

Deoxyribonucleic acid, 126–137
  errors, 137
  recombinant, 140
  replication, 128, 138
Deoxythymidine synthesis, 96
Development, nitrogen metabolism, 149
Dhurrin, from tyrosine, 85
α,γ-Diaminobutyrate, from aspartate, 73
L-α,ε-Diaminopimelate, amino acid
  metabolism, 89
meso-α,ε-Diaminopimelate, amino acid
  metabolism, 89
1,3-Diaminopropane, from ornithine, 87
γ,δ-Diaminovaleric acid, succinate cycle,
  95
N^2-(1,3-Dicarboxypropyl)ornithine
  synthesis, 120
Diguanide, from arginine, 87
Dihydroorotic acid, pyrimidine synthesis,
  96
2,3-Dihydropicolinic acid, amino acid
  metabolism, 89
Dihydrothymine, pyrimidine breakdown,
  100
Dihydrouracil, pyrimidine breakdown,
  100
α,β-Dihydroxyisovalerate, amino acid
  metabolism, 92
α,β-Dihydroxy-β-methylvalerate, amino
  acid metabolism, 89
3,4-Dihydroxyphenylalanine, from
  tyrosine, 85
Dimethylallyguanidine, from arginine, 87
Dinitrophenol, sulphate transport system,
  301
β-(3,5-Dioxo-1,2,4-oxazolidin-2-yl)-L-alanine,
  from serine, 94
γ,δ-Dioxovaleric acid, succinate cycle, 95
Dipeptidase, 313–315
DNA, *see* Deoxyribonucleic acid
DNP, *see* Dinitrophenol
Dolichol (lipids), 308
Dulcitol, leaf export, 361

E

Endonuclease, 130
Endoplasmic reticulum
  protein metabolism, 304
  protein synthesis, 309